Lecture Notes in Physics

Edited by J. Ehlers, München, K. Hepp, Zürich and
H. A. Weidenmüller, Heidelberg
Managing Editor: W. Beiglböck, Heidelberg

17

Strong Interaction Physics

International Summer Institute on
Theoretical Physics in Kaiserslautern 1972

Edited by W. Rühl and A. Vancura

Springer-Verlag

Berlin · Heidelberg · New York 1973

ISBN 3-540-06141-X Springer-Verlag Berlin · Heidelberg · New York
ISBN 0-387-06141-X Springer-Verlag New York · Heidelberg · Berlin

PREFACE

The International Summer Institute on Theoretical Physics in Kaiserslautern 1972 was the fifth in the series of summer schools organized by different universities in the Federal Republic of Germany, following Karlsruhe, Heidelberg, and Hamburg. It was devoted mainly to the strong interactions of elementary particles with emphasis on the theoretical investigation rather than phenomenology. Two major fields are covered by the lectures:

> Mathematical methods in analyzing scattering amplitudes and light cone singularities respectively conformal invariance in quantum field theory.

The present volume contains notes of sixteen lectures as they were prepared by the authors. The lectures by Prof. H. Lehmann (Hamburg) "Chiral invariant field theory and pion-pion scattering" and Prof. A. Tavkhelidze (Dubna) "On automodel asymptotic behaviour of deep inelastic form factors" are not included in the volume for technical reasons. The final preparation of the manuscripts for printing was in the hands of the editors, and all blame for misprints and mistakes of that sort should be cast on them.

The Institute took place on the campus of the very young University of Trier-Kaiserslautern in Kaiserslautern. On behalf of all participants we take great pleasure to thank the officials of this university for their cooperation and help before and during the Institute.

The Institute was sponsored by the Scientific Affairs Division of the North Atlantic Treaty Organization and generously supported by the Bundesministerium für Bildung und Wissenschaft in Bonn. Last but not least, special thanks are due to Mrs. von Aswegen from Kaiserslautern for typing the manuscripts. Without her painstaking effort in preparing the manuscripts we would have never been able to have it ready so soon.

October 1972

W. RÜHL
A. VANCURA

CONTENTS

<u>HIGH ENERGY EXPERIMENTS</u>

A. Minten

CERN, Genève

1.   <u>INTRODUCTION</u>

In this lecture we will report on the experimental progress observed during the last year <u>at high energy</u> accelerators. This energy range can be separated in this context into two regions:

(i)   <u>high energy</u>, well above the resonance region, ranging from ~ 10 to 70 GeV, experimenting with a variety of particles like $\pi^{\pm}$, $K^{\pm}$, p, $\bar{p}$, n, K^{o} and most recently with hyperons;

(ii) <u>very high energy</u>, up to 2000 GeV, until last year reserved to cosmic rays, now accessible with proton storage rings. This region is restricted to the study of pp collisions and it is limited by rates to the investigation of cross sections $\gtrsim 10^{-34} \, cm^2$.

In discussing the progress in the field we will put emphasis on experiments, i.e. measurements and their results, and we leave interpretations to other lectures. We cover, without any completeness, some elementary but representative processes which have been studied in the full energy region. These are:

- particle production (inclusive reactions);
- elastic scattering;
- total cross section.

2.   <u>INSTRUMENTS AND METHODS</u>

The novel instrument to study pp collisions at very high energies are the CERN <u>Intersecting Storage Rings</u> (ISR). Two almost circular rings intersect at 8 points with an intersection angle 14.8^{o}

(Fig. 1). Incident protons with equal energy E_p produce a system with (CM energy)^2 = s

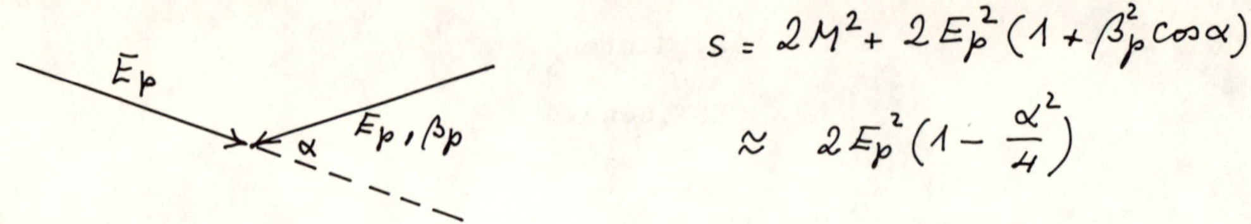

$$s = 2M^2 + 2E_p^2\left(1 + \beta_p^2 \cos\alpha\right)$$

$$\approx 2E_p^2\left(1 - \frac{\alpha^2}{4}\right)$$

The corresponding energy E_L for a proton from a conventional accelerator, producing the same s on a proton target at rest, is expressed by

$$s = 2M^2 + 2ME_L$$

The table below shows the standard ISR energies in the variables E_p, s and E_L :

					x)	
E_p	~ 10	15	22	26	31	GeV
s	~ 400	900	2000	2700	3800	GeV^2
E_L	~ 210	480	1050	1450	2000	GeV

x) acceleration in the ISR from 26 to 31 GeV

The circulating proton beam in the ISR, forms a band of 6 cm width and 0.5 cm height. If vertically adjusted, the beams intersect at the intersection regions, forming a "diamond" of about 45 cm length

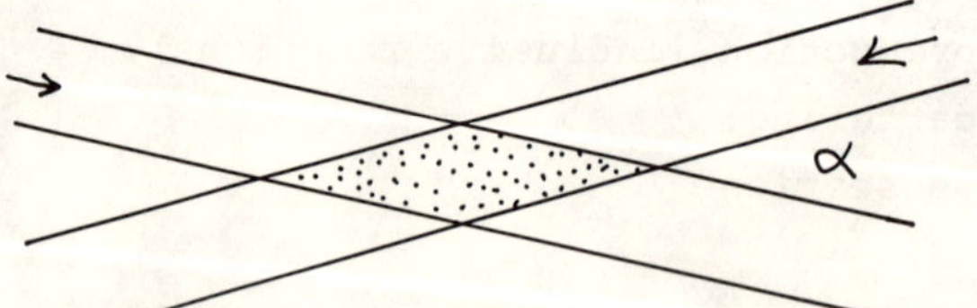

With N_i stored protons in each of the rings, the collision rate $\dot{N}$ is given by [1]

$$\dot{N} = n_1 \, n_2 \, \left|\vec{v}_{rel}\right| \, V \, \sigma$$

(n_1, n_2 beam densities $\left[\,cm^{-3}\,\right]$, v_{rel} relative velocity of beams

$[\text{cm sec}^{-1}]$, V interaction volume $[\text{cm}^3]$, σ cross section $[\text{cm}^2]$).
Expressed by convenient variables this is

$$\dot{N} = \frac{N_1 N_2}{(2\pi R)^2} \; \frac{c}{\tan \frac{\alpha}{2} \cdot h} \cdot \sigma$$

$$= L \cdot \sigma$$

(R = 150 m ISR radius, c light velocity, h effective beam height,
L luminosity $[\text{cm}^{-2} \text{ sec}^{-1}]$). The luminosity L

$$L \propto \frac{1}{h} \propto \int m_1(z) \, m_2(z) \, dz$$

must be experimentally determined, which limits the accuracy of cross
section measurements to 5 - 15% at present.

Relevant ISR parameters are :

N_i (max)	$2 \cdot 10^{14}$
(normal)	10^{14}
L	$10^{30} \text{cm}^{-2} \text{sec}^{-1}$
$\dot{N}$	$40\ 000 \text{ sec}^{-1}$

As we see, proton storage rings enable us to extend studies of proton-
proton interactions to very high energies, probably in the future up
to s $\simeq 10^5$ GeV^2 . On the other hand they are limited in interaction
rate and therefore in the smallest detectable cross section, which is
of the order of 10^{-34} cm^2, corresponding to about one event per day.
Expressing the interaction rate by the formula

$$\dot{N} = i_1 \cdot (\rho \ell)_2 \cdot \sigma$$

(i_1 $[\text{sec}^{-1}]$ beam current, ρ $[\text{cm}^{-3}]$ target density, ℓ $[\text{cm}]$ target length)
we reach in storage rings a 10^8 times higher current than in accelerat-
ors, but 10^{14} times lower target density, so that, in principle, 10^6
times lower cross sections can be studied at proton accelerators.

In addition, accelerators provide secondary beams and the
secondary interactions can be studied on secondary targets. Therefore a very large part of our future knowledge : rare particles and
reactions, $\pi, K, \bar{p}, \gamma$ induced processes etc., will come from

accelerators and will be at accelerator energies.

3.   <u>EXPERIMENTS</u>

   3.1 <u>Particle Production.</u>

The production of photons, pions, kaons and baryons has been studied during the first year at the CERN ISR[2-9]. Results can be compared with similar studies of pp collisions at lower energies mainly in bubble chambers[10].

Main investigations and results are :

i) <u>the invariant cross section</u> was measured

$$E\,\frac{d^3\sigma}{dp^3} = \frac{E}{\pi}\,\frac{d^3\sigma}{dp_\ell\,dp_t^2} = \frac{E}{p^2}\,\frac{d^2\sigma}{dp\,d\Omega} = f(p_\ell, p_t, s)$$

(E particle energy, p_ℓ (p_t) particle longitudinal (transverse) momentum, $d\Omega$ solid angle, f structure function). Alternative variables for p_ℓ are

$$x = p_\ell / p_\ell^{max} \qquad\qquad \text{reduced longitudinal momentum}$$

$$y = \frac{1}{2}\,\ell n\,\frac{E+p_\ell}{E-p_\ell} \xrightarrow[p_\ell \gg m]{} \ell n\,\frac{2\,p_\ell}{\sqrt{m^2 + p_t^2}} \qquad \text{rapidity}$$

Results (Fig.2) :

- at fixed p_ℓ the transverse momentum distribution is exponential[4,6]

$$E\,\frac{d^3\sigma}{dp^3} = A\,exp(-b\,p_t)$$

  with b approximatively independent of p_ℓ and s, but dependent on the nature of the particle. For pions one finds b $\approx$ 6.3 GeV^{-1}, for baryons values between 4 and 5 GeV^{-1}.

- at fixed p_t the x distribution for $0.1 \le x \le 0.4$ is found to be independent from the energy s within errors of $\pm$ 15% [7,8] (scaling). The x distribution for pions is approximately exponential and that for protons is roughly flat, with a rise towards x = 1 (leading proton).

  ii) <u>particle ratios</u> vary with x and s[3-8] (Fig.3)

- the π^+/π^- ratio varies from 2 at x = 0.4 to 1 at x = 0 (proton fragmentation vs. pion pair production) ;

- the $p/\bar{p}$ ratio decreases from about 100 at x = 0.4 to 1.5 $\pm$ 0.5 at x = 0 (leading protons vs. produced antiproton);

- the $\pi^-/\bar{p}$ ratio rises from about 0.01 at 20 GeV to 0.05 at 500 GeV and seems to be independent of s above that value (production threshold).

iii) differential cross sections for charged particles were measured[9]

$$\frac{d\sigma}{d\Omega} = \sum_{\pi k p} \int \frac{1}{E} f(p_\ell, p_t, s)\, p^2 dp$$

Results [9,10] :

-

$$\frac{d\sigma}{d\Omega}(\theta, s) = \sigma_{90°}(s)\, \frac{1}{\sin^2\theta} \qquad\qquad 30° \leq \theta \leq 90°$$

The $\sin^{-2}\theta$ angular dependence is equivalent with a structure function $f(p_t, s)$ independent of p_ℓ for x < 0.05 (cylindrical phase space) :

$$\frac{d\sigma}{d\Omega} = A(s) \int \frac{1}{E} e^{-b\,p_t}\, p^2 dp$$

with p_t = p sin θ , and E $\approx$ p

$$= \frac{A(s)}{\sin^2\theta} \int p_t\, e^{-b\,p_t}\, dp_t$$

$$= \frac{A(s)}{b^2}\, \frac{1}{\sin^2\theta} \ .$$

- the $90°$ cross section is <u>energy dependent</u> and rises from about 4.mb/sterad at 19 GeV to 7. $\pm$ 1. mb/sterad at 1500 GeV (<u>no</u> scaling at x = 0).

- transformation to a variable equivalent to the repidity y for photons and almost equivalent for pions

$$y = \ln \frac{2 p_\ell}{\sqrt{m^2 + p_t^2}} \quad\xrightarrow[m \ll p_t]{}\quad - \ln \tan \frac{\theta}{2} = \eta$$

transforms the $\sin^{-2}\theta$ distribution into a constant, since

$$\frac{d\Omega}{\sin^2\theta} = \frac{2\pi\, d\theta}{\sin\theta} = 2\pi\, d\left(\ln \tan \frac{\theta}{2}\right) = 2\pi\, d\eta$$

Experimental y distributions for pions and η distributions for
charged particles are constant at angles $> 30^{\circ}$ for all energies
above 250 GeV.

(iv) integration of the measured differential cross sections for all
charged particles permits the derivation of the charged mean multiplicity $\langle n_c \rangle$

$$\sum_{\pi K p} \int \frac{d^3\sigma}{dp^3} dp^3 = \int \frac{d\sigma}{d\Omega} d\Omega = \langle n_c \rangle \sigma_{in}$$

where $\sigma_{in} \simeq$ 32 mb is the total inelastic cross section. The
analysis was done combining data from different ISR experiments
in the y variable and integration[11] . Procedure and result are
given in Fig. 4.

3.2 Elastic Scattering

Important experiments provided recently detailed knowledge on the
behaviour of elastic scattering at relatively "low energy" (Fig. 5) :

-    the t-dependent structure of $\pi^{\pm}$, $K^{\pm}$, $p^{\pm}$ scattering on protons in
     the complete angular range up to 180° at 5 GeV[12];

-    the difference of particle/antiparticle scattering on protons at
     momentum transfers $|t| \leqslant 1.$ GeV^2 at 3.65 GeV revealing the "cross
     over" phenomenon[13];

-    the (t, s) dependent structure in pp scattering in the complete
     angular range up to 90° and at energies up to 24 GeV[14].

These experiments represent the most refined knowledge on elastic
cross sections at this moment. Therefore we want to present them here,
although they are outside the frame of our lecture. We now consider
the much more limited and less precise information at "high energies" :

(i) diffraction scattering has been measured for π^{-}, K^{-}, $\bar{p}$ on protons
    for $.1 \leqslant |t| < .4$ GeV^2 at 25 and 40 GeV[15]. Results are constant
    slope parameters b(s) (Fig.6).

$$\bar{\pi} p \qquad b = (8.5 \pm .2) \, GeV^{-2}$$
$$\bar{K} p \qquad (7.8 \pm .2)$$
$$\bar{p} p \qquad 11.5$$

(ii) pp scattering has been measured at the ISR for $|t| \leq 1.$ GeV^2 at 250, 500, 1000 and 1500 GeV[16,17]. The results indicate a change of slope at t = 0.1 GeV^2, with (Fig. 7)[17]

$$b = 11.6 - 12.9 \, GeV^{-2} \quad for \quad |t| < .1 \, GeV^2$$
$$b = 11.4 - 11.8 \, GeV^{-2} \quad for \quad |t| > .1 \, GeV^2.$$

The same experiment gives a value for the total pp cross section through the optical theorem

$$\frac{d\sigma}{dt} = A \, e^{bt} = \frac{\sigma_{tot}^2}{16\pi} \, (1 + \alpha^2) \, e^{bt}$$

where $\alpha = $ Re/Im of the scattering amplitude.

Result[18]:

$$\sigma_{tot} \, (1 + \alpha^2) = (38. \pm 1.5) \, mb$$

(iii) simultaneous measurement of Coulomb - and nuclear scattering give values for σ_{tot} and α with absolute calibration.

$$\frac{dN}{dt} = L \cdot Acc \left\{ \frac{4\pi e^4}{t^2} \, G^4(t) + \frac{\sigma_{tot}^2}{16\pi} \, (1 + \alpha^2) e^{bt} + \alpha \, C.I. \right\}$$

(L luminosity, Acc detector acceptance, G (t) proton form factor, C.I. Coulomb-nuclear interference term). Measurements at 250 and 500 GeV give[19]

$$\sigma_{tot} = (40 \pm 2) \, mb$$
$$b = 13. \, GeV^{-2}$$
$$\alpha = (-0.06 \pm 0.1)$$

(iv) Together with accelerator measurements up to 70 GeV[20] the above result may be interpreted as

$$\alpha = (-0.0 \pm 0.1)$$

independent of energy, at very high energy.

3.3 Total cross sections

(i) no direct measurement of the total pp cross section has been
performed at the ISR. Values obtained from elastic scattering
via the optical theorem indicate [18,19]

$$\sigma_{tot} = (39. \pm 2.)\,mb$$

within these large errors the pp total cross section seems to
remain constant at energies between 10 and 1500 GeV (Fig.8).

(ii) total cross sections for $\pi^{\pm}$, $K^{\pm}$ and $p^{\pm}$ on protons are
measured up to 70 GeV[21] (Fig.9). σ_{tot} of particles and anti-
particles converge with rising energy; logarithmic extrapolation is compatible with equal cross sections at about
500 GeV.

(iii) for the first time high energy total hyperon-proton cross
sections were measured in hyperon beams at the CERN PS.
Results :

$$\sigma_{tot}(\Lambda p) = (34.6 \pm 0.4)\ mb\ at\ 6 - 20\ GeV\ [22]$$

$$\sigma_{tot}(\Sigma^- p) = (34.9 \pm 1.2)\ mb\ at\ 19\ GeV\ [23]$$

4. CONCLUDING REMARKS

The CERN ISR have opened the region from 200 to 2000 GeV to
experiments. Accelerator studies reach a level of precision to see
the fine structure of cross sections. Hyperon beams become feasible.

The author thanks Profs. G. Hite, W. Rühl and A. Vančura for
their hospitality at the Institute and his collegues at CERN for many
discussions.

REFERENCES

1. W.C. Middelkoop and A. Schoch, CERN - AR /INT SG/63-40 (1963)

2. G. Giacomelli, VII Rencontre de Moriond, (1972)

3. J.C. Sens, Proceedings of the Int. Conf. on High Energy
    Collisions, Oxford (1972)
    M.G. Albrow, D.P. Barber, A. Bogaerts, B. Bosnjakovic,
    J.R. Brooks, A.B. Clegg, F.C. Erné, C.N.P. Gee, A.D. Kanaris,
    A. Lacourt, D.H. Locke, P.G. Murphy, A. Rudge, J.C. Sens and
    F. Van der Veen, Phys. Letters 40B, 136 (1972)

4. British Scandinavian Collaboration, Contribution to the Oxford
    Conference (1972)

5. Saclay-Strassbourg Collaboration, unpublished (1972)

6. G. Neuhofer, F. Niebergall, J. Penzias, M. Regler, W.Schmidt-
    Parzefall, K.R. Schubert, P.E. Schumacher, M. Steuer and
    K. Winter, Phys. Letters 38B, 51 (1972)

7. L.G. Ratner, R.J. Ellis, G. Vannini, B.A. Babcock, A.D. Krisch
    and J.B. Roberts, Phys. Rev. Letters 27, 68 (1971)

8. A. Bertin, P. Capiluppi, A. Cristallini, M.D'Agostini-Bruno,
    R.J. Ellis, G. Giacomelli, C. Maroni, F. Mercatali, A.M. Rossi
    and G. Vannini, Phys. Letters 38B, 260 (1972)

9. G. Barbiellini, M. Bozzo, P. Darriulat, G. Diambrini-Palazzi,
    G. de Zorzi, M. Holder, A. McFarland, G. Maderni, P. Mery,
    S. Orito, J. Pilcher, C. Rubbia, G. Sette, A. Staude, P. Strolin
    and K. Tittel, Phys. Letters 38B, 294 (1972)

    M. Breidenbach, G. Charpak. G. Coignet, D. Drijard, G. Fischer,
    G. Flügge, Ch. Gottfried, H. Grote, A. Minten, F. Sauli,
    M. Szeptycka and E.G.H. Williams, Phys. Letters 39B, 654 (1972)

10. H. Boggied, K.H. Hansen, and M. Su, Nucl. Physics B27, 1(1971)

    H.J. Muck, M. Schachter, F. Selonke, B. Wessels, V. Blobel,
    A. Brandt, G. Drews, H. Fesefeldt, B. Hellwig, D. Mönkemeyer,
    P. Söding, G.W. Brandenburg, H. Franz, P.Freund, D. Luers,
    W. Richter, Phys. Letters 39B, 303 (1972)

11. B. Breidenbach, G. Flügge, K.R. Schubert and E.G.H. Williams
    CERN/NP 72-6, unpublished (1972)

12.   V. Chabaud, A. Eide, P. Lehmann, A. Lundby, S. Mukhin,
      J. Myrheim, C. Baglin, P. Briandet, P. Fleury, P. Carlson,
      E. Johansson, M. Davier, V. Gracco, R. Morand and D. Treille,
      Phys. Rev. Letters 38B, 441, 445, 449 (1972)

13.   A.B. Wicklund, J. Ambats, D.S. Ayres, R. Diebold, A.F. Greene,
      S.L. Kramer, A. Lesnik, D.R. Rust, C.E.W. Ward and
      D.D. Yovanovitch,
      Contribution to the Oxford Conference (1972)

14.   J.V. Allaby, A.N. Diddens, R.W. Dobinson, A. Klovning, J. Litt,
      L.S. Rochester, K. Schlüpmann, A.M. Wetherell, U. Amaldi,
      R. Biancastelli, C. Bosio and G. Mathiae,
      Phys. Letters 34B, 431 (1971)

15.   Yu.M. Antipov, R. Busnello, G. Damgaard, M.N. Kienzle-Focacci,
      W. Kienzle, R. Klanner, L.G. Landsberg, A.A. Lebedev,
      C. Lechanoine, P. Lecomte, M. Martin, V. Roinishvili, R.D. Sard,
      F.A. Yotch and A. Weitsch, unpublished (1972)

16.   U. Amaldi, R. Biancastelli, C. Bosio, G. Matthiae,
      J.V. Allaby, W. Bartel, G. Cocconi, A.N. Diddens, R.W. Dobinson,
      V. Elings, J. Litt, L.S. Rochester, and A.M. Wetherell,
      Phys. Letters 36B, 504 (1971)

17.   G. Barbiellini, M. Bozzo, P. Darriulat, G. Diambrini Palazzi,
      G. De Zorzi, A. Fainberg, M.I. Ferrero, M. Holder, A. McFarland,
      G. Maderni, S. Orito, J. Pilcher, C. Rubbia, A. Santroni,
      G. Sette, A. Staude, P. Strolin and K. Tittel,
      Phys. Letters 39B, 663 (1972)

18.   G. Barbiellini, M. Bozzo, P. Darriulat, G. Diambrini Palazzi,
      G. de Zorzi, A. Fainberg, M.I. Ferrero, M. Holder, A. McFarland,
      G. Maderni, S. Orito, J. Pilcher, C. Rubbia, A. Santroni,
      G. Sette, A. Staude, P. Strolin and K. Tittel, unpublished (1972).

19.   CERN-Rome Collaboration, unpublished (1972)

20.   G.G. Beznogikh, A. Bujak, L.F. Kirillova, B.A. Morozov,
      V.A. Nikitin, P.V. Nomokonov, A. Sandacz, M.G. Shafranova,
      V.A. Sviridov, Truong Bien, V.I. Zayachki, N.K. Zhidkov,
      and L.S. Zolin, Phys. Letters 39b, 411 (1972)

21.   S.P. Denisov, S.V. Donskov, Yu.P. Gorin, A.I. Petrukhin,
      Yu D. Prokoshkin, D.A. Stoyanova, J.V. Allaby and G. Giacomelli,
      Phys. Letters 36B, 415 (1971)
      S.P. Denisov, Yu.P. Dmitrevski, S.V. Donskov, Yu. P.Gorin,

Yu. M. Melnik, A.I. Petrukhin, Yu. D. Prokoshkin, V.S. Seleznev,
R.S. Shuvalov, D.A. Stoyanova and L.M. Vasiljev,
Phys. Letters 36B, 528 (1971)

22. S. Gjesdal, G. Presser, P. Steffen, J. Steinberger, F. Vannucci,
H. Wahl, K. Kleinknecht, V. Lüth and G. Zech,
Phys. Letters 40B, 152 (1972)

23. J. Badier, R. Bland, J.C. Chollet, T. Devlin, J.M. Gaillard,
J. Lefrançois, B. Merkel, R. Meunier, J.P. Repellin, G. Sauvage,
unpublished (1972)

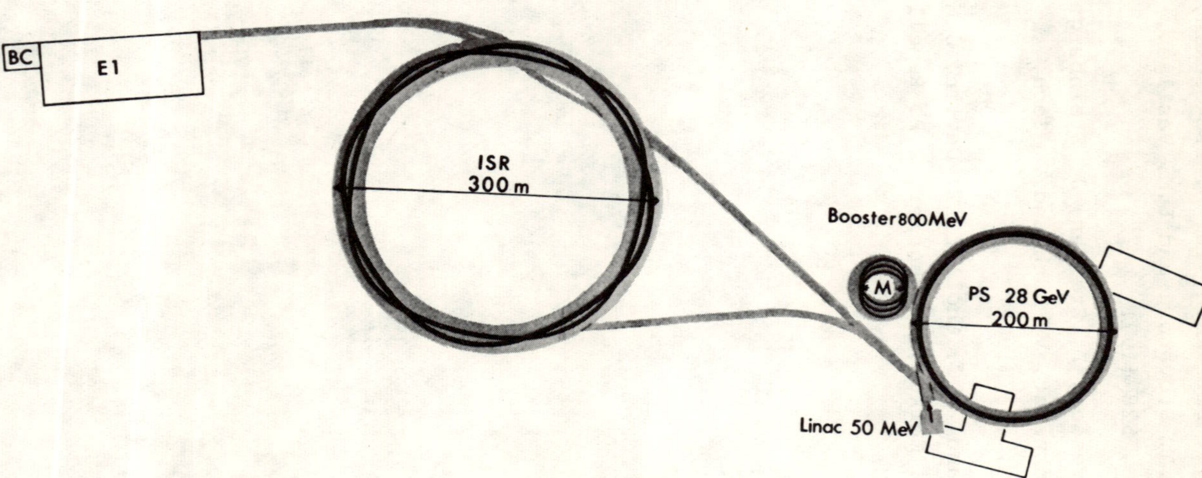

Fig. 1   Schematic layout of the CERN Intersecting Storage Rings

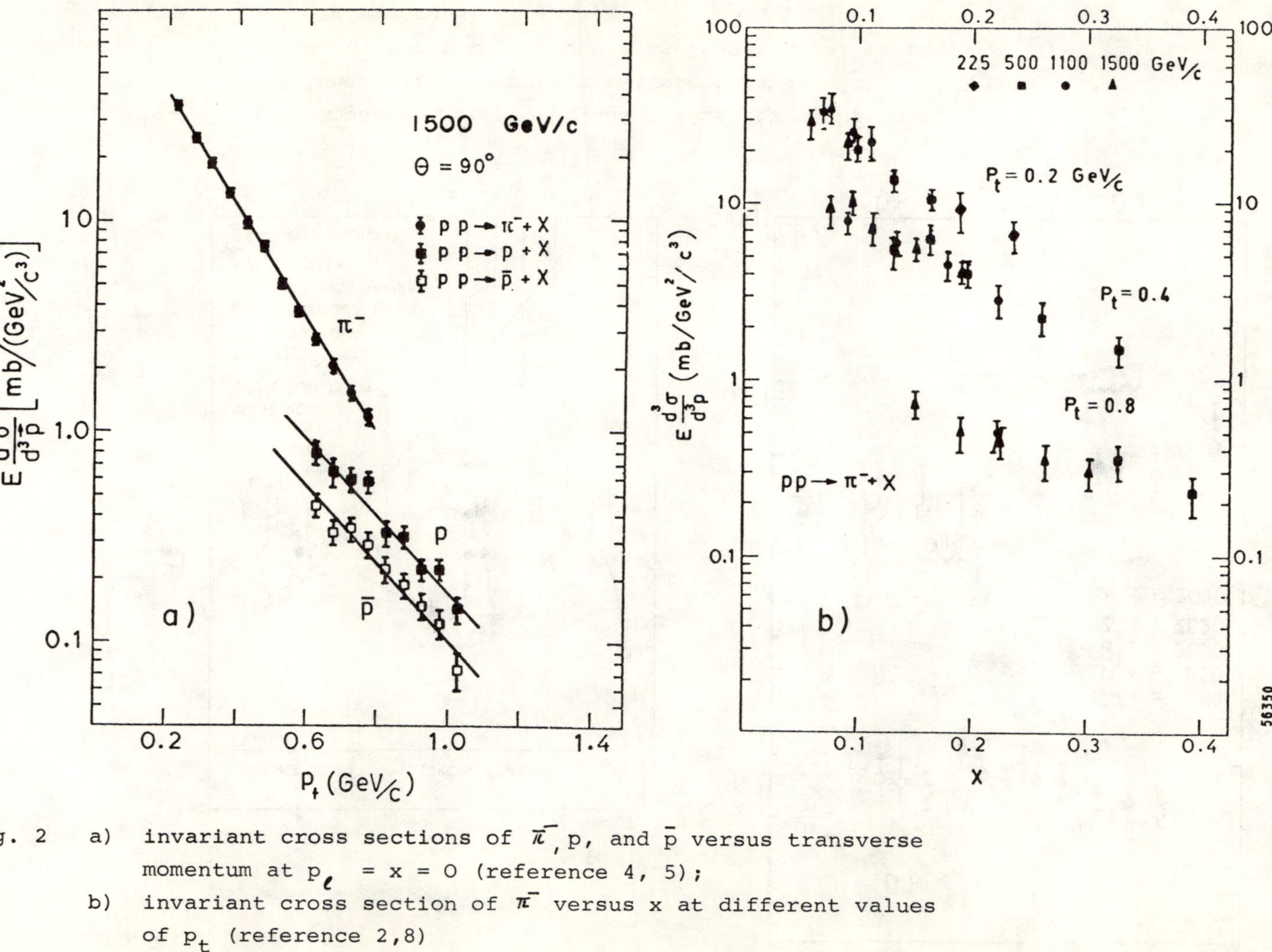

Fig. 2   a)   invariant cross sections of $\bar{\pi}^-$, p, and $\bar{p}$ versus transverse momentum at $p_\ell = x = 0$ (reference 4, 5);

   b)   invariant cross section of π^- versus x at different values of p_t (reference 2,8)

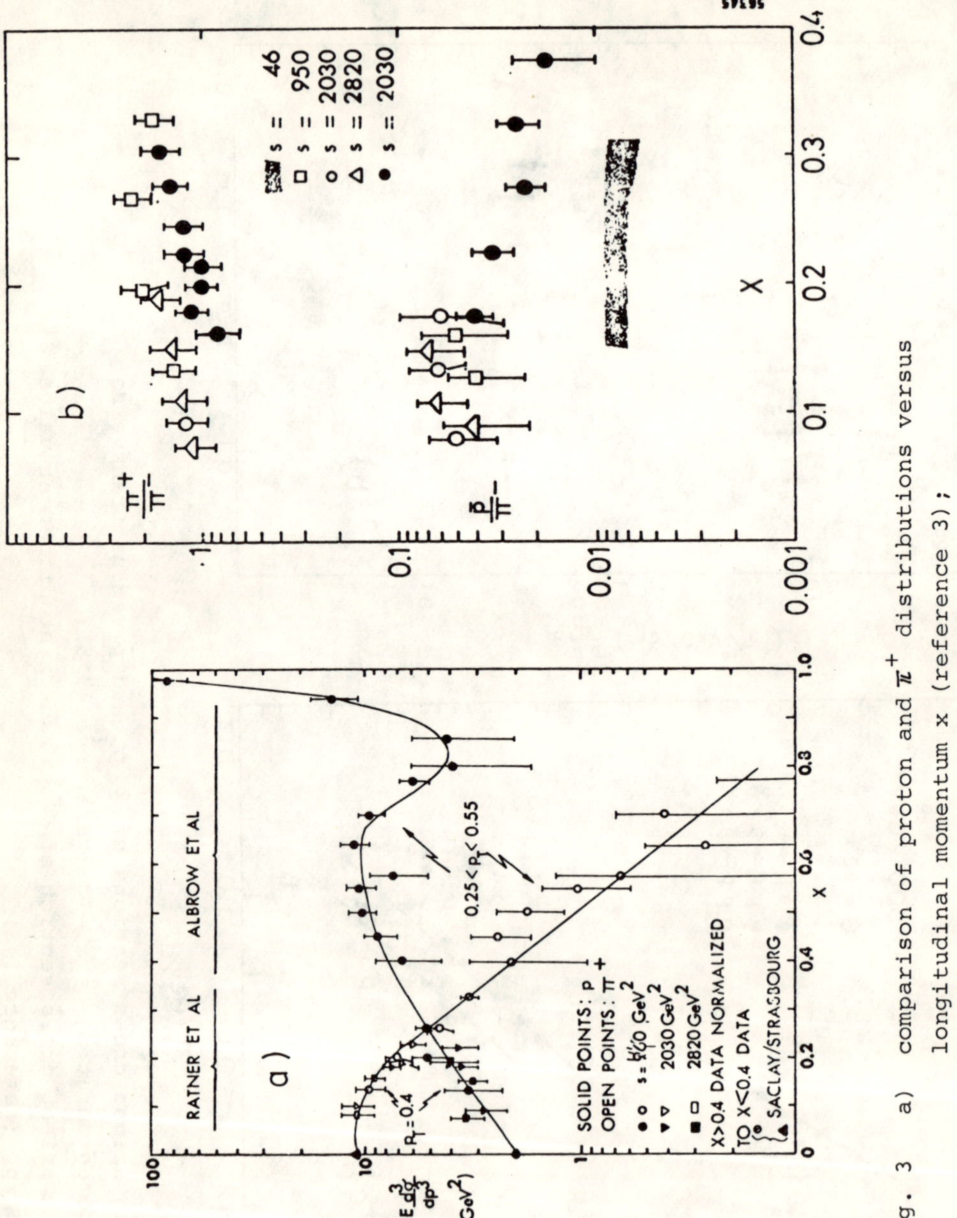

Fig. 3   a)   comparison of proton and $\bar{\pi}^+$ distributions versus longitudinal momentum x (reference 3);

      b) π^+/π^- and $\bar{p}/\pi^-$ ratio versus longitudinal momentum x (reference 3)

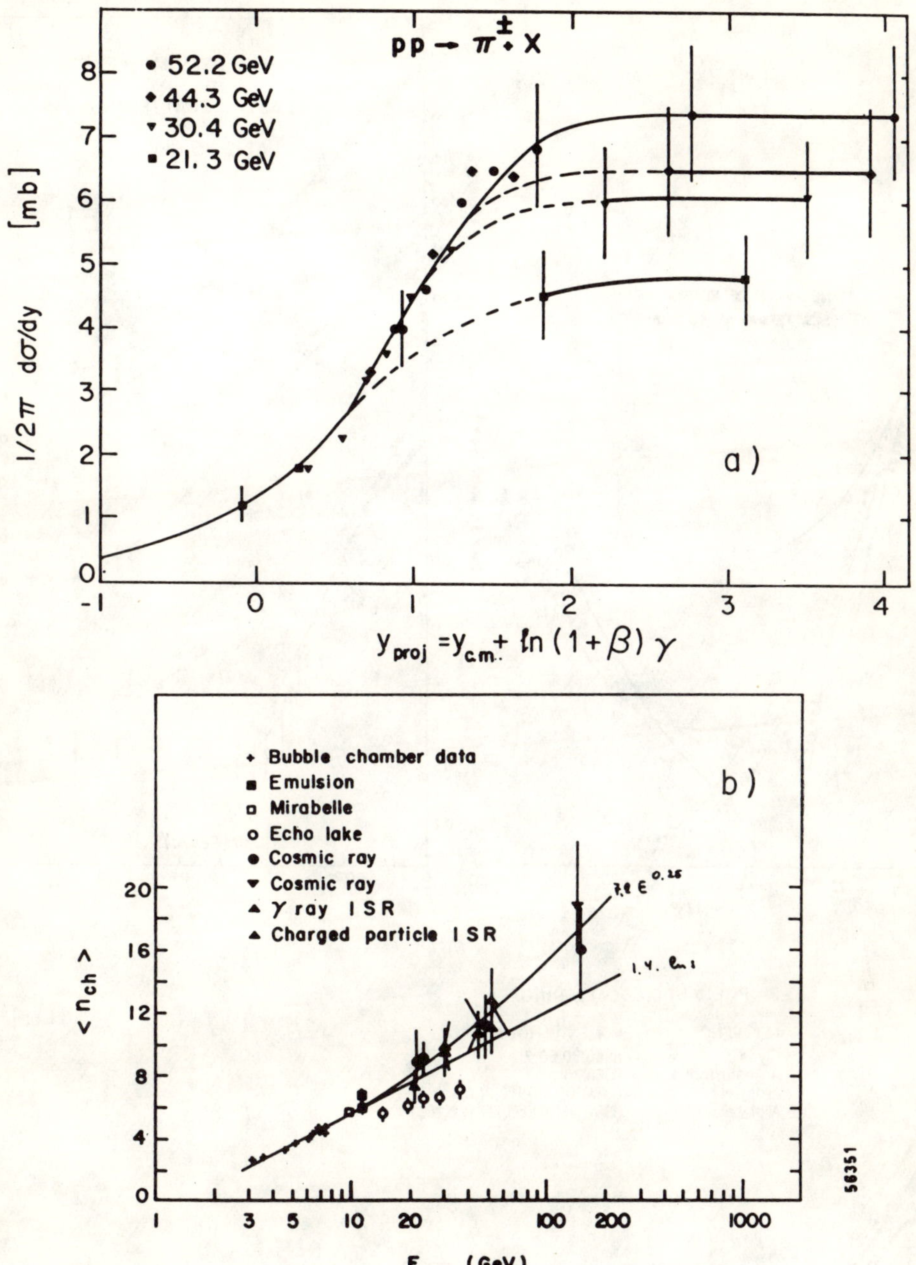

Fig. 4   a)   differential cross section versus repidity for different
              energies (reference 11);
         b)   average charge multiplicity from various experiments
              as a function of C.M. energy. The ISR values were
              obtained by integration of Fig.4a (reference 11)

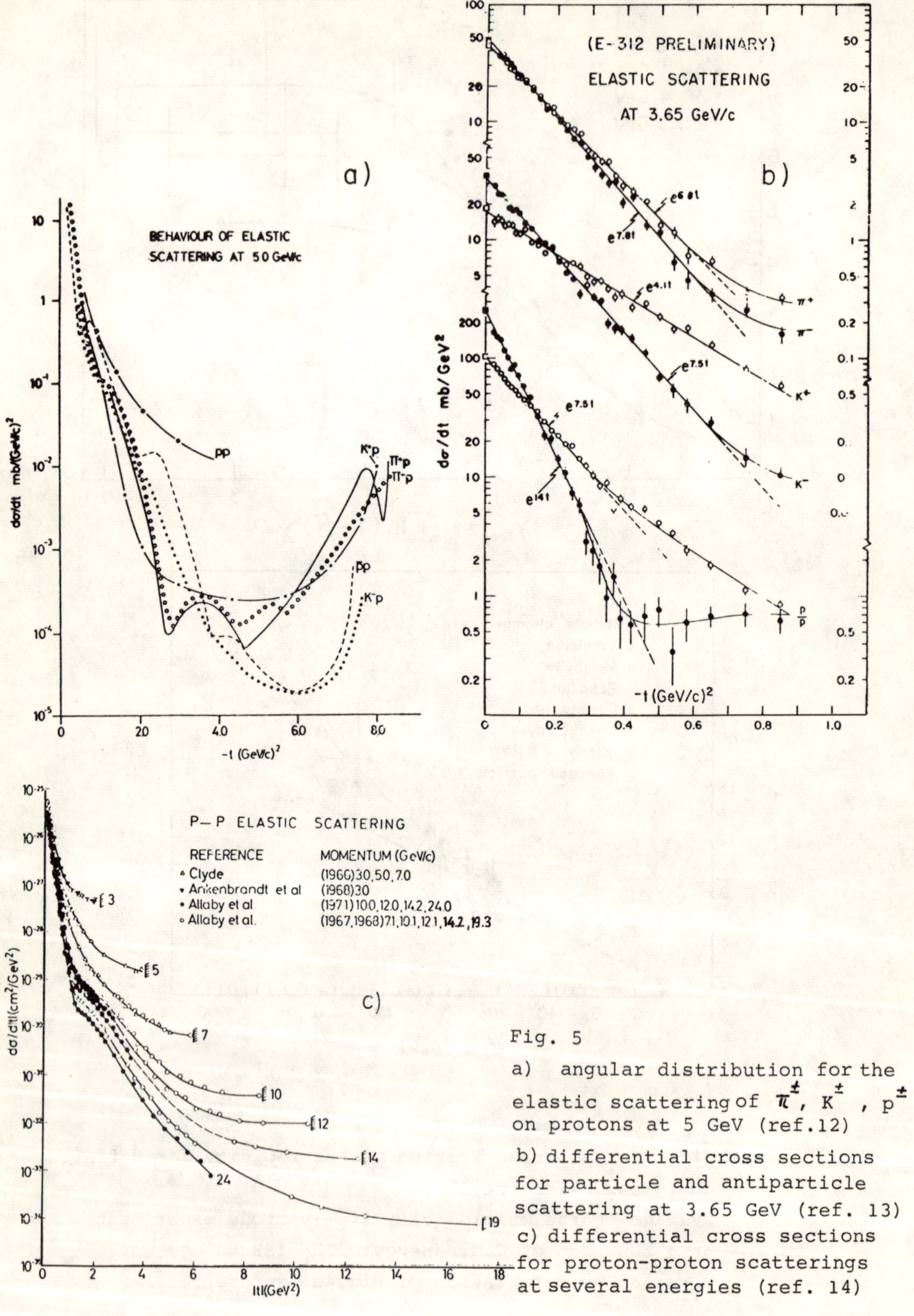

Fig. 5

a) angular distribution for the elastic scattering of $\pi^{\pm}$, $K^{\pm}$, $p^{\pm}$ on protons at 5 GeV (ref. 12)

b) differential cross sections for particle and antiparticle scattering at 3.65 GeV (ref. 13)

c) differential cross sections for proton-proton scatterings at several energies (ref. 14)

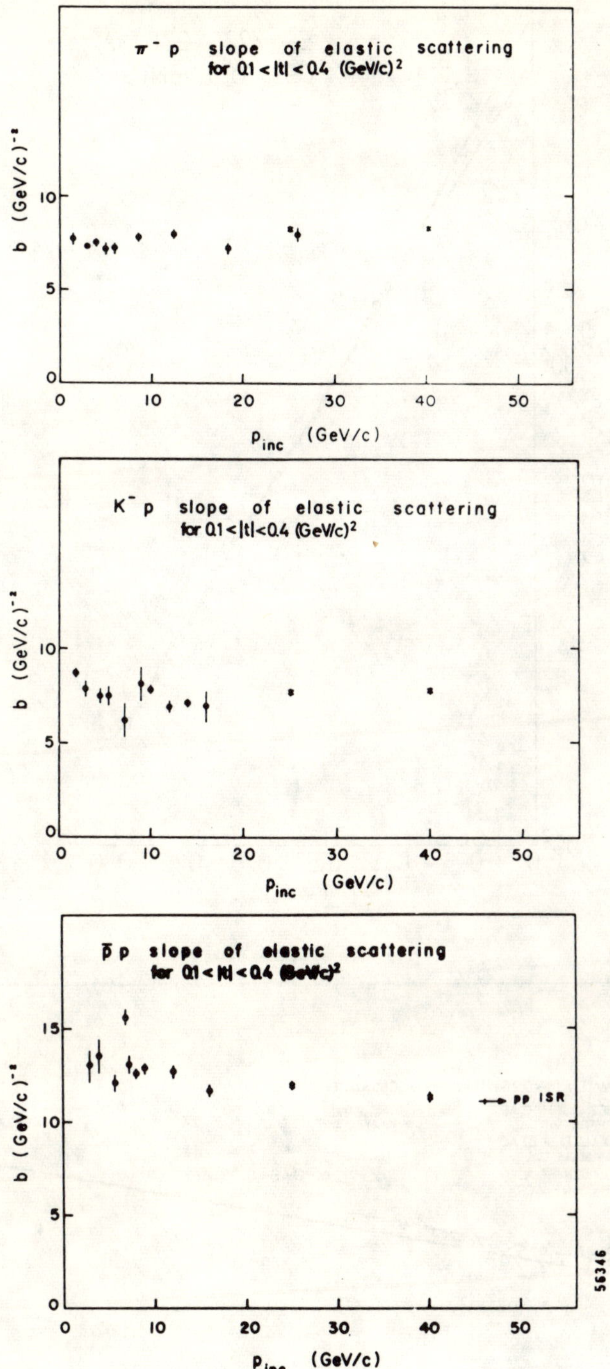

Fig. 6   Compilation of slope parameters b in elastic scattering as a
function of momentum of the incident π^-, K^- or $\bar{p}$ (ref. 15)

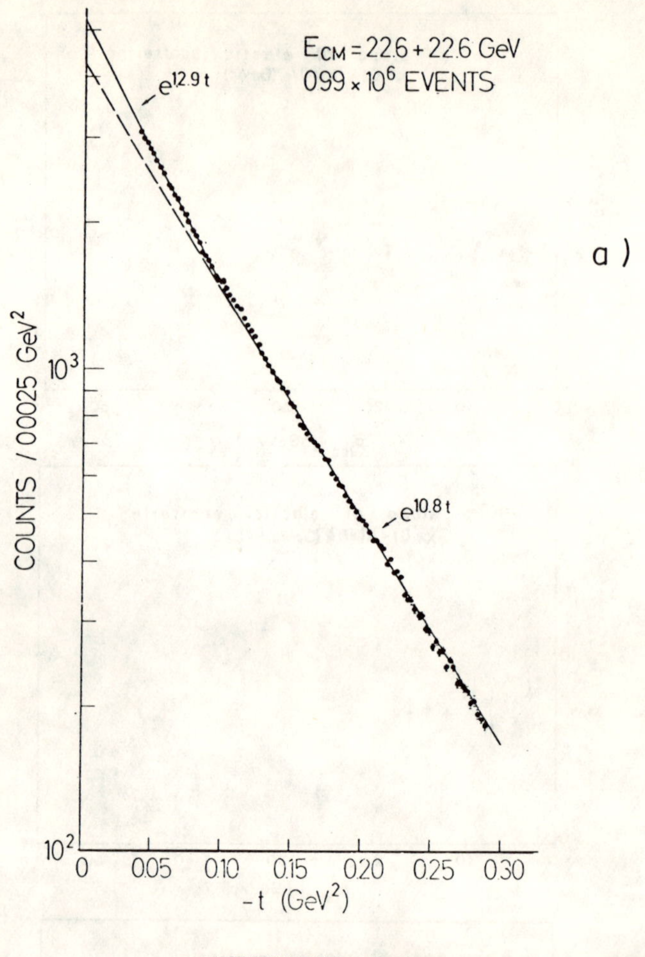

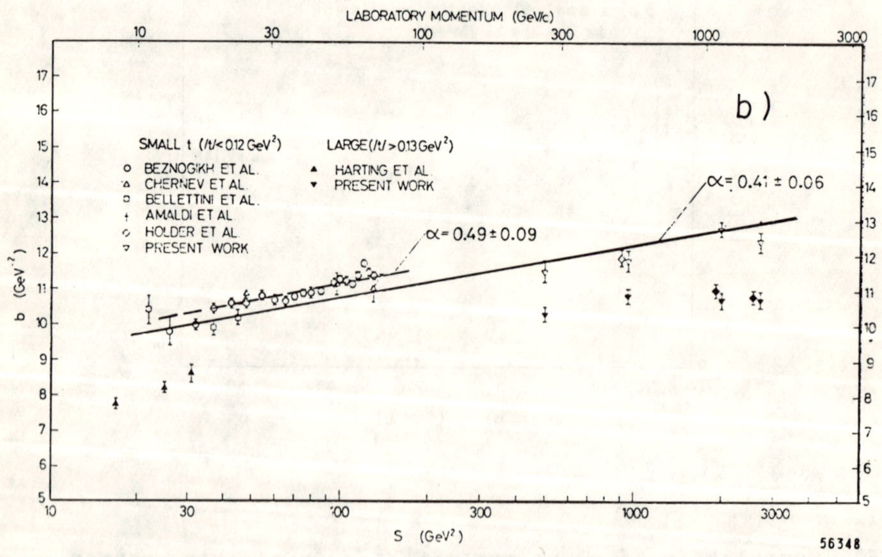

Fig. 7   a)   t-distribution in pp scattering at 1000 GeV (ref. 17)
b)   Compilation of slope parameters b in pp scattering
(ref. 17)

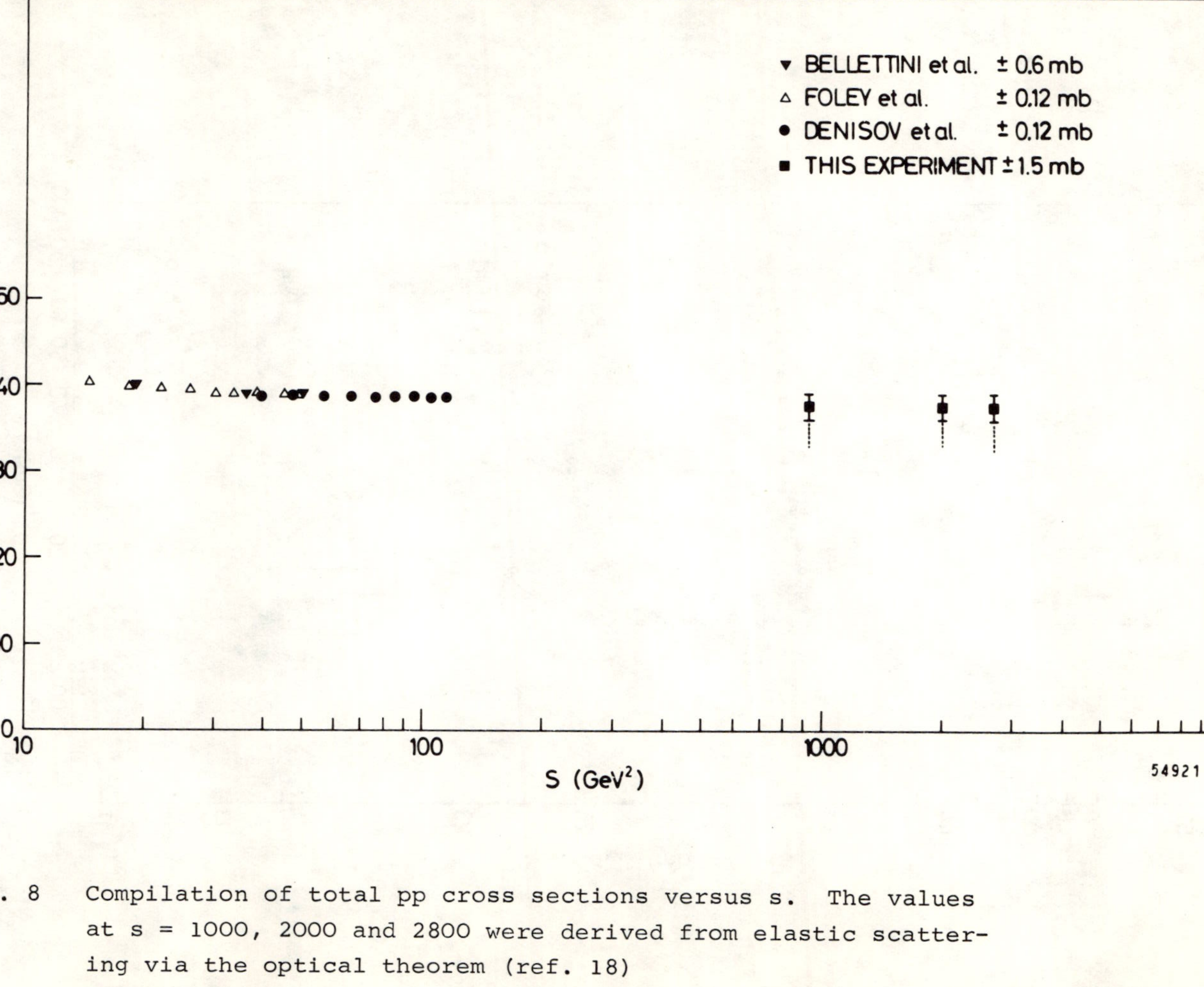

Fig. 8   Compilation of total pp cross sections versus s. The values
at s = 1000, 2000 and 2800 were derived from elastic scattering via the optical theorem (ref. 18)

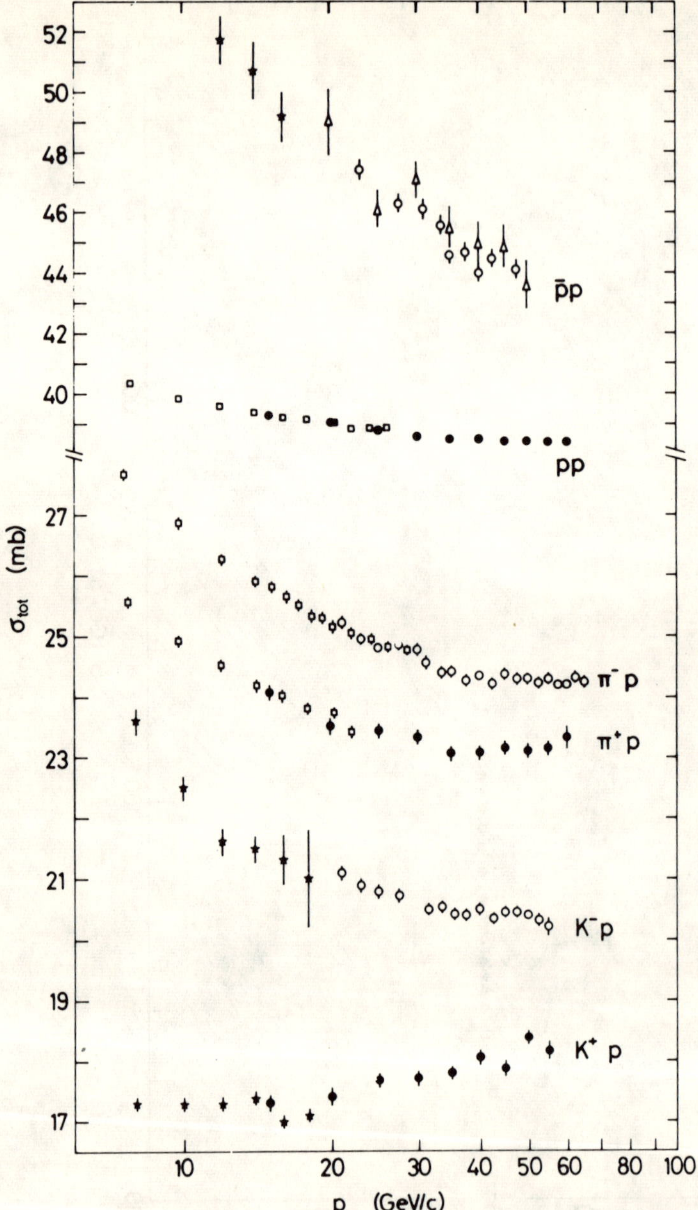

Fig. 9   Compilation of total cross section on protons versus
          incident momentum (ref. 21)

G. Höhler

Institut für Theoretische Kernphysik der
Universität Karlsruhe, Germany

1. INTRODUCTION

In these lectures we shall discuss several topics, which are related to the recent work of the Karlsruhe group.

There are two different approaches in πN phenomenology, which are not always clearly separated: amplitude analysis and the study of special models.

1.1 Amplitude Analysis

It is the aim of amplitude analysis to determine the different scattering amplitudes from the experimental data. This cannot be done without theoretical assumptions, since a common phase of the 4 amplitudes is not measurable in scattering experiments (except at very small angles via Coulomb interference and at $t=0$, using the optical theorem). In order to determine the remaining 7 numbers at each energy and angle in a unique way, one needs 8 measurements, since the relations are non-linear[1].

At present a complete set of data is available only at 6 GeV/c in the small t-range: $0.23 \leq |t| \leq 0.50$ $(GeV/c)^2$. Amplitude analysis of these data are treated in Refs.[2-4]. Since spin-rotation experiments are very difficult, further progress along these lines will be slow. Some of the general theoretical principles have already been used in the above version of amplitude analysis (Lorentz invariance, isospin

invariance, unitarity at t=0). As a next step one can try to exploit
analyticity. This leads to a rather accurate information on the forward amplitudes. But applications at t=0 are only possible, if the
results of phase shift analysis are taken into account. These questions
are treated in § 2.

1.2 Phase Shift Analysis[5-7]

Below 2 GeV/c one has a fairly complete knowledge of only
5 quantities: differential cross sections for elastic and charge-
exchange scattering and elastic polarizations. Nevertheless phase shift
analysts offer a result, from which the 4 complex-valued amplitudes are
completely determined.

The main additional input is the assumption that up to 2 GeV/c
all partial waves with angular momenta $\ell > 5$ are negligible. Since a
large number of solutions is found at intermediate energies, one starts
with a unique solution in the elastic region and calculates a unique
continuation to higher energies by imposing a "smoothness" condition
for the energy dependence of the partial waves (shortest path method).
In the CERN analysis[5] an important intermediate step is based on partial
wave dispersion relations, which follow from the assumption of Mandelstam analyticity. (However a success of the analysis is only a weak
argument in favour of this assumption).

A critical review of the present situation in phase shift
analysis has recently been given by F. Wagner[7].

1.3 Impact Parameter Representation

Phase shift analysis is well adapted to the low and intermediate
energy region, where nucleon resonances are described in a simple way,
since they are seen in a single partial wave, and the number of relevant
partial waves is rather small. Above 2-3 GeV/c all advantages are lost
and the partial waves give a very complicated description of the simple
t-dependence of the amplitudes.

A natural modification is the Fourier-Bessel integral representation of the amplitudes. The transforms are related to the partial waves and they depend on a parameter, which corresponds to the impact parameter in the classical limit. Since the parameter has the dimension of a length, this description is well suited for the formulation
of "geometrical" models of high energy scattering, which were investigated by many authors in recent years.

The impact parameter representation will be treated in § 3.

1.4 Test of Models

The determination of the amplitude is of course not the final aim. One would like to understand the "mechanism" of high energy scattering. This means that one wants to have a model, from which the main features of the amplitudes can be "derived". There are certainly many possibilities. Some are based on intuitive pictures, but the most satisfactory one would be a model, in which the amplitudes can be calculated from field theory. There is little hope that this can be achieved in the near future and one can expect that the study of phenomenological models will be an important intermediate step.

A very large number of models has been proposed in the literature and in almost all cases a "good agreement with the data" is reported. Even if one excludes those, which disagree with subsequent data, the number remains large and the question arises, which models should be taken seriously and which are not more than "numerology".

The best test is of course a comparison with the result of amplitude analysis. But one cannot hope that complete sets of measurements will be performed in large kinematical regions. Furthermore the application of analyticity is limited to t-values not far from t=0, since an analytic continuation is necessary in applications of fixed-t dispersion relations at t≠0. Even the highly developed method treated in Prof.Ciulli's lecture will not help to go to $|t|$ -values much larger than 0.5 $(GeV/c)^2$, at least with the present accuracy of the data.

Therefore one has to test the models also by comparing their results with the data. This is not always done in a very critical way and it might be useful to mention the following points.

i) Many models contain functions, which are restricted only weakly. This is true for instance for the residue functions in Regge pole models. Another example are the dual absorptive models, where the "definition" of the "Pomeron" and "Non-Pomeron" parts is to a large extent arbitrary. The authors start from "plausible" assumptions on these functions, leaving a few adjustable parameters, which are determined by a fit to the data. Of course the papers always show that a large number of data points was fitted with a few parameters and this is taken as evidence for the validity of the model.

We conclude that one should not only ask for the number of parameters but also for the number of _functions_, which are only weakly restricted in a model.

ii) A good fit to the existing data does not confirm all assumptions of a model, but only part of it. This follows from the fact that in large kinematical regions the differential cross sections and elastic polarizations are dominated by two amplitudes. The contributions of the others are comparable with the errors or even negligible.

We conclude that a critical test of a model should not be made as usual by a direct comparison with the data. Instead the fit should be shown for those _combinations of data_, which occur in amplitude analysis, for instance $(P_+ d\sigma_+/dt \pm P_- d\sigma_-/dt)/\sin\theta$ or $(P_0 d\sigma_0/dt)/\sin\theta$. It will be seen that "good" or "reasonable" fits to the P- and $d\sigma/dt$-data sometimes indicate a discrepancy, if the above combinations are plotted. Furthermore this method shows clearly, which part of the model has really been tested.

iii) Many authors tested their models by inserting their results into finite energy or continuous moment sum rules or they used these sum rules for the determination of parameters[8]. The sum rules are certainly useful tools, but one should notice that fixed-t dispersion relations are preferable, if one wants to include a _sensitive_ test of the compatibility between the input and the analyticity requirement.

There is no clear distinction between models and parametrizations of amplitudes in terms of quantities, which have no immediate physical interpretation. For instance the Regge pole fit by Barger and Phillips[8] was originally an attempt to determine the parameters of the high energy expansion. Then it turned out that its prediction disagrees with the data above 20 GeV/c. Nevertheless this fit is still useful as a parametrization of the amplitudes at 2-20 GeV/c. But one should notice that its "Regge poles" are "effective poles", which are not simply related to the true Regge poles (if these are really dominating at very high energies).

More details on these questions will be found in §§ 2 and 4.

2. πN AMPLITUDES AT SMALL MOMENTUM TRANSFER

A detailed review on this question has been given in my lecture at the International Seminar at Dubna in 1971[9]. Therefore I shall mainly treat results obtained during the last year. The notation is in the Appendix.

2.1 Forward Amplitudes $C^{\pm}$

The imaginary parts are given by the total cross sections

$$\mathrm{Im}\ C^{\pm} = k\sigma^{\pm} \equiv \frac{k}{2}\ (\sigma_- \pm \sigma_+). \tag{2.1}$$

σ^+ decreases from 27.8 mb at 5 GeV/c to 24.5 mb at 20 GeV/c and then it remains practically constant up to 60 GeV/c. This momentum dependence cannot be described by a P,P' Regge model fit. Other parametrizations have been proposed by many authors, but it seems that there is no model, which makes a reliable prediction for the energy dependence above 60 GeV/c.

Contrary to the optimistic statements of other authors we do not think that measurements of the real parts via Coulomb interference give strong restrictions on the total cross sections from the dispersion relation

$$\mathrm{Re}\ C^+ (\omega) = \mathrm{Re}\ C^+ (1) + \frac{4\pi f^2 k^2}{m(1-\omega_B^2)(\omega^2-\omega_B^2)} + k^2 \frac{2}{\pi} \int_0^\infty \frac{\sigma^+(k')dk'}{k'^2 - k^2} \tag{2.2}$$

where $\omega_B = 1/2m$. Unfortunately this relation is not a "crystal ball that can view the road to Asymptopia" [10], since it was shown that the data for the real parts in Ref. [11] essentially determine only an integral over the total cross sections [12].

Further information on σ^+ can be expected rather soon from experiments at Batavia. Since there are indications for a constancy of the total pp-cross section above 60 GeV/c, it is interesting to remember a possible consequence of constant total πN cross sections [12]: if σ^+ goes to a constant rapidly enough such that the integral converges, Lehmann's sum rule predicts [13] that Re C^+ goes to a constant in the high energy limit

$$\mathrm{Re}\ C^+(\infty) = \mathrm{Re}\ C^+ (1) + \frac{4\pi f^2}{m(1-\omega_B^2)} - \frac{2}{\pi} \int_0^\infty dk\left[\sigma^+(k) - \sigma^+(\infty)\right] \tag{2.3}$$

When Igi proposed the Regge pole P' in 1962, he started from the same sum rule. But his argument was not conclusive, since he assumed Re $C^+ (\infty) = 0$, without giving a justification [14].

Nowadays P' is supported mainly by the belief that the f-meson has an exchange degenerated trajectory. However, it is not possible to determine the P'-parameters in a reliable way from $\tilde{\pi}$ N scattering data, since there must be another term of comparable magnitude, which cancels the energy dependence of the P' contribution to σ^+ between 20 and 65 GeV/c. (This difficulty is ignored for instance in Ref.[15]).

The original analysis of the Coulomb interference data of Ref.[11] had several difficulties, mainly because of the parametrization and a normalization error of the cross sections. A reanalysis led to the result[16] that the test of causality is not as good as it was thought. It was pointed out that these data contain interesting information on the amplitudes, which can be derived, if the validity of the dispersion relation is assumed.

The detailed investigation of the dispersion relation for the charge-exchange forward amplitude[17] was recently continued by Dronkers and Kroll (Ref.[18]). There is a good compatibility between the dispersion relation, most of the charge-exchange forward cross sections and the total cross-sections up to 20 GeV/c and one obtains a result for the integral

$$J = \frac{1}{2\pi^2} \int_0^\infty \frac{dk}{\omega} \, \sigma^-(\omega) = -0.0485 \pm .0005,$$

(2.4)

where the error does not include uncertainties of the total cross sections below 20 GeV/c. Assuming a power law for $\sigma^-(k)$ above 20 GeV/c, one obtains from (2.4) practically agreement with the power law, which is known to be valid at 8-20 GeV/c. Unfortunately a direct fit to the Serpuchov data gives a markedly different value for the exponent (-0.31 $\pm$ .04 instead of -0.45). If the fit to the data at 20-60 GeV/c is inserted into the dispersion relation, one finds a large discrepancy (Ref.[9]), as long as only the above mentioned statistical errors of the total cross sections above 20 GeV/c are taken into account. However the agreement is restored, if one assumes that the total cross section data have to be corrected by the systematic errors, which were given by the authors.

2.2 Derivatives of the Amplitudes $C^\pm$ at t=0

Derivatives of fixed-t dispersion relations at t=0 are of interest[19], because information on $(\partial/\partial t)\,\mathrm{Im}\,C^+$ is available from the slopes of $d\sigma/dt$ at high energies and also from the Coulomb interference data [11]. It was found that the phase shifts lead to a systematic discrepancy above 1 GeV/c, presumably because the derivative depends strongly on high partial waves, some of which are neglected in phase shift analysis. The remarkable result that $\text{Re } C^+$ _increases_ in absolute value at small $|t|$ was confirmed in recent amplitude analyses [20].

The derivative is also of interest for the determination of the Σ-term in current algebra [21].

The situation is less favourable in the case of $(\partial/\partial t)C^-$, where information at high energies is not available. Jakob determined an average over $(\partial/\partial t)\text{Im}C^-$ in the GeV-range (mentioned in Ref. [9]), confirming the rapid t-dependence, which is indicated by the difference of the cross sections. (see Fig. 5 in § 4)

$$\frac{d\sigma_-}{dt} - \frac{d\sigma_+}{dt} \sim \text{Re}\,(C^+C^{-*}) \tag{2.5}$$

At present the experimental information on the "cross-over" at high energies is not much better than in the first detailed discussion of this effect [22]. Halzen and Michael[2] found an indication for an energy dependence of the cross-over zero, but this is not confirmed in an analysis of the most accurate high energy data [16]. The information following from an isospin bound will be treated in § 4. Accurate new measurements of the difference (2.5) at 3.6 - 6 GeV/c were recently completed at Argonne, but the results are not yet available.

2.3 The Amplitude B^- at t=0

In their careful analysis of the charge-exchange data Dronkers and Kroll[18] determined not only the forward cross sections but also the slopes of $d\sigma_0/dt$ at t=0. Combining this result with elastic polarization data they obtained B^- at t=0 for different energies. An attempt to test the compatibility with the forward dispersion relation and the phase shifts led to some difficulties at 2-3 GeV/c, where B^- has probably a similar resonance variation as it is known for C^- (Fig. 1).

2.4 The Amplitude A^+ - Helicity Conservation [23,4]

The determination of the amplitude A^+ is closely related to that of the s-channel flip amplitude F^0_{+-} (Appendix), since

$$A^+ \simeq 4\sqrt{\pi}\,m\,\left(2k\frac{F^0_{+-}}{\sqrt{-t}} - F^0_{++}\right) \tag{2.6}$$

and F^o_{++} is approximately known. Polarization data give the component of A^+ orthogonal to C^+ in the complex plane

$$A^+_\perp \simeq \frac{2\sqrt{2\pi}\, m\omega}{\sqrt{-t}} \frac{\sigma_+ P_+ + \sigma_- P_- - \sigma_o P_o}{\sqrt{\sigma_+ + \sigma_-}} \;,\quad \sigma \equiv \frac{d\sigma}{dt} \tag{2.7}$$

The component of F^o_{+-} orthogonal to F^o_{++} follows from (2.6), since F^o_{++} is practically parallel to C^+ in the t-range of interest.

Measurements of the spin-rotation parameters are necessary for the determination of the components $A^+_\parallel$ and $(F^o_{+-})_\parallel$. For theoretical discussions it is useful to introduce the parameters S and T (Appendix)

$$-S + iT = (R + iA)\exp(i\Theta_p) = \sqrt{R^2 + A^2}\,\exp(\Theta_p + \phi)i \;, \tag{2.8}$$

where Θ_p is the proton recoil angle in the lab. system. Since the measurements of P are more accurate as those of A and

$$P^2 + R^2 + A^2 = 1, \tag{2.9}$$

S and T are calculated from R and P

$$T = \sqrt{1-P^2}\,\sin(\Theta_p + \phi), \quad \cos\phi = R/\sqrt{1-P^2} \tag{2.10}$$

$$1 - S^2 = T^2 + P^2 = 4x^2(1 + x^2)^{-2}, \quad x = |F_{+-}|/|F_{++}|$$

Since we are interested in the isospin even combinations, we have to consider

$$2\sigma^+ T^+ = \sigma_+ T_+ + \sigma_- T_- - \sigma_o T_o \;,\; \text{etc.} \tag{2.11}$$

and the components of A^+ and F^o_{+-} follow from

$$A^+_\parallel \simeq 4\sqrt{\pi}\, m\sqrt{\sigma^+}\left(\frac{kT^+}{\sqrt{-t}} - 1\right), \quad (F^o_{+-})_\parallel \simeq \frac{\sqrt{\sigma^+}}{2} T^+ \tag{2.12}$$

The result for $A^+_\perp$ (Fig.2) was calculated from the new polarization data of Ref.[24]. It is considerably more accurate than earlier plots of this quantity[23,25]. The Barger-Phillips fit[8] is not reliable for A^+, since it predicts a change of sign near 15 GeV/c[25], whereas the data indicate that $A^+_\perp$ remains almost constant between 6 and 14 GeV/c.

Eq.(2.10) gives a result for x , which is consistent with a-symptotic s-channel helicity conservation (SHC), but x could as well go to a finite value and there is not even a strong argument against t-channel helicity conservation (Fig.3).

Barger and Halzen[25] claimed to have found a "resolution of the πN helicity conservation question from polarization data alone". This statement is misleading, since polarization data determine $(F^{O}_{+-})_{\perp}$, whereas $(F^{O}_{+-})_{\parallel}$ is relevant for SHC. P^{+} goes to zero at high energies, even if SHC is <u>not</u> fulfilled.

The same authors tried to prove from $\pi^{-}p$ data for the T-parameter at 6 GeV/c that SHC is approximately valid and t-channel helicity conservation is ruled out. However

i) from $T_{-} \approx 0$ one can only conclude that $(F^{O}_{+-}) \approx -(F^{1}_{+-})$ and the dominating T=1 amplitude is not negligible,

ii) A value at a single energy does not allow a conclusion on the asymptotic behaviour, one needs the trend of the energy dependence.

The point at 14 GeV/c in Fig. 3 was estimated from the R_{-}-data at 16 GeV/c, the polarization data at 14 GeV/c and the assumption of flip dominance

$$d\sigma_{o}/dt \approx |F^{1}_{+-}|^{2} \quad \text{at} \quad 0.15 \lesssim |t| \lesssim 0.30 \; (\text{GeV/c})^{2}. \quad (2.13)$$

In Ref.[23] it was shown that the conjecture $A^{+} \to 0$ at high energies was compatible with all data and that SHC is a consequence. This result was one of the arguments of Gilman et al.[26], when they proposed SHC for a more general class of reactions.

$A^{+} \to 0$ is attractive, since it has the consequence that the usual <u>subtraction is not required</u> in the dispersion relation. A comparison of $A^{+}(\nu = 0, t)$ as determined from the subtracted relation with the same amplitude as following from the known part (0-2 GeV/c) of the unsubtracted relation led to encouraging results[23]. We are investigating at present, whether the new data lead to serious objections against the unsubtracted relation.

In the Regge pole language $A^{+} \to 0$ means that P and P' are decoupled from A^{+}. But lower-lying trajectories must give an appreciable contribution, since $A^{+}_{\perp}$ is clearly different from zero up to 14 GeV/c. A decoupling of the f-trajectory should also be seen in backward dispersion relation. Two independent attempts to determine the J=2 $\pi\pi N\bar{N}-$ amplitude on the left hand cut led to somewhat different results[27].

Further details on the question of SHC can be found in Mueller's review article[28] and, as far as the most recent amplitude analyses are concerned, in Refs.[3,4].

2.5 Dispersion Relation Methods

Continuing earlier attempts in this direction[19,29] several authors[30-33] work at present on the problem to determine invariant amplitudes, which fulfill dispersion relations and are compatible with all data. The methods are different. Some authors use fixed-t dispersion relations. Other work with dispersion relations between modulus and phase and have to consider carefully the effect of the zeros of the amplitudes. Both methods are limited to rather small $|t| \lesssim 0.5 \, (GeV/c)^2$.

The results are very interesting, since they give the absolute phases of the amplitudes. In particular one can expect that $\arg C^+$ will be reliable and this allows to determine from experimental data the phases of all other amplitudes.

It is true that the different phase shift analyses essentially agree, even in their predictions for quantities, which have not yet been measured[34] *). But the methods contain possibly the same systematic error, therefore the independent test of the compatibility with analyticity is important, even below 2 GeV/c.

At 6 GeV/c and t = 0.2 $(GeV/c)^2$ Hecht et al.[32] found $\arg C^+ \approx 113^o$, which gives $|Re\ C^+/Im\ C^+| \approx 0.4$. The rather large real part should be a warning against the usual neglection of real part contributions to the diffraction peak.

For instance in Harari's two-component duality model[36] the dip in πN scattering cross sections near t = - 0.6 $(GeV/c)^2$ is attributed to the _imaginary part_ of the Non-Pomeron _no-flip amplitude_. However this solution of the "elastic puzzle" is doubtful. The dip-bump-structure is clearly seen at 1-3 GeV/c and goes over into a shoulder at higher momenta. At 1-2 GeV/c the amplitudes can be reconstructed from phase shifts and it turns out that the dip is mainly caused by the sharp rise of the _real part_ of the s-channel _flip amplitude_. Around t = - 1 $(GeV/c)^2$ $(Re\ F_{+-}^o)^2$ is the largest contribution to $d\sigma/dt$. (See Gig.6 in Ref.[37]).

It is interesting to study the k-dependence of $d\sigma/dt$ at this t-value. There is a rapid decrease $\sim k^{-2}$ up to about 7 GeV/c and then the curve flattens abruptly, indicating that another contribution becomes dominant (Fig.7 in Ref.[37]).

---

*) On the other hand new data sometimes disagree with the predictions[35]

Another aspect of the same phenomenon is the strange t-depend-
ence of the "effective α " in $\bar{\pi}N$ scattering (see Fig.9 in Ref.[38]). An
effective α for all data above a certain energy is only meaningful,
if one has checked that $d\sigma/dt$ follows a power law $\sim k^{2\alpha-2}$.

3.   IMPACT PARAMETER REPRESENTATION, EIKONALS, OPTICAL POTENTIALS

3.1 The Impact Parameter Representation

As discussed in the Introduction the partial wave expansion is
not suitable at high energies. In order to find another representation,
which is better adapted to the main features of diffraction scattering,
we insert the integral representation of the Legendre function

$$\int_0^\infty J_0\left(x\sin\tfrac{\theta}{2}\right) J_{2\ell+1}(x)\,dx = P_\ell(\cos\theta), \quad -1 < \cos\theta < 1 \tag{3.1}$$

$$= 0, \quad \cos\theta < -1$$

into the partial wave series (spinless case)

$$F(s,t) = q^{-1} \sum_{\ell=0}^\infty (2\ell+1)\, T_\ell(s)\, P_\ell(\cos\theta) \tag{3.2}$$

and obtain

$$F(s,t) = q\int_0^\infty J_0\left(\sqrt{-t}\,\beta\right) f(s,\beta)\, d\beta, \quad 0 > t > -4q^2$$

$$= 0, \quad t < -4q^2 \tag{3.3}$$

where $f(s,\beta)$ is given by a Neumann series

$$f(s,\beta) = (qb)^{-1} \sum_{\ell=0}^\infty (2\ell+1)\, T_\ell(s)\, J_{2\ell+1}(2bq), \quad \beta = b^2 \tag{3.4}$$

Inserting (3.3) into the projection formula

$$T_\ell(s) = \tfrac{1}{2}q \int_{-1}^{+1} F(s,t)\, P_\ell(z)\, dz = (4q)^{-1} \int_{-4q^2}^{0} F(s,t)\, P_\ell\left(1+\tfrac{t}{2q^2}\right) dt \tag{3.5}$$

and using

$$\int_0^1 P_\ell (1-2x^2) J_0(xy)\, x\, dx = \frac{1}{y} J_{2\ell+1}(y) \tag{3.6}$$

one finds the inverse of (3.4)

$$T_\ell(s) = \int_0^\infty f(s,\beta) J_{2\ell+1}(2\beta q)\, d(2\beta q) \tag{3.7}$$

The inverse of (3.3) reads

$$f(s,\beta) = (4q)^{-1} \int_{-4q^2}^0 F(s,t) J_0(\sqrt{-t}\beta)\, dt \tag{3.8}$$

T_ℓ is the dimensionless partial wave amplitude, which can be expressed by the real phase shift δ_ℓ and the absorption parameter η_ℓ

$$T_\ell(s) = (\eta_\ell \exp(2i\delta_\ell) - 1)/2i \tag{3.9}$$

The above formalism is due to Cottingham and Peierls[39], Predazzi[40] and Adachi et al.[41]. In the original proposal by Blanken-becler and Goldberger[42] the integral in (3.8) was extended to $-\infty$. This leads to a difficulty in applications, since one needs the scattering amplitude in the unphysical region $t < -4q^2$, where one has only very little information. Furthermore eq. (3.4) is not valid, which allows a simple calculation of the transform from phase shifts in the first case. Finally $f(s,\beta)$ is an entire function of β according to (3.8), but it has singularities (for instance at $\beta=0$), if the integral is extended to $-\infty$.

It is remarkable that at high energies the difference between the two transforms is small or negligible _in practice_, although the analytic structures are quite different.

In order to demonstrate that b has the physical meaning of an "impact parameter", many authors inserted the approximate relation

$$P_\ell(\cos\theta) \simeq J_0(b\sqrt{-t}) \tag{3.10}$$

where $b\sqrt{-t} = (2\ell+1)\sin\theta/2$, into the projection formula (3.5) and pointed out that the result agrees with the r.h.s. of (3.8), i.e.

$$T_\ell(s) \simeq f(s,\beta), \quad \text{if} \quad \sqrt{\beta} = b = (\ell+1/2)/q. \tag{3.11}$$

Since (3.10) is usually listed as an approximation for <u>large ℓ and small θ</u>, it is added that (3.11) is expected to be valid for <u>large ℓ and high energies</u>.

However a more detailed investigation[37] has shown that one has to distinguish carefully two different limits:

i) Eq.(3.11) becomes correct in the high energy limit at fixed b. This follows from the correction term

$$T_\ell(s) = f(s,\beta) - (2q)^{-2}\left\{ f' + 4\beta f'' + 4\beta^2 f'''/3 \right\} + O(q^{-4}). \tag{3.12}$$

ii) A general statement is not possible at fixed energy and large ℓ . If the usual assumption of an exponential t-dependence is made for the diffraction peak

$$\text{Im } F(s,t) = (q\sigma/4\pi)\, \exp\,(Bt/2), \tag{3.13}$$

where σ denotes the total cross section, it is easy to calculate the partial waves and the impact transforms and to check the approximation (3.11). It turns out that for kinematical conditions corresponding to pion-nucleon scattering at 2 GeV/c (3.11) is approximately <u>valid for small ℓ</u> up to about $\ell=5$ and it becomes <u>wrong for large ℓ</u> . This is discussed in more detail in Ref.[37].

3.2. <u>Eikonals and Optical Potentials</u>

The eikonal χ is defined in analogy to (3.9)

$$f(s,\beta) = \left\{\exp(2i\chi) - 1\right\}/2i \tag{3.14}$$

and the optical potential follows from

$$V(s,r) = -\frac{2q}{\pi} \int_{r^2}^{\infty} \frac{\partial \chi}{\partial \beta}(s,\beta)\, \frac{d\beta}{\sqrt{\beta - r^2}}\,, \quad \chi(s,\beta) = -\frac{1}{q}\int_{0}^{\infty} V(s,\sqrt{\beta+z^2})\, dz \tag{3.15}$$

The definition of the optical potential is modeled after approximate relations, which are valid in high energy small angle scattering. It differs from the potential in a Schrödinger or Klein-Gordon equation and its physical interpretation must be derived from (3.15). The concept of optical potentials (or quasipotentials) is to some extent analogous to that of a charge distribution, which is derived from the electromagnetic form factor of the nucleon.

The potential is a possibility to introduce the notion of a "distance" and therefore it might be of interest for the formulation of geometrical models.

3.3 Application to Pion-Nucleon Scattering

In general the introduction of spin leads to complications, but the above methods can easily be applied to the spin non-flip amplitude, since its partial wave expansion contains only the Legendre function, not its derivative

$$G(s,t) = q^{-1} \sum_{\ell=0}^{\infty} \left\{ (\ell+1) T_{\ell+}(s) + \ell T_{\ell-}(s) \right\} P_{\ell}(\cos\Theta) \qquad (3.16)$$

A detailed discussion of the transforms was given in Ref.[37], of which some results are shown in Fg.4 . This investigation is based on phase shifts[5] in the momentum interval 1-2 GeV/c. Since the main features of the diffraction peak are already clearly seen at these momenta,one can hope to find some information on the transforms of the diffraction amplitude, if one averages over the resonance structures. The trends of the momentum dependence are of interest for a comparison with assumptions, which have been made by other authors at higher energies.

The optical potential has a repulsive core in the real part and rather flat tails beyond 0.5 fermi in both real and imaginary parts at 2 GeV/c.

4. Isospin Bounds[43-45]

In principle isospin bounds are known already for a long time, but for some reason textbooks preferred to give only special cases and not the general formalism.

A simple treatment of the spin-dependent problem results, if the following combination of amplitudes is introduced (cf.Appendix)

$$K^{(\pm)} \equiv G \pm iH = (q/\sqrt{\pi}) \, e^{i\Theta/2} (F_{++} \pm iF_{+-}) \qquad (4.1)$$

The modulus of K is simply related to differential cross sections and polarizations

$$|K^{(\pm)}|^2 \equiv \Sigma^{(\pm)} = \frac{d\sigma}{d\Omega}(1 \pm P) \tag{4.2}$$

Because of isospin invariance the complex vectors for the reactions $\pi^{\pm}p \to \pi^{\pm}p$ and $\pi^- p \to \pi^0 n$ form the isospin triangle

$$K_+ = K_- + \sqrt{2}\, K_0 \tag{4.3}$$

where the upper index was omitted. The isospin bounds are given by the triangle inequalities

$$\sqrt{\Sigma_+} \leq \sqrt{\Sigma_-} + \sqrt{2\Sigma_0} \,, \quad \sqrt{\Sigma_-} \leq \sqrt{\Sigma_+} + \sqrt{2\Sigma_0} \,, \quad \sqrt{2\Sigma_0} \leq \sqrt{\Sigma_+} + \sqrt{\Sigma_-} \tag{4.4}$$

which are equivalent to

$$\left(\sqrt{\Sigma_+} - \sqrt{\Sigma_-}\right)^2 \leq 2\Sigma_0 \leq \left(\sqrt{\Sigma_+} + \sqrt{\Sigma_-}\right)^2 \tag{4.5}$$

and also to

$$\left(\Sigma_- - \Sigma_+\right)^2 \leq 4\Sigma_0(\Sigma_- + \Sigma_+ - \Sigma_0) \tag{4.6}$$

The bound (4.5) was recently investigated by Dass et al.[46]. Other authors used weaker bounds, for which polarization data are not needed. These bounds follow from the above ones, if Σ is replaced by $d\sigma/d\Omega$. They agree with them if $P_+ = P_-$.

We introduce the phase of the amplitudes $K_1 = |K_1|\exp i\phi_1$, $K_3 = |K_3|\exp i\phi_3$, where the lower index refers to isospin $T=1/2$ and $3/2$.

The relative phase $\Delta\phi$ and the absolute value of the ratio

$$K_1/K_3 = \rho \exp i\Delta\phi\,, \qquad \Delta\phi = \phi_1 - \phi_3 \tag{4.7}$$

can be calculated from $d\sigma/d\Omega$ and P data

$$\rho = \frac{3(\Sigma_0 + \Sigma_-) - \Sigma_+}{2\Sigma_+}\,, \quad \cos\Delta\phi = \frac{\Sigma_+ + 3\Sigma_- - 6\Sigma_0}{\{8\Sigma_+(3\Sigma_0 + 3\Sigma_- - \Sigma_+)\}^{\frac{1}{2}}} \tag{4.8}$$

The condition $\Delta\phi = 0$ or π is equivalent to the degeneracy of the isospin triangle and to the saturation of one of the isospin bounds. It is fulfilled along certain curves in the s,t plane, which will be called "$0,\pi$ phase contours". All zeros of K lead to a degeneracy of the isospin triangle and therefore they must lie on the $0, \pi$ phase contours.

A direct calculation of phase contours is not yet possible, since there are only a few P_O data. Therefore the contour lines were calculated from phase shifts and from the Barger-Phillips fit[44]. These lines have the interesting property that they are either closed loops or go from threshold to infinite energy. If a line goes to 0° or 180° in the plot belonging to $K^{(+)}$, it has a smooth continuation in the plot belonging to $K^{(-)}$ and vice versa.

Different applications of the contour plots were discussed in Ref.[44]. We shall mention here only that above 1.5 GeV/c there exists a $0, \pi$ phase contour line at approximately constant $t \simeq - 0.05$ $(GeV/c)^2$. This line is related to the large increase of the charge-exchange flip amplitude at $t=0$. It was shown in Ref.[45] that the saturation of the bound

$$\left(\sqrt{\Sigma_-} - \sqrt{\Sigma_+}\right)^2 \leqq 2 \, \Sigma_O \tag{4.9}$$

leads to a relation between $P_+ - P_-$ and $d\sigma_-/dt - d\sigma_+/dt$. At very small $|t|$ the experimental uncertainty is large for both quantities. A reasonable extrapolation of the $P_+ - P_-$ data to $t=0$ gives a strong restriction for the t-dependence of the cross section difference.

The special case of isospin bounds for backward amplitudes was treated in several papers[47,48]. There is a remarkable approximate saturation of the bound (4.9) in large energy intervals, which led the author of Ref.[48] to the speculation that there might be a "phase degeneracy" ($\Delta\phi = 0$ at 180°). But this is not true[43].

Further bounds are treated in the interesting review by S.M. Roy[49].

APPENDIX

Natural units $(\hbar = m_\pi = c = 1)$ are used unless stated otherwise.
m= nucleon mass, Θ = c.m. scattering angle, k, q = pion momentum in
the lab. and c.m. system respectively. $\omega^2 = 1+k^2$, $E^2 = m^2+q^2$, $s = W^2$
$t = -2q^2(1-\cos\Theta)$, $2m\omega = s-m^2-1$, $\nu = (s-u)/4m = \omega + t/4m$. $d\sigma/d\Omega$:
c.m.system.

<u>Spin no-flip and flip amplitudes</u>

$$M = G + i\,(\hat{n}\vec{\sigma})H$$

$$\frac{d\sigma}{d\Omega} = |G|^2 + |H|^2 \,, \qquad P\,\frac{d\sigma}{d\Omega} = 2\,\mathcal{I}m\,(GH^*)$$

<u>s-channel helicity no flip and flip amplitudes</u>

$$\frac{d\sigma}{dt} = |F_{++}|^2 + |F_{+-}|^2 \,, \qquad P\,\frac{d\sigma}{dt} = 2\,\mathcal{I}m\,(F_{++}F_{+-}^*)$$
$$= 2|F_{++}|\,(F_{+-})_\perp$$

$$S\,\frac{d\sigma}{dt} = |F_{++}|^2 - |F_{+-}|^2 \,, \qquad T\,\frac{d\sigma}{dt} = 2\,Re\,(F_{++}F_{+-}^*)$$
$$= 2|F_{++}|\,(F_{+-})_\parallel$$

$$(\sqrt{\pi}/q)\,G = F_{++}\cos\Theta/2 + F_{+-}\sin\Theta/2$$
$$(\sqrt{\pi}/q)\,H = -F_{++}\sin\Theta/2 + F_{+-}\cos\Theta/2$$

<u>Invariant amplitudes</u>

$$4\sqrt{\pi}\,qW\,F_{++} = m\cos\Theta/2\,(A+\omega B)$$
$$4\sqrt{\pi}\,qW\,F_{+-} = \sin\Theta/2\,\{EA + m(W-E)B\}$$

$$C \equiv A' = A + \frac{\nu}{1 - t/4m^2}\,B$$

REFERENCES

1.   N.W. Dean and Ping Lee, Phys. Rev. $\underline{D5}$, 2741 (1972)

2.   F. Halzen and C. Michael, Phys. Lett. $\underline{36B}$, 367 (1971)

3.   G. Cozzika et al., Phys. Lett. $\underline{40B}$, 281 (1972)

4.   G. Höhler, H.P. Jakob and F. Kaiser, preprint TKP 11/72, University of Karlsruhe

5.   S. Almehed and C. Lovelace, Nucl. Phys. $\underline{B40}$, 157 (1972)

6.   R. Ayed, P. Bareyre and Y. Lemoigne, Saclay preprint, submitted to the XVI International Conference on High Energy Physics, Batavia 1972

7.   F. Wagner, invited talk given at the meeting of the Studiengruppe at Plättig, Black Forest (June 1972), to be published

8.   V. Barger and R.J. Phillips, Phys.Rev. $\underline{187}$, 2210 (1969)

9.   G. Höhler in Proceedings of the International Seminar on Binary Reactions of Hadrons at High Energies (Dubna, 1971)

10.   S.J. Lindenbaum in: Pion-Nucleon Scattering. Ed.G.L. Shaw and D.Y. Wong, Wiley Interscience, New York, 1969

11.   K.J. Foley et al., Phys. Rev. $\underline{181}$, 1775 (1969)

12.   G. Höhler, F. Steiner and R. Strauss, Zeitschrift f.Physik $\underline{233}$, 430 (1970)

13.   H. Lehmann, Nucl. Phys. $\underline{29}$, 300 (1962)

14.   K. Igi, Phys. Rev. Lett. $\underline{9}$, 76 (1962)

15.   M. Davier, Phys.Lett. $\underline{40B}$, 369 (1972)

16.   G. Höhler and E. Krubasik, preprint and talk at the Titisee meeting 1971
        E. Krubasik, Nucl. Phys. $\underline{B44}$, 558 (1972)

17.   G. Höhler and R. Strauss, Zeitschrift f. Physik $\underline{240}$, 377 (1970)

18.   J. Dronkers and P.Kroll, preprint TKP 6/72, University of Karlsruhe, to appear in Nucl. Phys. B

19.   G. Höhler and H.P. Jakob, Nuovo Cim.Lett. $\underline{2}$, 485 (1971)

20.   C. Michael, Review Talk given at the Oxford Conference 1972, CERN preprint TH 1480

21.   G. Höhler, H.P. Jakob and R. Strauss, Phys.Lett. $\underline{35B}$,445(1971)
        H.P. Jakob, CERN preprint 1446 (1971)

22.   G. Höhler and N. Zovko, Zeitschrift f. Physik $\underline{181}$, 293(1964)

23.   G. Höhler and R. Strauss, Zeitschrift f.Physik $\underline{232}$, 205(1970)

24. M. Borghini et al., Phys.Lett. $\underline{31B}$, 405 (1970) and CERN preprint (Jan. 1972)

25. V. Barger and F. Halzen, Phys.Rev.Lett. $\underline{28}$, 194 (1972)

26. F.J. Gilman et al., Phys.Lett. $\underline{31B}$, 387 (1970)

27. J. Engels, Nucl.Phys. $\underline{B25}$, 141 (1970)
    R. Strauss, Thesis, University of Karlsruhe

28. H.J.W. Mueller, to be published in Particles and Fields, Ed.H.H. Aly, Dekker Pub. Co.

29. G. Höhler, H.Schlaile and R. Strauss, Zeitschrift f. Physik $\underline{229}$, 217 (1969)

30. J.A. McClure and L.E. Pitts, Phys.Rev. $\underline{D5}$, 109 (1972)

31. E. Pietarinen, Nordita preprint, submitted to Nucl.Phys.B.

32. G. Hecht, H.P. Jakob and P.Kroll, to be published

33. H. Schlaile and R. Strauss, private communication

34. F. Wagner, preprint MPI Munich, June 1972

35. J.E. Nelson et al., Berkeley preprint LBL-1027 (Sept.72) Charge-exchange Scattering between 1.0 and 2.4 GeV/c.

36. H.Harari, SLAC-PUB-914 and Invited Talk at International Conference on Duality and Symmetry in Hadron Physics, Tel-Aviv, 1971

37. G. Höhler and H.P. Jakob, preprint, Karlsruhe University(July 1972) Impact Parameter Expansion of πN Scattering Amplitudes TKP 9/72

38. G.C. Fox in High Energy Collision, Third International Conference in Stony Brook 1969. Gordon and Breach New York, p.367

39. W.N. Cottingham and R.F. Peierls, Phys.Rev.$\underline{137}$, B 147 (1965)

40. E. Predazzi, Ann. of Phys. $\underline{36}$, 228 (1966)

41. Adachi et al., Progress Theor.Phys. $\underline{35}$, 463 (1966), $\underline{35}$, 485(1966) $\underline{36}$, 745 (1966) and Suppl.Extra Number (1965), 316

42. R. Blankenbecler and M.L. Goldberger, Phys.Rev. $\underline{126}$, 766(1962)

43. G. Höhler and W. Schmidt, Phys.Lett. $\underline{38B}$, 237 (1972)

44. G. Höhler, F. Kaiser, H. Kohl and W. Schmidt, Nuovo Cim.Lett., in print

45. G. Höhler and H. Staudenmaier, preprint TKP 8/72, University of Karlsruhe

46. G.V. Dass et al., Phys.Lett. $\underline{36B}$, 339 (1971)

47. N.A. Törnquist et al., Nucl.Phys. $\underline{B33}$, 239 (1971) and earlier papers

48. R. Ayed et al., Submitted to the Amsterdam Intern.Conference, 1971

49. S.M. Roy, preprint, to be published in Physics Reports

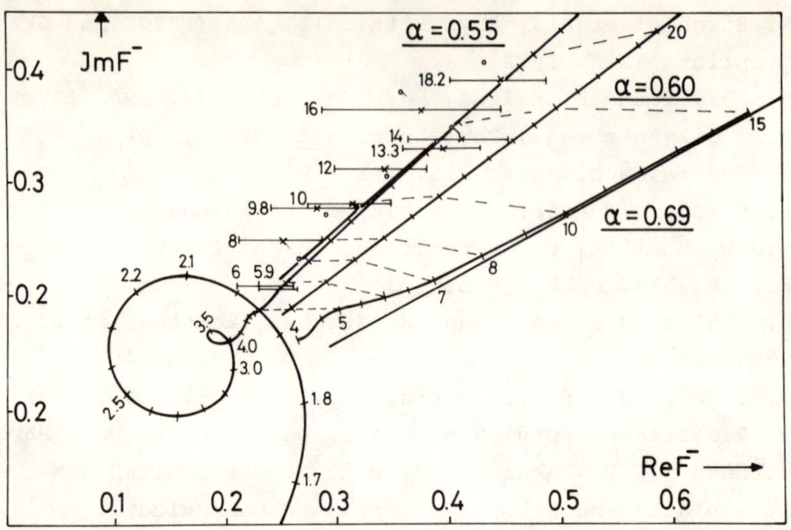

Fig. 1 The Charge-exchange Forward Amplitude. $F = C^-/4\tilde{\pi}$.
Experimental data are compared with predictions, assuming
different power laws at high energies. α = 0.69 is not
acceptable

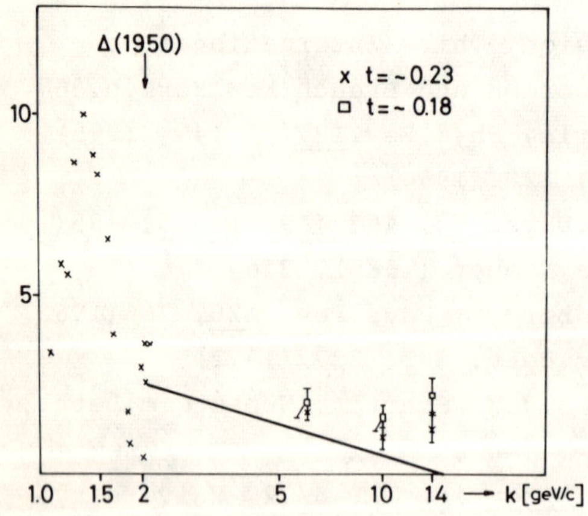

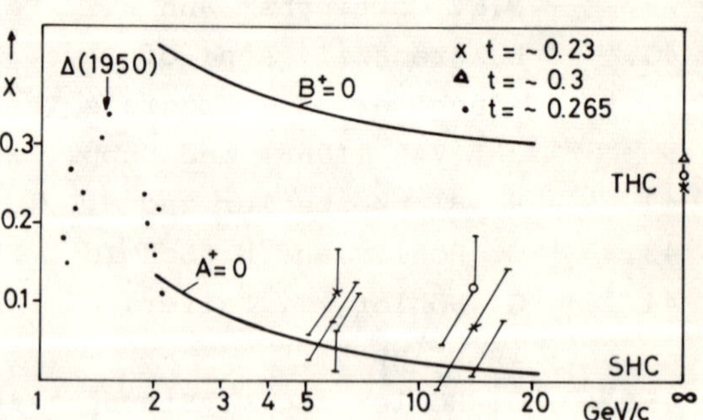

Fig. 2
$A^+_\perp$ as derived from the new data

Fig. 3
Flip/no flip ratio. Momentum
dependence from new data

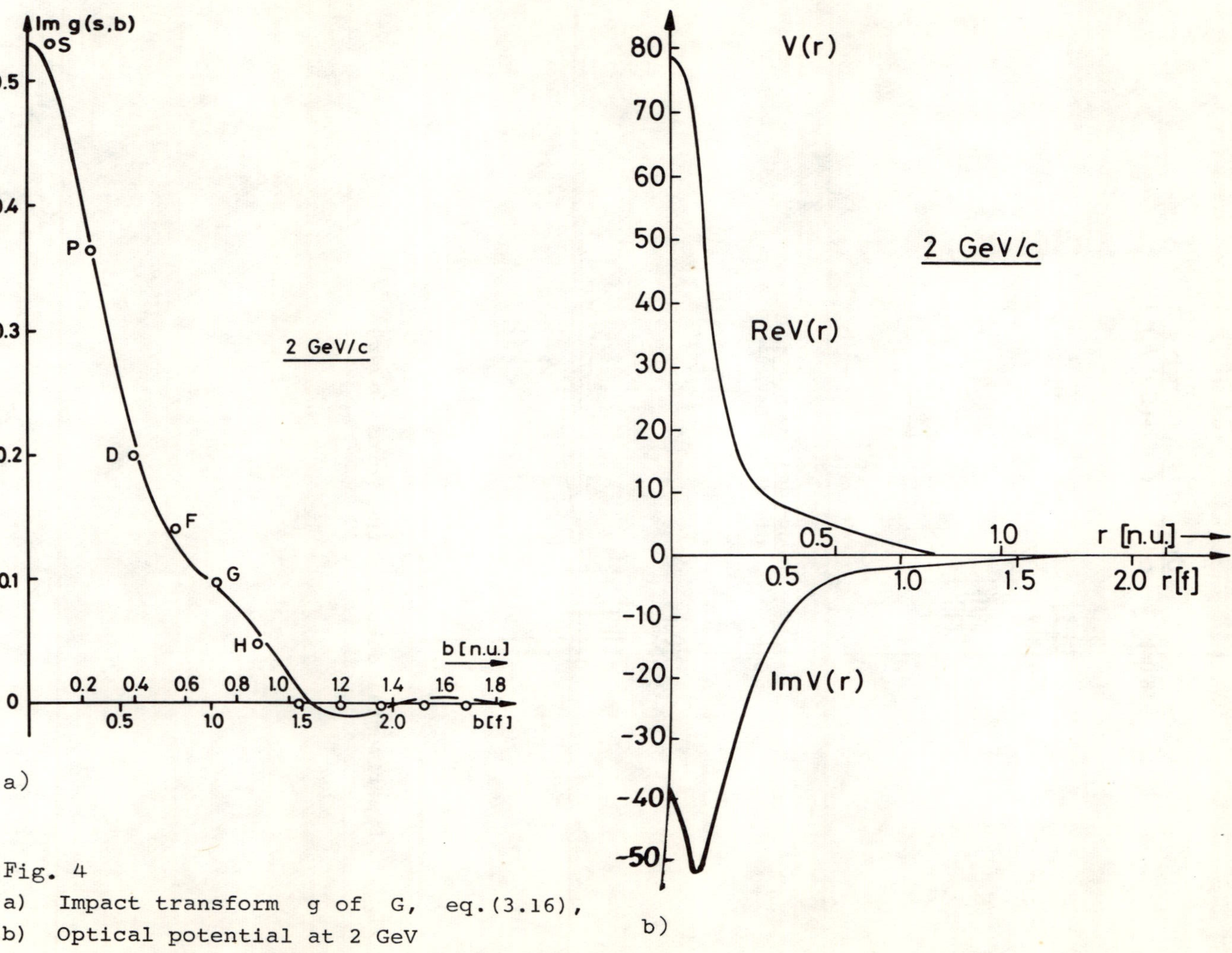

Fig. 4
a) Impact transform g of G, eq.(3.16),
b) Optical potential at 2 GeV

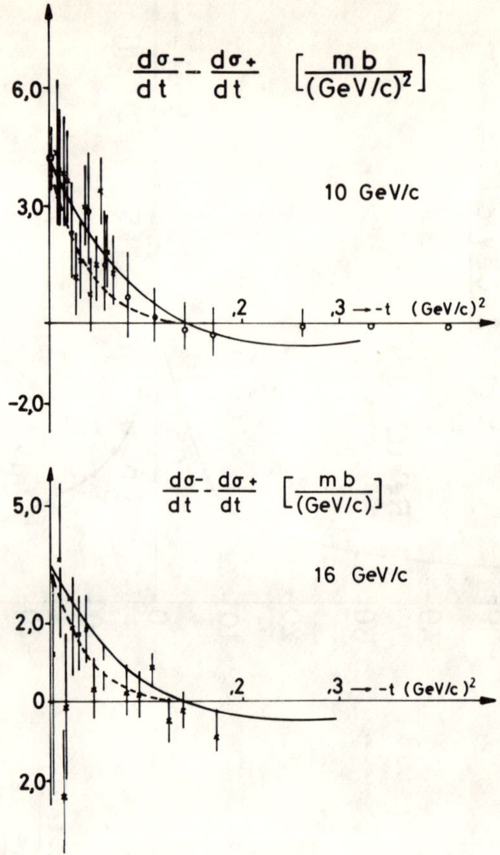

Fig. 5. Experimental data for the "cross over"

OPTIMIZATION OF COLLISION AMPLITUDES UNDER CONSTRAINTS

R.J. Eden

Cavendish Laboratory, Cambridge, England

1. ASSUMPTIONS AND OBJECTIVES

(a) Introduction

These lectures will describe methods for optimization under constraints. The methods will be illustrated by applications to collision amplitudes, both at asymptotic energies and at finite energies. The applications will include the use of integral constraints.

Starting from the Mandelstam representation, Froissart (1961) deduced that the total cross section $\sigma_{tot}(s)$ for the collision of two particles having c.m. energy $s^{\frac{1}{2}}$ must obey the following bound as $s \rightarrow \infty$,

$$\sigma_{tot}(s) \leq \text{constant} \left[\ln(s/s_0)\right]^2 \qquad (1.1)$$

where s_0 is a constant.

Since 1961 there have been many extensions of the idea of obtaining bounds on physical quantities from basic assumptions without any use of special theoretical models. These developments have been dominated by the work of Martin and his many associates and collaborators. It has been extensively surveyed in the review articles by Martin (1969) and by others listed in the list of references. These reviews contain detailed references to most original papers in this area of research, so these lecture notes will contain only minimal references to original papers.

Assumptions: The basic results of quantum field theory which are commonly used are:

      (i)   Unitarity

    (ii)   Analyticity in the Lehmann-Martin ellipse

   (iii)   Dispersion relations for collision amplitudes $F(s,t)$ at fixed momentum transfer in a limited range of t.

In addition it is useful to make phenomenological assumptions
about collision amplitudes, cross sections or scattering lengths.
These lead to bounds or constraints between different physical (or
analytically continued) quantities; examples will be given in later
lectures.

<u>Notation</u>: For equal mass particles with c.m. momentum k,

$$s = 4 \left(m^2 + k^2 \right) \quad , \qquad t = -2 k^2 \left(1 - \cos \vartheta \right) .$$

(1.2)

We shall also be discussing unequal mass collisions but shall not give
detailed formulae in those cases. The collision amplitude F (s,t) will
be normalised so that

$$\frac{d\sigma}{dt} = \frac{4\pi \, |F(s,t)|^2}{s \, (s - 4m^2)}$$

(1.3)

$$\sigma_{tot}(s) = \frac{4\pi}{k^2} \sum_{0}^{\infty} (2\ell+1) \, \text{Im} \, f_\ell(s) = \frac{4\pi}{k s^{1/2}} \, \text{Im} \, F(s, 0)$$

(1.4)

Within the Lehmann-Martin ellipse

$$F(s,t) = \frac{s^{1/2}}{k} \sum_{0}^{\infty} (2\ell+1) \, f_\ell(s) \, P_\ell(\cos\vartheta)$$

(1.5)

Unitarity requires that partial waves satisfy

$$|f_\ell(s)|^2 \leq \text{Im} \, f_\ell(s) \leq 1$$

(1.6)

Writing $f_\ell = r_\ell + i a_\ell$, the unitarity constraint can be written

$$a_\ell - a_\ell^2 - r_\ell^2 \geq 0$$

(1.7)

(b) <u>Objectives of Optimization</u>
    I shall indicate briefly some of the objectives that have been
achieved and return later to a discussion of further results.

<u>Dispersion Relations</u>. Using unitarity (1.6), the partial wave expansion
(1.5) and the properties of Legendre polynomials, Jin and Martin have
shown that the dispersion relation for F(s,t) at fixed t requires no

more than two subtractions,

$$|F(s,t)| \leq (s/s_0)^2, \quad (t < 4m^2).$$
(1.8)

<u>High energy bounds</u>. The Froissart bound (1.1) holds with the constant equal to $4\pi / (t_0 - \varepsilon)$, where $t_0 = 4m^2$ and ε may be arbitrarily small.

Integrated bounds have been obtained for averaged total cross sections by Common and Yndurain. These bounds will be discussed in section 4 of these lecture notes. Various bounds have been obtained on amplitudes and on differential cross-sections at fixed t. These bounds can be extended to inelastic reactions and to inclusive reactions as described in the review by Roy (1972).

<u>Absorptive parts of elastic amplitudes</u>
MacDowell and Martin and later Singh and Roy obtained a variety of bounds on ImF(s,t) at high energies. In section 3 some of these bounds will be considered as well as related bounds at finite energies.

<u>Bounds on crossed-channel amplitudes</u>
The integrated bounds of Common and Yndurain mentioned above, lead naturally to restictions on amplitudes in the crossed channel. For example, knowing N scattering results one can deduce restrictions on $N\bar{N} \longrightarrow \pi\pi$ amplitudes. These will be discussed in section 4 of these notes.

<u>Theorems on particle-antiparticle amplitudes</u>
The Pomeranchuk theorem has several variants; the following form is due to Martin. If F denotes a forward amplitude and

$$\lim_{s \to \infty} \frac{F(ab \to ab) - F(\bar{a}b \to \bar{a}b)}{s \, \ell n \, (s/s_0)} \to 0$$
(1.8)

and if the following limit exists,

$$\lim \left[\sigma_{tot}(ab) - \sigma_{tot}(\bar{a}b) \right]$$
(1.9)

then this limit (1.9) must be zero.

Numerous related theorems arise if it is assumed that (1.9) is not zero as was indicated by the early results from Serpukhov on total cross sections. Some of these theorems are discussed in the reviews by

Eden (1971) and Roy (1972).

2.   Optimization under Constraints
       References: M. Aoki (1971), M.R. Hestenes (1966) and M.B. Einhorn
              and R. Blankenbecler (1971).

(a) Formulation and terminology

Objective function: The function that we wish to optimize is called the objective function. Other names used include Lagrangian, criterion function, utility, cost or penalty functions. To be definite we will describe conditions required to minimise the objective function. This function depends on a set of real variables x_1, x_2, ... x_n, and it will be denoted by

$$f(x) = f(x_1, x_2 \ldots x_n) \tag{2.1}$$

Constraints: We consider equality constraints and inequality constraints. The equality constraints will be written

$$f_\alpha (x) = 0, \qquad \alpha = 1, 2, \ldots p. \tag{2.2}$$

The inequality constraints will be written

$$g_\beta (x) \geqslant 0, \qquad \beta = 1, 2, \ldots q. \tag{2.3}$$

Any point $x = (x_1, x_2, \ldots x_n)$ that satisfies the constraints is called a feasible point and the set of such points is called the feasible set S.

Tangent cone: the set of all (unit length) half lines h, originating at a point x_0 in S and tangent to a curve in S is called the tangent cone to S at x_0.

Increments: the increment of f along a small vector v through a feasible point x_0 is called the first differential of f and is denoted by

$$\delta f = f'(x_0, v) = \left(f'(x_0), v \right) = \sum \frac{\partial f}{\partial x_i} v_i \tag{2.4}$$

The second differential is defined by

$$f''(x_0, v) = \sum \frac{\partial^2 f}{\partial x_i \, \partial x_j} \, v_i \cdot v_j \tag{2.5}$$

Regular points of S (equality constraints only):
Let x_0 be a feasible point and let k be a unit vector satisfying

$$(f'_\alpha \, (x_0), k) \;=\; 0, \quad \text{for all} \tag{2.6}$$

If every k satisfying (2.8) lies in the tangent cone C at x_0, then x_0 is a regular point of S.

Normal points of S: If the gradients $f'(x_0)$ are linearly independent, x_0 is a normal point. Every normal point is a regular point (Hestenes 1966).

Inequality constraints: The definitions of normal points and regular points extend to inequality constraints if we divide these into ineffective or interior constraints β in I (x_0), and effective or boundary constraints β in B(x_0), defined by

$$I(x_0) \;=\; \left\{ \beta \mid g_\beta(x_0) > 0 \right\} \tag{2.7}$$

$$B(x_0) \;=\; \left\{ \beta \mid g_\beta(x_0) = 0 \right\} \tag{2.8}$$

(b) Minimization with Equality Constraints

The standard method of Lagrange multipliers determines all local minima that are regular points. It is summarised by the following two theorems:

Theorem 1

Let x_0 be a regular point of S and let x_0 be a local minimum of f(x) on S.
Then:

    (i)   there exist multipliers λ_α such that

$$L'(x_0) = \frac{\partial L}{\partial x_i} = 0, \quad i = 1, 2, \ldots n \tag{2.9}$$

where the auxiliary function L is defined by

$$L(x) \;=\; f + \sum \lambda_\alpha f_\alpha \; . \tag{2.10}$$

(ii)   for a minimum

$$L''(x_0, h) \;\geqslant\; 0 \qquad\qquad (2.11)$$

for all h in the tangent cone at x_0,

(iii) if x_0 is normal, the multipliers λ_α are unique.

Theorem 2

If (2.9) is satisfied and if $L''(x_0, h)$ is strictly positive for all h in the tangent cone at x_0, then x_0 is a local minimum of f(x).

(c)   <u>Interpretation of the Lagrange Multipliers</u>

The Lagrange multipliers can be interpreted as sensitivity coefficients with respect to small changes in the constraints. Thus if x^* denotes a local minimum of f(x), with equality constraints,

$$b_\alpha - f_\alpha(x) = 0 \;,\quad \alpha = 1, 2, \ldots p \qquad\qquad (2.12)$$

Then

$$\frac{\partial f}{\partial x_i^*} - \sum_\alpha \lambda_\alpha \frac{\partial f_\alpha}{\partial x_i^*} = 0 \qquad\qquad (2.13)$$

From (2.12) and (2.13) one finds

$$\frac{\partial f(x^*)}{\partial b_\gamma} = \lambda_\gamma \;. \qquad\qquad (2.14)$$

Thus λ_γ is the rate of change in the stationary value $f(x^*)$ with respect to small changes in the parameter b_γ in the constraints (2.12). In practice the sign of λ_γ may be intuitively obvious from this interpretation.

(d)   <u>Linear Programming</u>

Linear Programming problems are those in which both the objective function and the constraints are linear in the variables.

<u>Standard form</u>: Minimise the objective function

$$\phi(x) = \sum c_i x_i + d \qquad\qquad (2.15)$$

subject to

$$A x = b \quad , \quad x \geq 0 \qquad (2.16)$$

where A is an (m,n) matrix of rank m, (m<n) and

$$x^T = (x_1 \cdots x_m) \, , \quad b^T = (b_1 \cdots b_m) \, ; \qquad (2.17)$$

<u>Extreme points</u>: The feasible set S is the inside of a convex polyhedron (possibly extending to infinity). The extreme points of the linear objective function

$$\phi = \sum c_i x_i + d \qquad (2.18)$$

occur at the vertices of the polyhedron. Thus they are members of the set of vectors x in which (n-m) of the x_i are zero. Thus with m constraints only m of the x_i are non-zero.

<u>Lagrangian method</u>: The Lagrangian associated with the standard linear programming problem is

$$L(x,\lambda) = \sum c_i x_i + d + \sum_\alpha \lambda_\alpha [b_\alpha - (A x)_\alpha] \qquad (2.19)$$

subject to x $\geq$ 0.

(e) <u>Minimisation with Inequality Constraints</u>

Consider the minimisation of f(x) subject to the constraints

$$f_\alpha(x) = 0 \quad , \quad \alpha = 1, 2, \ldots p, \qquad (2.20)$$

$$g_\beta(x) \geq 0 \quad , \quad \beta = 1, 2, \ldots q. \qquad (2.21)$$

For any feasible x_0, let $I(x_0)$ be the set of indices β for which $g_\beta(x_0) > 0$ and $B(x_0)$ be those β for which $g_\beta(x_0) = 0$.

<u>Theorem 3</u>

Let x_0 be a regular point and a local minimum in the feasible set S. Then,

(i) There exist multipliers λ_α, and $\mu_\beta \leq 0$ such that

$$L'(x_0) = 0 \qquad (2.22)$$

where

$$L(x_o) = f + \sum \lambda_\alpha f_\alpha + \sum \mu_\beta g_\beta \tag{2.23}$$

(ii)   If β is in $I(x_o)$ we may choose $\mu_\beta = 0$.

(iii)   Let S_1 be the subset of S for which $g_\beta(x) = 0$ for all β in $B(x_o)$.

Then

$$L''(x_o,h) \geqslant 0 \tag{2.24}$$

for all h in the tangent cone of S_1 at x_o.

(iv)   If x_o is a normal point, the multipliers are unique.

Theorem 4

If (2.22) is satisfied and if (instead of (2.24))

$$L''(x_o,h) > 0 \tag{2.25}$$

for all h in the tangent cone S_1 at x_o, then x_o is a local minimum of $f(x)$.

Sometimes when L is linear in a coordinate, theorem 4 may not apply. Then one could use the following obvious theorem.

Theorem 5

If $f'(x_o,h) > 0$ for all h in the tangent cone at x_o, then x_o is a local minimum of $f(x)$.

3.   Collision Amplitudes under Constraints

(a) Linear Constraints at High Energies

For our first example to illustrate the general method we consider the problem discussed by Singh (1971b), but using the Lagrangian method as suggested by Hodgkinson (unpublished). Let $A(s,t) = ImF$ be the imaginary part of a scattering amplitude. Consider its maximum value in

$$0 < t \leq 4m^2 = t_1 \tag{3.1}$$

Define $w = 1 + 2t/(s-4)$, $w_1 = 1 + 2t_1/(s-4)$.   Then we have to maximise

$$A(s,t) = \sum_0^\infty (2\ell+1)\, a_\ell P_\ell(w) \tag{3.2}$$

subject to the constraints

$$a_\ell \geq 0, \text{ for } \ell = 0,1,2, \ldots \tag{3.3}$$

$$\sum_0^\infty (2\ell+1) a_\ell P_\ell(\omega_1) = (k/s^{\frac{1}{2}}) A(s,t_1) = A_1 \tag{3.4}$$

(where $A_1 < (s/s_0)^2$ follows from the Jin-Martin bound),

$$\sum_0^\infty (2\ell+1) a_\ell = \frac{k^2 \sigma(tot)}{4\pi} = (k/s^{\frac{1}{2}}) A(s,0) = A_0. \tag{3.5}$$

The method of linear programming tells us that the maximum of $A(s,t)$ will occur when all a_ℓ are zero except for two, a_p and a_q say. It does not at once inform us that $p+1 = q$. This can be seen from the Lagrangian for the constrained problem.

$$L = \sum (2\ell+1) a_\ell P_\ell(\omega) + \alpha \left[A_0 - \sum (2\ell+1) a_\ell \right]$$

$$+ \beta \left[A_1 - \sum (2\ell+1) a_\ell P_\ell(\omega_1) \right]$$

$$+ \sum (2\ell+1) \lambda_\ell a_\ell \tag{3.6}$$

It is obvious that max $A(s,t)$ will increase if either A_0 or A_1 is increased, hence $\alpha > 0$ and $\beta > 0$. Also since λ_ℓ are inequality multipliers $\lambda_\ell \geq 0$ for a maximum (from theorem 3). The differential of L is zero at a maximum,

$$\frac{1}{(2\ell+1)} \frac{\partial L}{\partial a_\ell} = 0 = P_\ell(w) - \alpha - \beta P_\ell(w_1) + \lambda_\ell. \tag{3.7}$$

Let I denote the interior subset and B_0 the boundary subset of solutions of (3.7),

$$I : a_\ell > 0, \quad \lambda_\ell = 0, \quad P_\ell(w) = \alpha + \beta P_\ell(w_1) \tag{3.8}$$

$$B_0 : a_\ell = 0, \quad \lambda_\ell = \alpha + \beta P_\ell(w_1) - P_\ell(w) > 0 \tag{3.9}$$

From (3.8) and (3.9) it is obvious that a_ℓ is zero except for a_p, a_{p+1}.

The value of p can be found from the constraints (3.4) and (3.5). For large values of s, p is also large and one can use

$$P_p(w_1) \quad \sim \quad P_{p+1}(w_1) \quad \sim \quad I_o(y_1) \sim \frac{\exp(y_1)}{(2\pi y_1)^{\frac{1}{2}}} \qquad (3.10)$$

where $y_1 = L(t_1/k^2)^{\frac{1}{2}}$. This gives

$$p \leq \frac{1}{4} s^{\frac{1}{2}} \ell n \left[s/(s_o^2 \sigma_{tot}) \right] . \qquad (3.11)$$

Substituting in (3.2) one obtains Singh's result, namely

$$\frac{A(s,t)}{A_o} \leq \frac{A^{max}(s,t)}{A_o} \leq I_o(y) . \qquad (3.12)$$

The above method extends readily to the use of non-linear unitarity
$a_\ell - a_\ell^2 > 0$ instead of the linear form (3.3).

(b) <u>Phenomenological Constraints at Finite Energy</u>

Singh and Roy (1970) bounded $A(s,t)$ given by (3.2) in the region
$t < 0$, using the partial wave series for σ(total) and for σ(elastic)
as constraints. In the region $|t| < 0.1$ $(GeV/c)^2$, they found

$$max \left| \frac{A(s,t)}{A(s,0)} \right|^2 \approx \left(\frac{d\sigma}{dt} \right) \Big/ \left(\frac{d\sigma}{dt} \right)_{t=0} \qquad (3.13)$$

At high energies the Singh-Roy bound given by the left hand side of
(3.13) exceeds the data by only about 10% in $|t| < 0.1$ $(GeV/c)^2$, but
for larger $|t|$ it differs from the data by more than an order of
magnitude.

Savit, Einhorn and Blankenbecler (1971) imposed the additional
constraint that partial waves decrease monotonically $a_{\ell+2} \leq a_\ell$.
However, they found, only a slight improvement was obtained.

Jacobs et al (1970) imposed instead, an additional constraint
that fixed a phenomenological value of $A(s,t)$ at a physical value of
$t \approx -0.1$ $(GeV/c)^2$. This was found to extend the region in which max
$A(s,t)$ was close to the data in the sense of (3.13). For the constraint
at larger values of $|t|$ oscillations are induced in max $A(s,t)$ and its
agreement with the data becomes poorer.

Hahn and Hodgkinson (1971) impose an additional
constraint at $t > 0$ instead of $t < 0$. Thus they use (3.4) with
Jin-Martin bound replaced by a phenomenological value of A_1 .

This value is obtained by extrapolating from experimental data in $t < 0$ to $t = t_1 = 4m^2$, using

$$A(s,t) = A_0 \exp\left[\tfrac{1}{2}bt + \tfrac{1}{2}ct^2\right] \tag{3.14}$$

The resulting auxiliary objective function (or Lagrangian) is given by (3.2) and

$$\begin{aligned}
L = \; & A(s,t) + \alpha\left[A_0 - \sum(2\ell+1)\,a_\ell\right] \\
& + \beta\left[A_1 - \sum(2\ell+1)\,a_\ell P_\ell(w_1)\right] \\
& + \frac{1}{2a}\left[\sigma_{\ell\ell} - \sum(2\ell+1)\,(a_\ell^2+r_\ell^2)\right] \\
& + \sum(2\ell+1)\,\lambda_\ell\,(a_\ell-a_\ell^2-r_\ell^2)
\end{aligned} \tag{3.15}$$

For a maximum of $A(s,t)$, $r_\ell = 0$ and

$$P_\ell(z) - \alpha - \beta P_\ell(w_1) - (a_\ell/a) + \lambda_\ell(1-2a_\ell) = 0, \tag{3.16}$$

$$\lambda_\ell \geqslant 0, \quad (1/a) + 2\lambda_\ell \geqslant 0. \tag{3.17}$$

where $-1 < z < 1$ and $w_1 > 1$.
There are three classes of ℓ values for the solutions of these equations,

$$\ell \in I \; : \; \lambda_\ell = 0, \; (a_\ell/a) = P_\ell(z) - \alpha - \beta P_\ell(w_1).$$

$$\ell \in B_0 \; : \; a_\ell = 0, \; \lambda_\ell = \alpha + \beta P_\ell(w_1) - P_\ell(z) \geqslant 0 \; .$$

$$\ell \in B_1 \; : \; a_\ell = 1, \; \lambda_\ell = P_\ell(z) - (1/a) - \alpha - \beta P_\ell(w_1). \tag{3.18}$$

The parameters are evaluated from the equality constraints that cause each of the square brackets in (3.15) to be zero. The solutions for a_ℓ give a partial wave profile for $A^{max}(s,t)$, that depends on both s and t.

The normalised bound U is defined by

$$U(s,t) = A^{max}(s,t)/A_0 \tag{3.19}$$

One can similarly obtain a lower bound,

$$L(s,t) = A^{min}(s,t)/A_0 \tag{3.20}$$

The resulting upper and lower bounds obtained by Hahn and Hodgkinson are compared with experiment in Fig. 1, which also shows the Singh-Roy bound, in pion-nucleon scattering. It is remarkable how close $U(s,t)$ comes to the data, when one recalls that it involves only one constraint (at $t = 4m^2$) additional to σ_{tot} and σ_{el} . It will of course be noted that the use of the ratio (3.19) compensates partly for the fact $U(s,t)$ involves only the imaginary part of the amplitude.

4.   Integral Constraints

(a) Bounds on Averaged $\pi\pi$ Cross Sections

Reference: Common and Yndurain (1971), Roy (1972) and Steven (1972).

We shall consider $\pi^0\pi^0$ scattering, using units with $m_\pi = 1$. From the Froissart-Gribov formula for partial waves in the t channel, evaluated in the threshold limit $t = 4$, we obtain the scattering length α_2^t . This gives

$$a = \frac{15\pi}{8}\,\alpha_2^t = \int_4^\infty \frac{A(s,4)}{2s^3}\,ds \tag{4.1}$$

In terms of s channel partial waves,

$$A(s,t) = 2(s^{1/2}/k) \sum (2\ell+1)\, a_\ell(s)\, P_\ell(z) \tag{4.2}$$

$$\sigma_{tot}(s) = 2\,(\frac{4\pi}{k^2}) \sum (2\ell+1)\, a_\ell \tag{4.3}$$

where the factor 2 in (4.2) and (4.3) comes from the identity of the pions.

We will assume that the scattering length α_2^t is known, for example by considering the effect of $\pi\pi$ scattering on πN dispersion relations using experimental πN data (Morgan and Shaw 1969). We then maximise the averaged total cross-section $\bar\sigma_{tot}$ defined by

$$\Sigma_T = \frac{\bar\sigma_{tot}(s_1,s_2)}{32\pi} = \int_{s_1}^{s_2} \sum (2\ell+1)\, a_\ell(s)\, \frac{q(s)\,ds}{(s-4)} \tag{4.4}$$

where $q(s)$ is a chosen weight function, and $a_\ell(s)$ are subject to the constraint (4.1) with (4.2), and the unitarity constraint

$$a_\ell - a_\ell^2 - r_\ell^2 \geq 0. \tag{4.5}$$

The auxiliary objective function is

$$L = \sum_T + D(s_1,s_2) \left[a - \int_4^\infty (ks^{5/2})^{-1} \sum_\ell (2\ell+1) a_\ell P_\ell(w)\, ds \right]$$

$$+ \sum (2\ell+1) \int_4^\infty (2ks^{5/2})^{-1} \lambda_\ell(s,s_1,s_2)(a_\ell - a_\ell^2 - r_\ell^2)\, ds \qquad (4.6)$$

where $w = 1 + 8/(s-4)$.

From the general theory considered earlier, for a maximum,

$$\lambda_\ell(s,s_1,s_2) \geqslant 0 \qquad (4.7)$$

$$\lambda_\ell = 0 \quad \text{if} \quad 0 < a_\ell(s) < 1 \qquad (4.8)$$

Taking functional derivatives of L, we obtain

$$\frac{\partial L}{\partial r_\ell} = 0 = \lambda_\ell(s,s_1,s_2) r_\ell(s) \qquad (4.9)$$

$$\frac{\partial L}{\partial a_\ell} = 0 = \frac{\theta(s-s_1)\,\theta(s_2-s)\,q(s)}{(s-4)} - \frac{D(s_1,s_2)\,P_\ell(w)}{k\,s^{5/2}} +$$

$$+ \frac{\lambda_\ell(s,s_1,s_2)(1-2a_\ell)}{2k\,s^{5/2}} \qquad (4.10)$$

In general when $\lambda_\ell = 0$ the solutions for a_ℓ and r_ℓ are indeterminate, but solutions exist to (4.10) only for discrete values of s. The contribution of the corresponding values of a_ℓ can make no contribution to $\bar\sigma_{tot}$ so they can be ignored. Exceptions to this general rule occur only for certain choices of weight function. For example, Roy (1972) considers

$$q(s) = C(k/s^{5/2}). \qquad (4.11)$$

Then (4.10) can be satisfied for $\ell = 0$ for all s by choice of D. The problem reduces in this case to optimising $\bar\sigma_{tot}$ when only $a_0(s)$ is non-zero. One finds that if the scattering length α_2^t is not too large, one obtains the bound

$$\frac{15\pi}{8}\alpha_2^t > \frac{1}{8\pi} \int_{s_1}^{s_2} \sigma_{tot}(s) \frac{kds}{s^{5/2}} \qquad (4.12)$$

Roy (1972) finds that bounds of this type are approached to within a factor 2 by currently accepted scattering lengths and cross-sections.

The corresponding bounds for more general weight functions have been evaluated by Steven(1972). The method follows from (4.10) ignoring the class I in which $0 < a_\varrho < 1$, and considering only the classes $B_o : a_\varrho = 0$, and $B_1 : a_\varrho = 1$.

Blankenbecler and Savit (1972) have introduced a modification to the above method by assuming the $\tilde\pi\tilde\pi$ amplitude and its partial waves to be known for energies less than a known constant C. Above this energy they assume a particular functional form for σ_{tot} and they take $\sigma_{e\varrho} \le \frac{1}{2}\sigma_{tot}$ for energy greater than C. These constraints lead to a lower bound on the scattering length. The requirement that this lower bound is less than the experimental value then sets a constraint on the assumed functional form for σ_{tot}. In the particular case where σ_{tot} is taken to be a constant equal to $\sigma(\infty)$ above $s = 25\ m_{\tilde\pi}^2$, they find that $\sigma(\infty)$ is less than about 40 mb, which compares well with the factorization estimate of 15-20 mb.

(b) <u>Pion-Nucleon Amplitudes</u>

Reference: Common and Yndurain (1971 and 1972),
Kolanowski and Lukaszuk (1972), Steven (1972).

The above techniques for bounding averaged cross-sections have been extended by the above authors to pion-nucleon scattering. They also invert the argument to obtain bounds on the amplitudes in the crossed channel. Thus (4.12) provides a lower bound on the $\tilde\pi\tilde\pi$ scattering length if the total cross-section is assumed to be known, for example by putting in phenomenological $\tilde\pi\tilde\pi$ resonances.

In the case of $\tilde\pi$N scattering the total cross-sections are known from experiment. Then the formalism leads to bounds in the crossed channel, namely

$$\tilde\pi\tilde\pi \rightarrow N\overline{N} \tag{4.13}$$

Common and Yndurain (1972) introduce the additional feature that the annihilation cross-section $N\overline{N} \rightarrow \tilde\pi\tilde\pi$ is constrained via unitarity by the $\tilde\pi\tilde\pi$ elastic amplitude,

$$\left| f_\varrho (N\overline{N} \rightarrow \tilde\pi\tilde\pi)\right|^2 \le \mathrm{Imf}_\varrho (\tilde\pi\tilde\pi) - \left| f_\varrho (\tilde\pi\tilde\pi)\right|^2 \tag{4.14}$$

This observation gives an immediate gain of a factor 4 in the analogue of the Froissart bound for the total annihilation cross-section

$\sigma(N\bar{N} \to \pi\pi)$. It can also be used in two further ways. One way
is to obtain a local bound at finite energies on the annihilation
cross-section when $\sigma_{tot}(\pi\pi)$ is assumed to be known. The latter
constrains the partial wave series involving $\mathrm{Im}f_{\ell}(\pi\pi)$ and via (4.14)
it constrains the partial wave expansion of $\sigma(N\bar{N} \to \pi\pi)$. The local
value of $\sigma_{tot}(\pi\pi)$ may be taken from a Regge model since the energy
for $N\bar{N}$ annihilation exceeds 2 GeV.

The constraint (4.14) is most relevant at low energies near the
annihilation threshold. In this region (above 2 GeV) the bound may be
used to test models for extrapolation of experimental values for
$\sigma(N\bar{N} \to \pi\pi)$. These experimental values are obtained at energies of
several GeV where the bounds are rather weak. At lower energies the
bounds provide a cut-off to the region where a Regge model extrapolation for $\sigma(N\bar{N} \to \pi\pi)$ might be applicable. The results of Common
and Yndurain are summarised in Fig. 2 which is adapted from their (1972)
paper.

Finally Common and Yndurain use the Froissart-Gribov expression
for the $\pi\pi$ scattering length to constrain the $\pi\pi$ amplitude as in
(4.1) and (4.2). This then leads via (4.14) to constraints on the
energy-averaged partial wave expansion of $\sigma(N\bar{N} \to \pi\pi)$. These energy
averaged bounds are not very tight presumably because (4.14) is a weak
constraint at most energies, since it is clear from experimental results
and Regge models that $f_{\ell}(N\bar{N} \to \pi\pi)$ should tend to zero as the energy
increases.

(c) <u>Other applications of optimization theory</u>

Within particle physics as further experimental evidence becomes
available there will be more scope for studying more inequalities based
on phenomenological constraints. This should apply particularly to
inequalities involving spin parameters in two-body reactions and to
inequalities involving multiparticle production and inclusive reactions.
From a more theoretical viewpoint it is   valuable to use
inequalities to limit the effects on bootstrap calculations of unknown
couplings to inelastic processes (Ciulli et al. (1972)).

Outside particle physics optimization theory forms an important
part of many studies of complex systems whether in operational research,
systems analysis or control theory. One of the central problems in its
use for environmental or social applications arises from the conflict
between desirable objective functions. This conflict, coupled with
the multitude of influences, the paucity of data and the varying time
scales of observation of change, leads to problems of such magnitude

that the particle theorist may pause to reflect on the simplicity
of his own problems of understanding the fundamental laws of physics.

REFERENCES

1.   Review Articles on High Energy Bounds

R.J. Eden: Reviews of Modern Physics $\underline{43}$, 15 (1971)
M.B. Einhorn and R. Blankenbecler: Annals of Physics $\underline{67}$,
                                    480, 1971.
T. Kinoshita: Lectures at Boulder Summer School (1966).
A. Martin: Scattering Theory: Unitarity, Analyticity and
            Crossing (Springer Verlag, Berlin, 1969).
S.M. Roy: High Energy Theorems (Saclay preprint to be
           published in Physics Reports 1972).
V. Singh: Fields and Quanta $\underline{1}$, 151 (1971a).
F.J. Yndurain: Rigorous Constraints, Bounds and Relations for
               Scattering Amplitudes ,Rev.Mod.Phys.(1972).

2.   Introductory Books on Optimization Theory

M. Aoki: Optimization Techniques (Macmillan 1971).
M.R. Hestenes: Calculus of Variations and Optimal Control
                Theory (Wiley 1966).

3.   Other Articles

R. Blankenbecler and R. Savit: SLAC preprint (1972).
A.K. Common and F.J. Yndurain, Nucl.Phys. $\underline{B26}$, 167 and ibid.
$\underline{34B}$, 509 (1971).
A.K. Common and F.J. Yndurain, preprint University
of Madrid (1972), to appear in Nuclear Physics.
S. Ciulli et al., Bucharest preprint (1972) Tautologies
and Optimization of N/D Equations.
M.A. Jacobs et al. Phys. Rev. $\underline{D2}$, 1970 (1970).
B.D. Hahn and D.P. Hodgkinson (Cavendish Laboratory, Cambridge
preprint to appear in Nuclear Physics 1972).

M. Kolanowski and L. Lukaszuk, preprint (1972) from
Institute of Theoretical Physics, Warsaw, (to appear in
Nucl. Phys. B.).
D. Morgan and D. Shaw, Nucl. Phys. $\underline{B10}$, 261 (1969).
R. Savit, R. Blankenbecler and M.B. Einhorn: J. Math.Phys.$\underline{12}$,
2092 (1971).
V. Singh, Phys. Rev. $\underline{26}$, 530 (1971b).
V. Singh and S.M. Roy, Phys. Rev. $\underline{D1}$, 2368 (1970).
A. Steven, preprint Cavendish Laboratory, Cambridge (1972).

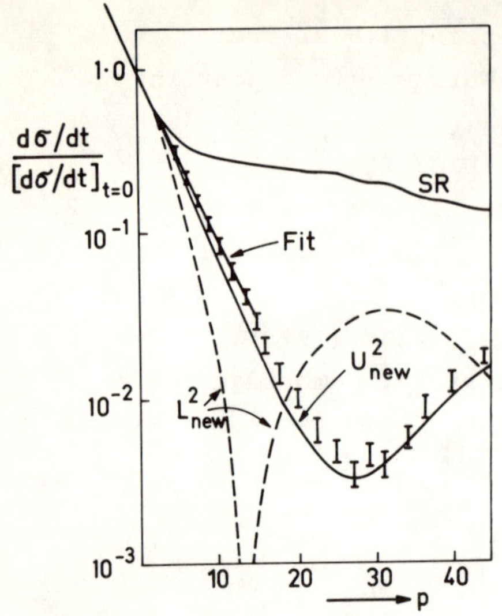

Comparison with experiments (for $\pi^- p$ elastic at 2 GeV/c) of the Hahn-Hodgkinson semi-phenomenological bounds U = max ImF s,t), and L = min ImF(s,t). The parameter ρ is proportional to t, the range shown being about 1 (GeV/c)^2. Their fit to the data that led to the constraint at $4m_\pi^2$ is also shown. The Singh-Roy bound SR follows from unitarity and σ_{el} ,σ_{tot} only

Fig. 1

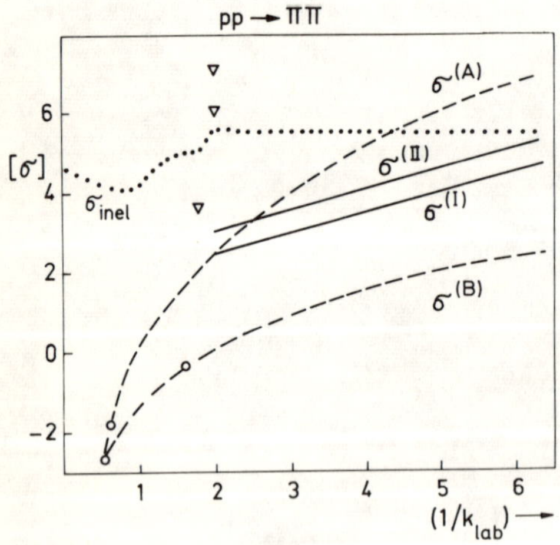

The bounds of Common and Yndurain, compared with two extrapolations (A) and (B) of experimental values of $(p\bar{p} \to \pi\pi)$. The broken lines denote the extrapolations. The continuous lines denote the bounds using two different Regge parametrisations I and II for the $\pi\pi$ amplitude. The dotted line denotes the total inelastic cross section for $p\bar{p}$. The triangles denote integrated bounds using different scattering lengths, the top one being that normally accepted

Fig. 2

A THEORETICAL INVESTIGATION OF

PHASE-SHIFT ANALYSIS

D. Atkinson

Institute for Theoretical Physics, P.O. Box 800, Groningen

I think that most people are surprised by the inadequacy of
phase-shift analysis when it is first explained to them. Let us recall the problem in the simplest situation, that of an elastic scattering process, with no competition from inelastic channels, and with no
spin or isospin. One wants to find a complex amplitude, $M(x)$, where
x is the cosine of the scattering angle, with the following partial-
wave expansion:

$$M(x) = \sum_{\ell=0}^{\infty} (2\ell+1)\, P_\ell(x)\, e^{i\delta_\ell} \sin \delta_\ell \tag{1}$$

where the phase-shifts, δ_ℓ, are to be real. One knows only $\left| M(x) \right|$,
which is essentially the square-root of the differential scattering
cross-section. It is not at all obvious that the δ_ℓ are determined
uniquely by $\left| M(x) \right|$.

I want first to describe some results of Newton[1] and Martin[2]
that pertain to this elastic case, together with some improvements
that Johnson, Warnock and I[3] have recently obtained. Then I will
discuss the extension of this analysis to energies above the inelastic
threshold. We will see that a continuum ambiguity is to be expected
in this case, a fact that has been illustrated by Bowcock and Hodgson[4]
in a specific example. Finally, I propose to touch on the possibility
of a numerical investigation of the ambiguities in actual cases of
experimental interest[5], and I will indicate the complications that
spin and isospin introduce.

Let us write

$$M(x) = g(x) \exp\left[i\phi(x) \right],\tag{2}$$

where $g = |M|$, which is to be supposed known, and ϕ is the phase.
If we could determine ϕ, the problem of phase-shift analysis would
have been solved. Below the inelastic threshold, the reality of the
phase-shifts in (1) is equivalent the following relation, the elastic
unitarity condition:

$$\text{Im } M(x) = \frac{1}{4\pi} \int_{-1}^{1} dy \int_{0}^{2\pi} dw \; M^*(y)\, M(z),\tag{3}$$

where

$$z = xy + \cos w \left[(1-x^2)(1-y^2) \right]^{1/2}.\tag{4}$$

This may be rewritten

$$g(x) \sin \phi(x) = \frac{1}{4\pi} \iint dy \; dw \; g(y)g(z) \cos\left[\phi(y) - \phi(z) \right],\tag{5}$$

or, for short,

$$\phi(x) = A[\phi;x] = \sin^{-1} B[\phi;x],\tag{6}$$

where

$$B[\phi;x] = \frac{\iint dy \, dw \, g(y)g(z) \cos\left[\phi(y) - \phi(z) \right]}{4\pi \, g(x)}\tag{7}$$

We can regard eq. (6) as a nonlinear equation for ϕ, given g. For a
given g, there is in general no guarantee either of the existence or
uniqueness of a solution of eq. (6). As Newton[1] pointed out, we can
easily construct a class of functions, g, for which there is no solution,
ϕ. For when $x = 1$, $z = y$, so that (6) becomes

$$\sin\phi(1) = \frac{\int_{-1}^{1} dy \, [g(y)]^2}{2 g(1)}.\tag{8}$$

If g is such that the right-hand side here is greater than unity,
then there is no real solution, ϕ. On the other hand, for some other
g's, there may be more than one solution. Crichton[6] showed, as late
as 1966, that there is a non-trivial ambiguity in the simple case
where only S, P and D waves are non-vanishing. Any amplitude with an
expansion

$$M(x) = \sum_{\ell=0}^{\infty} (2\ell+1) P_\ell(x) \left[e^{2i\delta_\ell} - 1 \right]/2i,\tag{9}$$

with real phase-shifts, δ_ℓ , will satisfy equ.(6), with the appropriate $g(x)$. We can restrict δ_ℓ to lie in $[0,\pi]$, since M is unchanged if δ_ℓ is replaced by $\delta_\ell \pm n\pi$, n = 1, 2, 3, ... If the sign of every δ_ℓ is changed, then M is changed into its complex conjugate and g = |M| is unchanged. This is called the trivial ambiguity. Crichton's example is not trivial: he showed that, for any δ_2 between about $12\frac{1}{2}$ deg. and 24 deg., there are <u>two</u> possible choices of δ_0 and δ_1, such that g is the same, but ϕ is different.

Following Martin, we define a functional of g:

$$\sin\mu = \sup_{-1 \leq x \leq 1} \frac{\iint dy\,dw\, g(y)\,g(z)}{4\pi\, g(x)} . \qquad (10)$$

This quantity is by no means forced to be less than unity: with the help of Dr.N. Mehta I recently worked out that it is about 3.3 for the Crichton example, with $\delta_2 = 20°$. As we shall see, there is the possibility of numerical analysis when $\sin\mu \geq 1$; but for the moment we restrict our attention to the case $\sin\mu < 1$. It is only here that we can prove our fixed-point theorems.

Let me indicate first the outlines of our version of a proof that, if $\sin\mu \leq 0.62$, then there is one, and only one phase, ϕ ,that satisfies the unitarity equation (6). Let S be the space of all functions, f(x), that are continuous on $-1 \leq x \leq 1$, and associate the following norm with each f:

$$\| f \| = \sup_{-1 \leq x \leq 1} |f(x)| . \qquad (11)$$

Let T be the set of functions, ϕ, contained in S, such that $0 \leq \phi(x) \leq \mu$, $-1 \leq x \leq 1$. We suppose that g(x) is continuous, and we set

$$\phi'(x) = A[\phi, x] . \qquad (12)$$

Then it is easy to see that $\phi'(x)$ is continuous, and $\phi'(x) \geq 0$. Moreover,

$$\phi'(x) \leq \sin^{-1}\left[\frac{\iint dy\,dw\, g(y)\,g(z)}{4\pi\, g(x)}\right] \leq \mu \qquad (13)$$

so ϕ' belongs to T if ϕ does, for any $\sin\mu < 1$. We say that T is mapped into itself by the operator A.

To apply the Contraction Mapping Principle, we have now to show that, if ϕ_1 and ϕ_2 are any two functions belonging to T,

$$\| A(\phi_1) - A(\phi_2)\| \le \beta \|\phi_1 - \phi_2\| \, , \tag{14}$$

with $\beta < 1$. In this case, we can iterate A:

$$\phi^{(n+1)}(x) = A\left[\phi^{(n)};x\right], \tag{15}$$

where the starting point, $\phi^{(O)}$, can be any function belonging to T, for instance $\phi^{(O)}(x) \equiv 0$. The iteration is guaranteed to converge, since

$$\|\phi^{(n+1)} - \phi^{(n)}\| \le \beta \|\phi^{(n)} - \phi^{(n-1)}\| \le \beta^n \|\phi^{(1)} - \phi^{(O)}\|, \tag{16}$$

from which it follows easily that $\{\phi_n\}$ is a Cauchy sequence.

A neat way of obtaining a sufficient condition for contraction is to use the following, operator version of the Mean Value Theorem:

$$\| A(\phi_1) - A(\phi_2)\| \le \sup_{0 < t < 1} \| A'(t\phi_1 + (1-t)\phi_2)\| \cdot \|\phi_1 - \phi_2\| \, , \tag{17}$$

where A' is the Fréchet differential of the operator A, and is continuous. Now the set, T, is convex, i.e. if ϕ_1 and ϕ_2 belong to T, then so does $t\phi_1 + (1-t)\phi_2$, $0 \le t \le 1$. Accordingly, it is enough to require $\|A'(\phi)\| < 1$, for all ϕ belonging to T.

Let us rewrite eq. (7) as

$$B(\phi;x) = \iint dy \ dz \ H(x,y,z) \ \cos\left[\phi(y) - \phi(z)\right] \, , \tag{18}$$

where

$$H(x,y,z) = \frac{g(y)g(z)}{2\pi g(x)} \, k^{-\frac{1}{2}}(x,y,z), \tag{19}$$

$$k(x,y,z) = 1 + 2xyz - x^2 - y^2 - z^2. \tag{20}$$

We have changed to the integration variables y and z. The region of integration is the interior of an ellipse, corresponding to the condition $k(x,y,z) \ge 0$. The Fréchet derivative of A is then

$$A'(\phi)\delta\phi = -\left[1 - B^2(\phi)\right]^{-\frac{1}{2}}\iint dy \ dz \ H(x,y,z) \sin \left[\phi(y) - \phi(z)\right]\left[\delta\phi(y) - \delta\phi(z)\right] \, . \tag{21}$$

Now $B(\phi;x) \le \sin\mu$, $\sin\left[\phi(y) - \phi(z)\right] \le \sin\mu$, and $\iint dy \ dz \ H(x,y,z) \le \sin\mu$, so

$$\|A'(\phi)\delta\phi\| \leq \frac{2\sin^2\mu}{[1-\sin^2\mu]^{1/2}} \|\delta\phi\| \qquad (22)$$

i.e.

$$\|A'(\phi)\| \leq 2\sin\mu\tan\mu . \qquad (23)$$

A sufficient condition for a contraction is that the right-hand side be less than unity, which means $\sin\mu \lesssim 0.62$.

Martin[2] showed that there cannot be more than one solution if $\sin\mu \lesssim 0.79$, and we[3] have in fact obtained a contraction mapping in this domain by using the weighted L^2 norm:

$$\|\phi\| = \left[\frac{1}{2} \int_{-1}^{1} dx\, g(x)\, \phi^2(x) \right]^{1/2} . \qquad (24)$$

Hence an iteration can be again used. The algebra is a little involved, but one finds eventually that

$$\|A'(\phi)\delta\phi\| \leq \frac{2\sin^2\mu}{[1-\sin^2\mu\cos^2\mu]^{1/2}} \|\delta\phi\| \qquad (25)$$

with the L^2 norm, and the condition that this coefficient be less than unity is $\sin\mu \lesssim 0.79$.

One can use Schauder's theorem to show that there is at least one solution of $\phi = A(\phi)$, if $\sin\mu < 1$. If A maps a closed, convex subset, F, of a Banach space, continuously into a compact subset of itself, then Schauder's theorem guarantees the existence of at least one fixed point in F. The usual difficulty in applying the theorem to an infinite-dimensional system, like $\phi = A(\phi)$, is to find a set, F, such that A(F) is compact. According to the Ascoli theorem, this is equivalent to requiring that the family of functions, A(F), be uniformly bounded and equicontinuous. Newton[1], in fact, tried to use the set T, with the topology of the sup norm, eq. (11), but this proof is not correct, since one can show explicitly[3] that the set A(T) is not equicontinuous.

Martin[2] overcame this difficulty by taking F to be the closure of the convex hull of the set A(T). He showed that A(F) is compact, although he used a partial-wave expansion, the convergence of which needs to be handled carefully. We[3] have a more elementary proof that does not use a partial-wave series, and does not indulge in convex hulls. One can prove directly that

$$\iint dy\, dz \left[k^{-\frac{1}{2}}(x_1,y,z) - k^{-\frac{1}{2}}(x_2,y,z) \right] g(y)g(z)\, \cos[\phi(y)-\phi(z)] \leq$$

66

$$\leq M \left| \frac{x_1 - x_2}{1 - x_2^2} \right|^{\alpha}, \tag{26}$$

where M is constant, $1 \geq x_2^2 \geq x_1^2$, and $0 < \alpha < \frac{1}{2}$. The proof is a little tricky, since the domain of integration is different for x_1 and x_2.

It follows from (26) and the condition $\sin \mu < 1$, that

$$\left| A(\phi; x_1) - A(\phi; x_2) \right| \leq M \left| \frac{x_1 - x_2}{1 - x_2^2} \right|^{\alpha} + M \left| g(x_1) - g(x_2) \right|. \tag{27}$$

Accordingly, if g(x) is continuous, then the family $A(\phi; x)$ is equi-continuous, except at the end-points $x = \pm 1$. It is this difficulty at the end-points that invalidated Newton's proof, but we may avoid the difficulty by the following trick: Set

$$\psi(x) = (1 - x^2)^{\frac{1}{2}} \phi(x), \tag{28}$$

and
$$C(\phi; x) = (1 - x^2)^{\frac{1}{2}} A(\phi; x), \tag{29}$$

so that (6) becomes

$$\psi = C(\psi). \tag{30}$$

For F of Schauder's theorem, we use the set of continuous functions, $\psi(x)$, such that $0 \leq \psi(x) \leq \mu (1 - x^2)^{1/2}$, which of course corresponds to $0 \leq \phi(x) \leq \mu$. It is possible to show that C maps F continuously into itself, and

$$\left| C(\psi; x_1) - C(\psi; x_2) \right| \leq \left| (1 - x_2^2)^{1/2} \left[A(\phi; x_1) - A(\phi; x_2) \right] \right| +$$

$$+ \left| \left[(1 - x_2^2)^{1/2} - (1 - x_1^2)^{1/2} \right] A(\phi; x_2) \right| \leq$$

$$\leq M \left| x_1 - x_2 \right|^{\alpha} + M \left| g(x_1) - g(x_2) \right|, \tag{31}$$

so C(F) is compact, and the Schauder result applies.

Now let us consider the physically more interesting case when there is competition from inelastic channels. We write

$$\sin \phi = B(\phi) + I(\eta), \tag{32}$$

$$I(\eta) = \left[g(x) \right]^{-1} \sum_{\ell=0}^{\infty} (2\ell + 1) \frac{1 - \eta_\ell^2}{4} P_\ell(x), \tag{33}$$

where $0 \leq \eta_\ell \leq 1$. If we assume that g and I are known, continuous functions, then Schauder's theorem applies if $\sin \mu < 1$, where now

$$\sin \mu = \sup_{-1 \leq x \leq 1} \left[\iint dy\ dz\ H(x,y,z) + |I(\eta;x)| \right] \ . \quad (34)$$

For a contraction mapping, we define

$$\sin \mu' = \sup \iint dy\ dz\ H(x,y,z), \quad (35)$$

and we can repeat the manipulations with the sup norm as follows:

$$\phi = A(\phi;\eta) = \sin^{-1}\left[B(\phi) + I(\eta) \right] \quad (36)$$

$$A_\phi(\phi;n)\delta\phi = \left\{ 1 - \left[B(\phi) + I(\eta) \right]^2 \right\}^{-\frac{1}{2}} B'(\phi)\delta\phi \quad (37)$$

$$\| A_\phi(\phi;\eta) \| \leq 2 \sin \mu' \tan \mu \ . \quad (38)$$

So if $2 \sin \mu' \tan \mu < 1$, we have a contracting mapping, which means that there is a locally unique solution, ϕ, for a given g and I. If we keep g fixed, but change I continuously, in such a way that (32) is still respected, it follows that ϕ will change continuously. This is easy to see, since, if we change I by changing the η_ℓ, ϕ will change as follows:

$$\delta\phi = A_\phi\ \delta\phi + A_\eta\ \delta\eta. \quad (39)$$

since $\| A_\phi \| < 1$, this equation has a unique solution (as we can see by using the contracting mapping again in the space of continuous functions) which we may write as

$$\delta\phi = \left[1 - A_\phi \right]^{-1} \delta\eta. \quad (40)$$

This result in fact is an example of the use of the Implicit Function Theorem.

The basic idea of phase-shift analysis is to determine ϕ and I, given g, since this is equivalent to a knowledge of all the (complex) phase-shifts. We have seen that, in the contraction domain at least, there is a continuum ambiguity, a phenomenon that has been recently illustrated by a specific example of Bowcock et al[4]. Indeed, this consequence of the Implicit Function Theorem holds in the neighbourhood of any solution for which A_ϕ is not singular. We can expect that, for each set of phase-shifts that fit the cross-section, there will be a whole continuum of alternative sets, and the boundaries of the patches of candidate phase-shifts will be defined, for each partial-wave, by a subset of the points for which A is singular, and by the unitarity circle itself. Hence an interesting possibility would be to take the results of a phase-shift analysis and to find nearby solutions that correspond to the same cross-section but different inelasticities. This could be done in principle by interacting eq. (7); but it would probably be preferable to use the Newton-Kantorovitch method, which uses the Fréchet differential (21). One could then see in practical cases how important (or otherwise) the ambiguity might be.

Spin and isospin complicate the situation, but the picture is basically similar: in the elastic region there is a domain where the contraction mapping theorem can be used to demonstrate uniqueness; and in the inelastic region, a continuum ambiguity exists. For example, in $\pi\pi$ scattering, there are five different cross-sections $\pi^+\pi^+ \to \pi^+\pi^+$; $\pi^+\pi^0 \to \pi^+\pi^0$; $\pi^+\pi^- \to \pi^+\pi^-$; $\pi^+\pi^- \to \pi^0\pi^0$; $\pi^0\pi^0 \to \pi^0\pi^0$, which could be measured (only in principle, of course!). These measurements would suffice to enable $g_I(x)$, $I = 0,1,2,$ to be calculated, where I is the isospin. In the elastic region, within the contraction domain, the phases, $\phi_I(x)$, would then be determined, which means that there would be two constraints between the five cross-sections. In the inelastic contractive region, we have 3 inelastic terms, 3 unitarity equations and 5 cross-sections, so that only one inelastic term would be independently variable. There would be a continuum ambiguity, but it would not be independent for each isospin state.

As another example, consider π^+p scattering, where we have the spin complication, but can ignore isospin. Here there are two unitarity conditions, corresponding to the two helicity states, and four measurable quantities, $\frac{d\sigma}{d\Omega}$, the polarization, P, and the parameters S and T (or R and A). However, aside from a sign ambiguity, only three of these quantities are independent. In the contractive, elastic domain, there will be one constraint, which should allow S and T to be calculated from $\frac{d\sigma}{d\Omega}$ and P. In the inelastic region, there would again be a continuum ambiguity, even if $\frac{d\sigma}{d\Omega}$, P, S and T were measured[7].

REFERENCES

1. R.G. Newton, J. Math. Phys. $\underline{9}$, 2050 (1968).

2. A. Martin, Nuovo Cim. $\underline{59A}$, 131 (1969).

3. D. Atkinson, P.W. Johnson, and R.L. Warnock, Determination of the Scattering Amplitude from the Differential Cross-Section and Unitarity", Groningen preprint (1972); to be published in Comm. Math. Phys.

4. J.E. Bowcock and D.C. Hodgson, Birmingham preprint (1971).

5. R.L. Warnock, New Methods for Numerical Solution of Nonlinear Equations, with Suggested Applications in S Matrix Theory, Bonn preprint (1972).

6. J.H. Crichton, Nuovo Cim. $\underline{45A}$, 256 (1966).

7. See a forthcoming paper by D. Atkinson, G.Mahoux and F.J. Yndurain.

<u>STABILITY PROBLEMS IN ANALYTIC</u>

<u>CONTINUATION</u>

Sorin Ciulli

Institute for Atomic Physics

Bucharest, Romania

1.   <u>INTRODUCTION</u>

1.1. <u>Ill Posed Problems</u>

In elementary particle physics, information is available only along some finite regions of the energy (or momentum transfer) cut complex plane and the problem one is faced with is to extract from this incomplete, error affected knowledge, information concerning the reactions in crossed channels or at energies outside the initial range. This is in general an ill posed mathematical problem, in the sense that minute changes (errors) in the input data may provide uncontrollable responses in the output.

Dispersion relations, the first and most common way of performing analytic extrapolations in particle physics, are far from being exempted from this general evil, as Cauchy integrals are stable only if the integrand is known over the w h o l e contour. Neither are the "purely theoretical" dynamical schemes as they usually include at least at one step a process of analytic continuation.

In early times the situation concerning the stability of the analytic continuation was a bit confused among physicists, as every one of them knew that an analytic function is uniquely determined by its values given along some arbitrary continuum. Nevertheless - and this point seems to have been ignored - as soon as one admits of the thinnest error corridor around these data values, one is already able to slip inside it amplitudes differing among themselves as much as one might wish. A familiar example is provided by the Laplace equation,where although the solution is granted to be locally uniquely determined by the values of the function and its derivative along an open curve, one never uses the latter as initial data, since the solution would be discontinuous (i.e. infinitely sensitive) with respect to them. This sort of problems are all instances of inversion of the symbolic equatio n Af = h, where   h   is a given input and A is a c o n t i n u o u s operator on some functional space (for instance the operation of restricting an analytic function to a part of its boundary). Uniqueness of its inverse by no means implies the unique determination of f in practical situations, as, if A^{-1} is discontinuous, arbitrarily small variations in h will cause uncontrollable changes in f .

1.2.   <u>STABILIZING LEVERS</u>

If one restricts in a convenient way the set $\mathcal{F}$ of f's in which the solution is sought, the restricted inverse $A^{-1}\big|_{\mathcal{F}}$ might be continuous and stability with respect to changes of h, restored. For instance it turns out to be sufficient to ask that "the admissible" f's make up a compact. There are different conditions ("stabilizing levers") which do the job and it is most important to choose those of them which are most relevant for physical predictions. As an example consider the analytic continuation off an open contour to the points of a domain D; a first way to compactify $\mathcal{F}$ is to require a Hölder condition for f(z)

$$|f(z_1) - f(z_2)| < K|z_1-z_2|^{\alpha} \quad ; \quad z_1, z_2, \in \Gamma_2 \qquad (1.2.1)$$

on the boundary Γ_2 of D. If f(z) is also bounded by some constant M

$$|f(z)| < M \qquad z \in \Gamma_2 \qquad (1.2.2)$$

then, by the Arzelà theorem, the set $\mathcal{F}$ is compact and the extrapolation is stable up to and including the cut Γ_2.

On the other hand, as (1.2.2) implies, for analytic function, the Hölder condition (1.2.1) for every closed region $\overline{D'}$ interior to D, the boundness condition (1.2.2) alone represents a complete stabilizing lever for interior points.

The actual values of these stabilizing parameters might be thought of as controlling in some way the opening of the range of predictions at points outside the experimental reach. For instance, for a given error channel width ε (z) around the data function h(z), high values of M in (1.2.2) might allow too many possible amplitudes differing among them by too large amounts, so that the predictions are

insignificant, whereas a too low value for M might go to the other
extreme where there are no amplitudes left at all (for $M < M_o$). It is
important to note that this happens not only in analytic continuation,

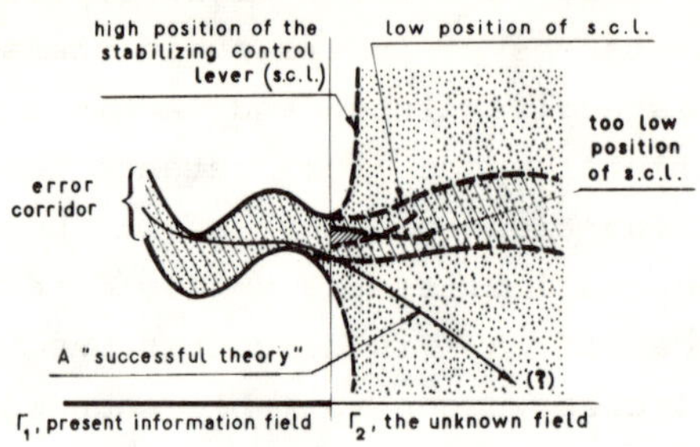

Fig. 1

but in every branch of science where theoretical predictions are made
outside the experimental range.

Although in these lectures only the Hölder and the boundedness
condition will be used, we would like to stress that the actual choice
of the kind of stabilizing condition is a very serious scientific problem of great importance for an Optimal Science Strategy. Ideally the
stabilizing parameters should be themselves within experimental reach
in order to allow one to set up yes-or-no experiments; we shall try
to make these things clearer in section 5.4 where it will be shown
that, if a precise bound M_{true} is known for the amplitude, one can
decide on the existence or nonexistence of its singularities.

1.3. THE TOTAL APPROACH

Once the stabilizing parameter was chosen and a precise value
was assigned to it, one could, at least in principle, construct the
whole set $\mathcal{F}$ of admissible amplitudes f(z). Taking then the center
f(z) of the values of all f(z) $\in \mathcal{F}$ at every point z $\in$ D, one gets
thus the absolute optimal extrapolation. This problem was solved up to
this ultimate stage only for interior points under the stabilizing
condition (1.2.2) and will be exhaustively treated in chapter 4. In
general however, this total program is bound to be a very hard mathematical task. Then we might be less ambitious and proceed along the lines
of a set of mathematically equivalent (tautological) theories and
optimize inside this set. We would like to work this out in a bit more
detail in order to emphasize the clear cut difference between this new,
"tautologic approach" and previous, "total", one.

1.4. <u>THE TAUTOLOGY APPROACH</u>

Physics handbooks provide us with theories (methods) of constructing a certain mathematical being, provided the input data are
"correct" in a certain sense, i.e. obey certain conditions. In this
case one can conceive a whole set of logically equivalent (tautological)
theories yielding exactly the same result, but which might produce
uncontrollably different results when fed with "inaccurate" data.
To have a simple but instructive example, let's consider the continuation of an analytic function f(z) from its boundary values f(t) via
the method of dispersion relations:

$$f(z) = \frac{1}{2\pi i} \int_{\Gamma_1} \frac{f(t)}{t-z}\, dt \qquad\qquad (1.4.1)$$

Exactly the same result is obtained with a weighted integral

$$f(z) = \frac{1}{2\pi i\; g_z(z)} \int_{\Gamma_1} \frac{f(t)\, g_z(t)}{t-z}\, dt \qquad\qquad (1.4.2)$$

where $g_z(z')$ is any function holomorphic in z' and arbitrary in z.
(The various possible g's make up the set of tautologies for dispersion
relations). Now if the "correct" boundary data f(t) on Γ_1 are replaced by an approximate data function h(t),

$$\left| h(t) - f(t) \right| < \varepsilon(t) \qquad\qquad t \in \Gamma_1 \qquad\qquad (1.4.3)$$

the symnetry of the integrals (1.2.1) and (1.2.1) is broken and there
appears the meaningful mathematical problem to find that tautology
(that $g_z(z')$) which produces at a certain point z a value least affected by the input inaccuracy. One can show (see chapter 3) that the
optimal weighted integral (1.4.2) can be found and it is a Poisson
weighted dispersion relation (P.W.D.R).

Similar problems, usually leading to extremal problems of
functional analysis can be solved also for other physical problems
where the tautology group can be explicitly exhibited. For instance
one could optimize dynamical N/D equations as will be done in section
7.2. Here "inaccurate" data (breaking the tautology symmetry) mean
left hand cut data which are not limiting values of unitary amplitudes.

2.   <u>OPTIMIZATION OF POLYNOMIAL EXTRAPOLATIONS</u>

       2.1. <u>Maximal Convergence and Optimal Conformal Mapping</u>

Dispersion relations compute the scattering amplitude at interior points starting from its boundary values. The converse, i.e. going to the cu$_2$ from data along some region Γ_1 inside the analyticity domain D can be dealt with, in principle, by expanding the amplitude f(z) in an appropriately chosen series still convergent **on** the cuts. The coefficients of the expansion are to be determined from the data function h(z) on Γ_1 :

$$\left| h(z) - f(z) \right| \, \Gamma_1 \; < \; \mathcal{E}(z) \qquad\qquad (2.1.1)$$

Historically, such expansions were first used [1,2], in connection with Chew-Goldberger-Low-Nambu integral equations where an extrapolation of the nearly forward scattering amplitude (Γ_1 reduces to the point $z \equiv \cos\theta = 1$) to the rest of the physical region ($-1 < z < 1$) is needed in order to express the imaginary part of the forward scattering amplitude and its cosine derivatives in terms of themselves, via unitarity. A power expansion

$$f(z) = \sum_{o}^{\infty} a_n \, w(z)^n \qquad\qquad (2.1.2)$$

was used, where the function

$$w(z) = \frac{\sqrt{u+1} - \sqrt{u-1}}{\sqrt{u+1} + \sqrt{u-1}} \;, \qquad u = (\frac{a^2-1}{z-1} - 1)/a \qquad (2.1.3)$$

is chosen so as to map the cuts $(-\infty, -a)$ and (a, ∞) onto the unit circle and the point z=1 into the origin.

Almost at the same time, Frazer [3] had the idea to make use of the quick convergence of this series to store the available information about f(z) in its few first coefficients (new effective range formulae [3]; see also [4-5]). Using the maximally convergent polynomials to be described below, Cutkosky pushed further this idea in his modified phase shift analysis (see section 2.4) and succeeded to express the cosine dependence of the amplitude in terms of very few coefficients at energies where hundreds of usual partial waves are needed.

The first to use these series up to the cuts (to the double spectral function) was Lovelace [6], who hoped that they will converge there like Fourier series; indeed, since the "worse" singularities are always confined to the first thresholds, they could be removed by suitable factors [7] so that the convergence of the Fourier series can always be ensured (see also [8]). The same method was also used in the energy plane by Atkinson [9] who expanded in terms of w's Hamilton's discrepancy function [10-13] defined as the difference between the actual amplitude and its right hand cut dispersion integral, and summarized hence the effects of the crossed channel cuts on the amplitude. As it is well known, the rate of convergence of a Taylor series depends in an essential way on the point around which the expansion is performed. Now in most physical extrapolations we do not expand around a point, but rather "around" some curve Γ_1, so that the Taylor series looses its relevance and we have to resort [14-17] to the theory of maximally converging polynomials.

The following theorem can be found in most texbooks (see Appendix A of [16] , or, better, [18]): Let Γ_1 be some simply connected open curve and $\zeta(z)$ a test conformal mapping which brings Γ_1 onto the unit circle $|\zeta| = 1$ and the point at infinity into itself. If R is the distance from the origin $\zeta = 0$ to the nearest singularity of $f(z)$ in the ζ-plane, then for every $R' < R$ there exist ("maximal converging") polynomials $P_n(z)$ such that for every z, (such that $|\zeta(z)| < R$) we have:

$$\left| f(z) - P_n(z) \right| \le M \left(|\zeta(z)| \big/ R' \right)^n \tag{2.1.4}$$

but there are none such that (2.1.4) hold also for R" s greater than R.

As usual, it is assumed that the positions of the singularities of the scattering amplitudes are known. The strategy gets then clear ($^{14-17}$): we shall leave the z plane and look for that conformal mapping of it, $W(z)$, for which the "maximally converging" polynomials in $W(z)$ have a highest convergence ratio.

To this end, let $W_a(z)$ and $W_b(z)$ map respectively the domains D_a and D_b ("falsely" cut along Γ_1) into the ring $1 < |W_a| < R_a$ and $1 < |W_b| < R_b$; here Γ_1 is mapped onto the unit circle, the test transformation $\zeta(W_{a,b})$ being hence already included in W. Suppose that the first singularities of $f(z)$ lie on the outer boundaries of both Γ_a and Γ_b and that D_a is included in D_b. Then we shall show that

$$\frac{|W_b(z)|}{R_b} \le \frac{|W_a(z)|}{R_a} \qquad (z \in D_a \subset D_b) \tag{2.1.5}$$

which simply means that the optimal convergence rate is obtained in terms of that conformal mapping $W(z)$ which maps the whole cut plane of $f(z)$ into the ring

$$\left| W_{(z\,\in\,\Gamma_1)} \right| = 1, \qquad \left| W_{(z\,\in\,\Gamma_2)} \right| = R \qquad\qquad (2.1.6)$$

<u>Proof of (2.1.5)</u>: Let us [16] compare the harmonic functions

$$O_a(z) = \ln \left\{ \left| W_a(z) \right| \Big/ R_a \right\}$$

$$O_b(z) = \ln \left\{ \left| W_{b(z)} \right| \Big/ R_b \right\} \qquad\qquad (2.1.7)$$

both on Γ_a and Γ_1. Since $O_b(z)$ is negative on Γ_1 and zero on Γ_b

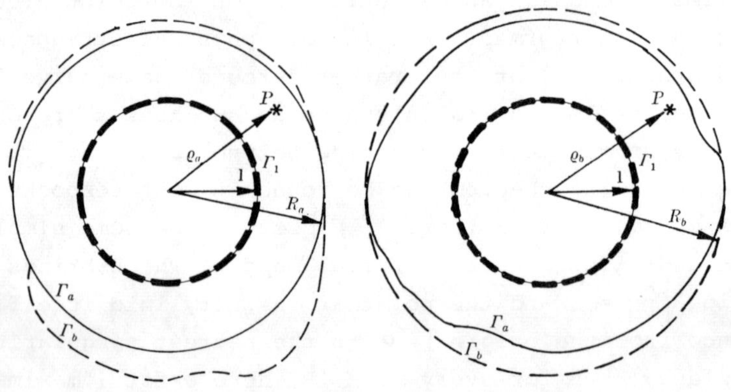

Fig. 2

and since $\Gamma_a \subset \overline{D_b}$, we have $O_b\,(z \in \Gamma_a) \leqslant 0$ and hence, (as $O_a\,(z \in \Gamma_a) = 0$),

$$O_b(z) \leq O_a(z) \quad , \qquad \text{for} \qquad z \in \Gamma_a \quad . \qquad\qquad (2.1.8)$$

On Γ_1 we have to resort to the less trivial principle of Groetzsch (see appendix B of [16]) which asserts that if $D_a \subset D_b$ then $R_a \leq R_b$. Hence

$$O_b\,(z \in \Gamma_1) = -\ln R_b \leqslant -\ln R_a = O_a\,(z \in \Gamma_1) \quad . \qquad\qquad (2.1.9)$$

Now, both $O_a(z)$ and $O_b(z)$ being harmonic, the ineq. (2.1.8) and (2.1.9) remain valid in every interior point $z \in D_a$, which proves (2.1.5).

The explicit form of the optimal mapping (2.1.6) can easily be derived; for instance, if the cut Γ_2 is symmetric ($\Gamma_2(-\infty,\ -1/k)$, and $(1/k,\infty)$, $k < 1$) and if the physical region Γ_1 extends from -1 to $+1$, one gets [16]

$$W(z) = i \exp\left\{ - i\,\pi\,u(z)/2\,K \right\},$$

$$u(z) = \int_0^z \frac{dt}{\sqrt{(1-t^2)\,(1-k^2t^2)}} \quad , \qquad K = \int_0^1 \frac{dt}{\sqrt{(1-t^2)\,(1-k^2t^2)}} \quad ,$$

the outer radius of the ring ($\left| W \right|_{r_2} = R$) being

$$R = \exp\left\{\tau K'/2K\right\} \quad \text{with} \quad iK' = \int_1^{1/k} \frac{dt}{\sqrt{(1-t^2)(1-k^2t^2)}} \; . \tag{2.1.10}$$

Since Γ_1 is not really a cut, $f(z)$ may be continued up to $|w| = 1/R$ using the Schwartz reflection principle on Γ_1 $\left(f(1/w^*) = f^*(w)\right)$; furthermore, if $f(z)$ is of the real type one has

$$f(w) = f^*(w^*) \quad \text{and thus, also } f(1/w) = f(w) \tag{2.1.11}$$

Hence $f(w)$ can be expanded in a (quickly converging) Laurent series of the form

$$f(w) = a_0 + a_1 \; (w + w^{-1}) + a_2 \; (w^2 + w^{-2}) + \dots \tag{2.1.12}$$

where all coefficents a_0, a_1,... are real.
It is obvious that instead (2.1.12) one could equally well use [14]a
Legendre expansion in the variable $z = 1/2(w + w^{-1})$ which maps the ring (2.1.16) into a "Lehmann" ellipse of foci ± 1, the theory remaining otherwise unchanged.

Note: At very high energies the mapping of Deo and Parida should be used [19] , as it emphasizes the now important physical region around $\cos\theta = 1$.

2.2. Hölder's Condition and Stability to the Cuts

In the preceding section we have proven that the partial sums

$$P_{f,n}(z) = a_0 + \sum_{k=1}^{n} a_k \; (w^k(z) + w^{-k}(z))$$

of the Laurent series in the w-plane converge to $f(z)$ the quickest when $w(z)$ is the optimal mapping (2.1.9).

Since the amplitude $f(z)$ <u>is unknown</u>, <u>so are</u> the exact polynomials $P_{f,n}$, the unique polynomials at hand being those $P_{h,n}$ found by the best (Tchebycheff) approximation of the data function $h(z)$ on Γ_1. Studying stability means to estimate the extent to which the extrapolation in terms of $P_{h,n}(z)$ approximates the unknown amplitude $f(z)$ at points outside Γ_1 [16].

If the p-th derivative of the amplitude is subjected, on the cut Γ_2, to the Hölder condition

$$\left| f^{(p)}(z_1) - f^{(p)}(z_2) \right| < M |z_1 - z_2|^{\alpha} \; ; \quad z_1, z_2 \in \Gamma_2 \tag{2.2.1}$$

then it is known from the Fourier series theory that

$$\left| f(z) - P_{f,n}(z) \right| \leq \bar{\eta}_n \frac{|w|^n}{R^n} \quad , \quad \bar{\eta}_n = A \ln[n/n^{p+\alpha}] \tag{2.2.2}$$

for every $1 \leq |w| \leq R$, so that on the physical region Γ_1 we get

$$\left| f(z) - P_{f,n}(z) \right|_{\Gamma_1} \leq \bar{\eta}_n/R^n \tag{2.2.3}$$

The <u>mere existence</u> of the unknown polynomials $P_{f,n}$ is enough to ensure

$$\left| h(z) - P_{h,n}(z) \right|_{\Gamma_1} < \varepsilon + \bar{\eta}_n/R^n \tag{2.2.4}$$

as from (1.4.3) and (2.2.3) one gets

$$\left| h(z) - P_{f,n}(z) \right|_{\Gamma_1} < \varepsilon + \bar{\eta}_n/R^n \tag{2.2.5}$$

and, $P_{h,n}$ being by construction the best polynomial approximation to $h(z)$ on Γ_1, the bound (2.2.4) has to be <u>at least as good</u> as the bound (2.2.5). Applying again the Schwarz inequality one gets ($|w|_{\Gamma_1} \equiv 1$) :

$$\left| \frac{P_{f,h}(z) - P_{h,n}(z)}{w^n(z)} \right|_{\Gamma_1} < 2 \left(\varepsilon + \bar{\eta}_n/R^n \right) \tag{2.2.6}$$

Now, by the maximum modulus theorem [1], since the l.h.s. of (2.2.6) is holomorphic for $|w| \geq 1$, the same holds also on Γ_2; thus $\left(|w|_{\Gamma_2} \equiv R \right)$

$$\left| P_{f,n}(z) - P_{h,n}(z) \right|_{\Gamma_2} < 2(\varepsilon R^n + \bar{\eta}_n) \tag{2.2.7}$$

which combined with (2.2.2) (on Γ_2!) <u>yields</u> the stability condition on Γ_2 [16] :

$$\left| f(z) - P_{h,n}(z) \right|_{\Gamma_2} \leq 2\varepsilon R^n + 3\bar{\eta}_n \equiv E_n \tag{2.2.8}$$

(For interior points one has simply to multiply E_n with $|w|^n/R^n$).
For small ε, E_n decreases first with n, until it reaches the minimum E_c at the critical $n = n_c$:

$$E_c = \frac{3}{n_c^{p+\alpha}} \left\{ \frac{(p+\alpha)\ln n_c - 1}{n_c \ln R} + \ln n_c \right\} , \tag{2.2.9}$$

$$n_c \geq \frac{\ln\left[3(p+\alpha)A/2\varepsilon \ln R \right]}{\ln\left[R \, 1{,}44^{(p+\alpha)} \right]} .$$

---

[1] For L^2 norm problems, this argument is no longer valid but we refer the reader to ref. [17].

Taking more terms, although $h(z)$ is better approximated by $P_{h,n}$ on Γ_1, the error function E_n on Γ_2 blows up and the extrapolation looses any physical meaning. Nevertheless one should note that for $\varepsilon \to 0$, $n_c \to \infty$ and hence $E_c \to 0$, which proves the stability of the whole procedure.

2.3. Hamilton's Discrepancy Method

The severe lack of space makes impossible a description here of the otherwise particularly interesting physical applications of the extrapolation procedures; we shall however, be humbly happy if these lines could merely serve as a guide to the papers quoted at the References.

An important class of papers [10-13] and [20-28] is connected to the discrepancy function first introduced by Hamilton and Spearman [10]

$$\overline{\Xi}(z) = A(z) - \frac{1}{\pi} \int_{r.h.cut} dz' \; \frac{\mathrm{Im}\; A(z')}{z' - z} \quad , \tag{2.3.1}$$

which obviously has only a left hand cut (Γ_2) and is known on Γ_1, which is part of its holomorphy domain. This method allows one to pass from a cut-to-cut extrapolation problem to the already solved interior points to the cut extrapolation.

As it is known, the $\pi\pi$ phases are connected to the l.h.cut of the very well known πN ones, and much effort has been devoted to their evaluation. Following the classical papers of Hamilton et al., [10-13], Nielsen, Petersen and Pietarinen [20], using CERN's latest πN P- and D-phases, obtained a model independent evaluation of the $T=0$, S- and D-waves, as well as the δ_1^1 phase. This program was continued by Elvekjaer, Nielsen and Oades [21-23], using discrepancies for the πN P- and D-waves, with CERN phases above 270 MeV and the revised phase shifts of [24] below 270 MeV, while in a recent paper [25] Elvekjaer, taking into account also the πN S-wave succeeded to solve the δ_0^0 up-down ambiguity in favour of the between-down solution.

For an excellent review of the new techniques applied to the analysis of πN scattering we refer the reader to the papers of J.Hamilton [26,27].

The discrepancy function has also been used for interior points extrapolation (also in the energy complex plane) to evaluate coupling constants. Thus, recently, Y.A.Chao and E.Pietarinen [28] obtained the best estimate so far of the residues of the important $K^- p$ channel poles.

2.4. <u>Cutkosky's Modified Phase Shift Analysis</u>

Optimal polynomials were successfully used not only in the energy plane, but, beginning with Cutkosky and Deo [14-15] in the cosine plane, and this in two ways: first for the determination of the residues of the cosine poles of K^-p amplitude, secondly to devise an impressively efficient way [29-30] of storing information on the cosine dependence of the amplitude (in the physical region), replacing the conventional phase shift analysis [1].

In order to determine the residues, Cutkosky and Deo checked the convergence of the $\mathcal{W}$-polynomials <u>in the physical region</u> for functions obtained by subtracting from the amplitude pole terms $g/(c-c_o)$ with various g's: when g fell upon its real value, a clear dip was seen in the number of polynomials necessary to fit the data to a given accuracy. If instead of optimal polynomials one uses other functions, the dip gets insignificant : this was a beautiful check of the theoretical prediction that the ratio of rates $1/\rho^n$ and $1/R^n$ of convergence when the pole is present/absent, is greatest with the optimal polynomials.

An impressive example of the efficiency of optimal polynomial extrapolation for information storage is furnished by the paper [29] of Yung-An Chao : computing the first nine $\mathcal{W}$-parameters for the NN amplitude from the first nine phase shifts, he obtained the next five ones ($^1G_4$, ε_4, $^3H_4$, $^3H_5$ and $^3H_6$) almost perfectly between 200 and 400 MeV. The point is that actually at these energies, only 9 or 10 parameters are essential as it is also found by using Cutkosky's modified χ^2-test (by mean of CTF functions, see Section 5.3) : starting with the tenth $\mathcal{W}$-coefficient, Cutkosky's χ^2 function goes flat, as theoretically expected.

A phase shift analysis for the K^+p scattering could be found in [30]. Also in connection with information storage (this time in the energy plane), we refer to reader to a paper of Cutkosky and Shih [31] which uses a $\mathcal{W}$-expansion for D(s) in the N/D formulation of the partial waves, ready for extrapolations. Finally we should like to mention a somewhat older paper[32] of G.Steinbrecher, which used a <u>double</u> $\mathcal{W}$-expansion (both in s and t) ; computations just under way show that crossing and unitarity are well fulfilled even in second order.

------

[1] It can be shown that whereas the number of significant partial waves goes up like $\sqrt{s}$, the corresponding number of $\mathcal{W}$-expansion coefficients goes up only as log s.

3.   WEIGHTED DISPERSION RELATIONS

3.1. Hölder Condition (again) and Cut to Cut Extrapolation

When the physical region Γ_1 is a part of the cuts Γ which might even be a boundary of the analyticity domain [33], polynomial extrapolations certainly fail and one is naturally led to use dispersion relations weighted with functions which damp down the influence of the $\Gamma_2 = \Gamma \setminus \Gamma_1$ cut .

The conditions

$$\left| f(z) - h(z) \right|\, \Gamma_1 < \varepsilon(z) \tag{3.1.1}$$

$$\left| f(z) \right|\, \Gamma_2 < M \tag{3.1.2}$$

are sufficient for stability at interior points, and it will be proven in (3.3) that (for constant [1] $\varepsilon(z) = \varepsilon$) the optimal weight is of the form

$$C_0(\xi) = \exp \left\{ - \lambda (\omega(\xi) + i\, \widetilde{\omega}(\xi)) \right\} \tag{3.1.3a}$$

where $\omega(\xi)$ is harmonic in D, <u>zero</u> on Γ_1 and <u>one</u> on Γ_2 and $\widetilde{\omega}(\xi)$ is its harmonic conjugate, so that

$$\left| C_0(\xi) \right|\, \Gamma_1 \equiv 1, \qquad \left| C_0(\xi) \right|\, \Gamma_2 \equiv e^{-\lambda} \tag{3.1.4}$$

Here $\omega(\xi) + i\,\widetilde{\omega}(\xi) = 1/2 + i/\pi \,.\, \ln \left\{ (1+i\xi)/(1-i\xi) \right\}$ (3.1.5) $\xi(z)$ being that mapping which brings D (for simple connected D's) onto the unit disk and Γ_1 (Γ_2) onto the right (left) semicircle.

To get a feel of how the C-function works, notice that for <u>interior</u> points, the integral over the unknown part Γ_2 in

$$f(\xi) = \frac{1}{2\pi i\, C_0(\xi)} \int_{\Gamma_1} \frac{C_0(\xi')d\xi'}{\xi' - \xi}\, f(\xi') + \frac{1}{2\pi i\, C_0(\xi)} \int_{\Gamma_2} \frac{C_0(\xi')d\xi'}{\xi' - \xi}\, f(\xi') \tag{3.1.6}$$

is damped by a factor $\exp \left\{ -\lambda(1 - \omega(\xi)) \right\}$, since $\omega(\xi)$, being harmonic, is strictly less than 1. Nevertheless λ cannot be taken too large, as it would enhance too much the noise due to the errors (3.1.1) in the first term; for optimization at interior points, see section 3.3.

---

[1]   In the variable error case the optimal weight is $C(z) = C_0(z) \cdot C_1(z)$

with

$$C_1(\xi) = \exp \left\{ 1/(2\pi) \cdot \int_{-\pi/2}^{\pi/2} \frac{e^{i\theta'} + \xi}{e^{i\theta'} - \xi}\, \ln\left(\frac{\varepsilon(z)}{\varepsilon} \right) d\theta' \right\} \tag{3.1.3b}$$

Unfortunately, the effect of this damping factor vanishes if ξ goes to Γ_2, as $\omega|_{\Gamma_2} \equiv 1$, but if f(z) is smooth enough, the quick oscillatory behaviour of the C-function is expected to diminish the second integral considerably. To make things more transparent, we shall treat here the simple example when the cuts Γ_1 and Γ_2 are disjoint; for more "physical" cases we refer the reader to the original paper[34].

Using again the old mapping $\mathcal{W}(z)$ (2.1.9) which brings Γ_1 and Γ_2 onto the circles of radii 1 and R respectively, and taking $C_o(\omega) = \mathcal{W}^{-n}$ (indeed, $|C_o|_{\Gamma_1} = 1$ and $|C_o|_{\Gamma_2} = 1/R^n$), we write

$$f(\mathcal{W}) = \frac{\mathcal{W}^n}{2\pi i} \int_{\Gamma_1} \frac{f(\mathcal{W}')}{\mathcal{W}'^n(\mathcal{W}'-\mathcal{W})} d\mathcal{W}' + \frac{\mathcal{W}^n}{2\pi i} \int_{\Gamma_2} \frac{f(\mathcal{W}')}{\mathcal{W}'^n(\mathcal{W}'-\mathcal{W})} d\mathcal{W}' \qquad (3.1.7)$$

so that the extrapolated amplitude $\check{h}(z)$, constructed solely from the data function h(z) on Γ_1 reads

$$\check{h}(\mathcal{W}) \equiv \frac{\mathcal{W}^n}{2\pi i} \int_{\Gamma_1} \frac{f(\mathcal{W}')}{\mathcal{W}'^n(\mathcal{W}'-\mathcal{W})} d\mathcal{W}'. \qquad (3.1.8)$$

Let us estimate how much $\check{h}$ departs from the true amplitude $f(\mathcal{W})$:

$$\left| f(\mathcal{W}) - \check{h}(\mathcal{W}) \right|_{\Gamma_2} \equiv \left| \frac{\mathcal{W}^n}{2\pi i} \int_{\Gamma_1} \frac{(f(\mathcal{W}') - h(\mathcal{W}'))}{\mathcal{W}'^n(\mathcal{W}'-\mathcal{W})} d\mathcal{W}' + \right.$$

$$\qquad (3.1.9)$$

$$\left. + \frac{\mathcal{W}^n}{2\pi i} \int_{\Gamma_2} \frac{f(\mathcal{W}')}{\mathcal{W}'^n(\mathcal{W}'-\mathcal{W})} d\mathcal{W}' \right|_{\Gamma_2} \leq \varepsilon R^n K + \left| \frac{\mathcal{W}^n}{2\pi i} \int_{\Gamma_2} \frac{f(\mathcal{W}')}{\mathcal{W}'^n(\mathcal{W}'-\mathcal{W})} d\mathcal{W}' \right|_{\Gamma_2}$$

Here $K = 1/(2\pi) \cdot \int_{\Gamma_1} |\mathcal{W}'-\mathcal{W}|^{-1} |d\mathcal{W}'|$; then, since $(\mathcal{W}'-\mathcal{W})^{-1} \cdot \mathcal{W}^n/\mathcal{W}'^n \equiv$

$$\equiv \mathcal{W}'^{-1} \sum_{k=0}^{\infty} (\mathcal{W}/\mathcal{W}')^{n+k},$$ the integral of the r.h.s. of (3.1.9) is nothing but the Taylor remainder $\left| f_+(\mathcal{W}) - \sum_{k=0}^{n-1} a_k \mathcal{W}^k \right|$ which, if f(z) is subjected to a Hölder condition (2.2.1), is bounded by

$$\eta_n = A \frac{\ln n}{n^{p+\alpha}}. \qquad (3.1.10)$$

Just like in section 2.2, there is a critical n which minimizes the error-bound $E_n = K \varepsilon R^n + \eta_n$ and hence the stability of the extrapolation (3.1.8) is proven. Unfortunately although all this matter was known [34] by the time of the 1969-Lund Conference, we have not succeeded so far to find a "total approach" for boundary points extrapolations as we did (see section 4) for interior points[1].

3.2. <u>Pišút and Prešnajder's Extrapolation "In The Average"</u>

One can get stability on the cut Γ_2 not for point estimates as in section 3.1. but for the average $\int_\gamma f.g$ of the amplitude with a weight function g over some (small) part $\gamma \subset \Gamma_2$, and this without any smoothnes requirements[37].

If g (z) is holomorphic in D (and <u>real</u> on γ) one has

$$\int_{\Gamma_1} hg \;+\; \int_\gamma fg \;=\; \int_{\Gamma_1} (h-f)g \;-\; \int_{\Gamma_2 \backslash \gamma} fg \quad . \qquad (3.2.1)$$

Dividing by the normalization $N = \int_\gamma g$ one gets

$$\left| \frac{1}{N} \int_{\Gamma_1} hg \;+\; \frac{1}{N} \int_\gamma fg \right| \;<\; \frac{1}{N} \varepsilon \left(\sup_{\Gamma_1} g \right) \ell_{\Gamma_1} \;+\; \frac{1}{N} M \left(\sup_{\Gamma_2 \backslash \gamma} g \right) \ell_{\Gamma_2} \;\equiv\; E_g \qquad (3.2.2)$$

To find an optimal g, they work with D mapped onto the upper unit semicirle with γ along the diameter so that g(z) (real on γ) can be extended to the whole unit circle; one then notices that for fixed $\sup_{\Gamma_1} g$ and $\sup_{\Gamma_2} g$ the maximum of N (and hence, the minimum of E_g) is reached when g has constant modulus on Γ_1 and $\Gamma_2 \backslash \gamma$ respectively and has no zeroes in D. It is obvious that g depends then solely on the parameter $c = \sup_{\Gamma_2 \backslash \gamma} g \;-\; \sup_{\Gamma_1} g$, the minimum of E_g with respect to c being then found on a computer.

The remarkable result one gets is the equality between the average $\frac{1}{N} \int_\gamma fg_{opt.}$ and the data function integral $-\frac{1}{N} \cdot \int hg_{opt.}$ within the error $E_{min}(\varepsilon)$, which could be shown[37] to vanish with ε , and all that without any Hölder but only with the boundedness condition (3.1.2).

---

[1]

These items are under active consideration by Prof.J. Fischer's group, in Prague[35]. See also the pioneering work of Bowcock and John[36].

3.3. <u>Poisson Weighted Dispersion Relation (PWDR)</u>

Among the various tautological kernels of the weighted dispersion relations (1.4.2) one also finds the Poisson one[1] $id(\mathcal{G}(z,z') + i \mathcal{X}(z,z')) \equiv \frac{\partial \mathcal{G}}{\partial n'}$ ds' = real (!) : Indeed, an analytic function is represented by a pair of harmonic functions and hence it is "reproduced" by the Poisson kernel. Following[38] we shall now prove that among all tautologic dispersion relations, the least error bound is obtained with Poisson kernels weighted by a C_0-function (3.1.3) with

$$\widetilde{f} \equiv C_0 f, \quad \widetilde{h} \equiv C_0 h \quad \text{on} \quad \Gamma_1 \text{ and } \widetilde{h} \equiv 0 \quad \text{on} \quad \Gamma_2. \tag{3.3.1a}$$

Both conditions (3.1.1) and (3.1.2) take the simple, symmetric form:

$$\left| \widetilde{f}(z) - \widetilde{h}(z) \right|_{\Gamma_1 + \Gamma_2} < \varepsilon \tag{3.3.1b}$$

Let $\widehat{h}(z)$ be the extrapolated function one gets using the P.W.D.R.

$$\widehat{h}(z) = \frac{1}{2\pi C_0(z)} \int_{\Gamma_1} h(z') C_0(z') \, i \, d\left(\mathcal{G}(z,z') + i \mathcal{X}(z,z') \right) \equiv \frac{\widehat{\widetilde{h}}(z)}{C_0(z)} \tag{3.3.2}$$

Since $\widehat{\widetilde{f}} = \frac{1}{2\pi} \int_{\Gamma_1 + \Gamma_2} \widetilde{f} \, i \, d(\mathcal{G} + i\mathcal{X})$ and $\widehat{\widetilde{h}} = \frac{1}{2\pi} \int_{\Gamma_1 + \Gamma_2} \widetilde{h} \, i \, d(\mathcal{G} + i\mathcal{X}),$

owing to (3.3.1b) and to the reality of the Poisson kernel, one gets

$$\left| \widehat{\widetilde{h}}(z) - \widehat{\widetilde{f}}(z) \right| = \left| \frac{1}{2\pi} \int_{\Gamma_1 + \Gamma_2} (\widetilde{h} - \widetilde{f}) \frac{\partial \mathcal{G}}{\partial n} \, ds \right| \leq \frac{\varepsilon}{2\pi} \int_{\Gamma_1 + \Gamma_2} \frac{\partial \mathcal{G}}{\partial n} \, ds \equiv \mathcal{E} \tag{3.3.3}$$

as, the number 1 being itself a harmonic function, it is reproduced by last Poisson integral of (3.3.3). Now, if instead of $\widehat{h}(z)$ one works with any other weighted dispersion relation

$$\widehat{\widetilde{h}}_{[g]} \equiv \frac{1}{2\pi g_z(z)} \int_{\Gamma_1 + \Gamma_2} \widetilde{h}(z') g_z(z') \frac{\partial \mathcal{G}}{\partial n} \, ds \tag{3.3.4}$$

($g_z(z')$ being holomorphic in z' and arbitrary in z) one gets instead of (3.3.3)

---

[1] Here $\mathcal{G}(z,z')$ is the Green function of the domain and $\mathcal{X}(z,z')$ is its harmonic conjugate.

$$\left|\hat{\tilde{h}}_{[g]}(z) - \tilde{f}(z)\right| \leq \varepsilon \; \frac{\mathcal{Y}_z(z)}{|g_z(z)|} \qquad with \qquad \mathcal{Y}_{z_o}(z) \equiv \frac{1}{2\pi}\int |g_{z_o}(z')| \frac{\partial g}{\partial n} \qquad (3.3.5)$$

Now $\mathcal{Y}_{z_o}(z)$ is by construction harmonic in z and has the boundary

values $\left|g_{z_o}(z)\right|_{z \in \Gamma}$. Then, since $g_{z_o}(z)$ is holomorphic in D,

the modulus $\left|g_{z_o}(z)\right|$ is subharmonic, the same being valid also for

the function $\left|g_{z_o}^o(z)\right| - \mathcal{Y}_{z_o}(z)$,

which, moreover, has zero valued boundary values. Hence, for interior

points we have $\left|g_{z_o}(z)\right| - \mathcal{Y}_{z_o}(z) < 0$ i.e.

$$\varepsilon \; \frac{\mathcal{Y}_z(z)}{|g_z(z)|} > \varepsilon \qquad (3.3.6)$$

which proves[38] the optimality of the P W D R extrapolation (3.3.2)

for interior points.

4.   Complete Solution For Extrapolation To Interior Points
4.1. The Constant $\varepsilon_o[h,M/\varepsilon]$.

As we have seen, among all possible dispersion relations, the
Poisson Weighted ones are least sensitive to the errors of the input
data. But dispersion relations do not exhaust all methods of analytic
continuation: in what follows we shall find[40] the absolute optimum
one can reach with analytic extrapolations by constructing effectively
the whole set of amplitudes compatible with(3.1.1) and (3.1.2).
Starting again from the standard form (3.3.1) we note that it may
happen that there be no analytic functions f(z) at all satisfying it;
indeed, if $\tilde{h}_1(e^{i\theta})$ and $\tilde{h}_2(e^{i\theta})$ are, respectively, the positive and
negative Fourier frequency parts of the weighted data function
$\tilde{h}(e^{i\theta})$ putting

$$\tilde{f}(\xi) = \tilde{h}_1(\xi) - \tilde{\chi}_1(\xi) \qquad (4.1.1)$$

the holomorphic function $\tilde{\chi}_1(\xi)$ would have to approximate $\tilde{h}_2(e^{i\theta})$
(which is not analytically extendable to D), better than

$$\left\|\chi_1 + \tilde{h}_2\right\|_{L^\infty(\Gamma)} \leq \varepsilon \qquad (4.1.2)$$

But the L^∞ norm of $\tilde{\chi}_1 + \tilde{h}_2$ cannot be made as small as one
likes as it has to be greater, at least, than the L^2 norm of $\tilde{h}_2$.

Its smallest possible value $\varepsilon_o[\tilde{h}] \equiv \varepsilon_o[\ell, M/\varepsilon]$ depends indeed solely on the negative Fourier coefficients c_{-1}, c_{-2},... of $\tilde{h}(e^{i\theta})$, being[40] the norm of the Hankel matrix

$$\varepsilon_o[\ell, M/\varepsilon] = \left\Vert \begin{pmatrix} c_{-1} & c_{-2} & c_{-3} \\ c_{-2} & c_{-3} & \cdots \\ c_{-3} & \cdots & \cdots \end{pmatrix} \right\Vert . \tag{4.1.3}$$

For $\varepsilon < \varepsilon_o[h, M/\varepsilon]$ there are no holomorphic function $\tilde{\chi}_1$ satisfying (4.1.2) and hence no holomorphic amplitudes $f(z)$ satisfying (3.1.1) and (3.1.2).

Remark: In contradistinction to the L^∞-problem, the optimal solution for the L^2-problem is straightforward[40], namely, as $\Vert \tilde{f} - \tilde{h} \Vert_{L^2}^2 = \Vert \tilde{f} - \tilde{h}_1 \Vert_{L^2}^2 + \Vert \tilde{h}_2 \Vert_{L^2}^2$ the L^2 optimal solution is $\check{f} \equiv \tilde{h}_1/C$. Unfortunately, the L^2 method has been the only one applied so far in physical problems.

4.2. The Whole Set Of Admissible Amplitudes

Let us restrict ourselves to a finite number N of terms in two Fourier expansion of $\tilde{h}_2(e^{i\theta})$. Then, multiplying the l.h.s of (4.1.2) by $e^{iN\theta}/\varepsilon$ one gets[40]

$$\left| \psi_o(\xi) \right|_\Gamma < 1 \tag{4.2.1}$$

$$\psi_o(\xi) = \frac{1}{\varepsilon}\left\{ c_{-N} + c_{-N+1}\xi + \ldots + c_{-1}\xi^{N-1} + \xi^N \chi_1(\xi) \right\} . \tag{4.2.2}$$

The problem of finding such a bounded function (4.2.1) with pre-assigned first N Taylor coefficients, can be solved in a recurrent way. Putting

$$\psi_k(\xi) = \sum_{i=o}^{\infty} \psi_{k_i} \xi^i \tag{4.2.3}$$

and

$$\psi_{k+1}(\xi) = \frac{1}{\xi} \frac{\psi_k(\xi) - \psi_{k_o}}{1 - \psi_{k_o}^* \psi_k(\xi)} \tag{4.2.4}$$

the bound (4.2.1) is preserved ($|\psi_{k+1}(\xi)| \leq 1$) owing to the special Blaschke form of eq. (4.2.4) but the number of preassigned coefficients

ψ_{ki} decreases by <u>one</u> with each new function $\psi_{k+1}(\zeta)$. Hence $\psi_o(\zeta)$ and thus also $\tilde{\gamma}_1(\zeta)$ and $f(\zeta)$ can be expressed in terms of an unity bounded, otherwise <u>arbitrary</u> function $\psi_m(\zeta)\left[\text{if } \varepsilon = \varepsilon_o, \ \psi_N(\zeta) \equiv 0\right]$.

Hence, one could show[40] that (for $\varepsilon > \varepsilon_o$) the set of values of the different admissible function $f(\zeta)$ fills, for every $\zeta \in D$, densely a circle of center $\hat{f}(\zeta)$ and radius $\gamma(\zeta)$, depending on ε (and vanishing for $\varepsilon = \varepsilon_o$ when a unique function $f_o(\zeta)$ survives). Obviously, $\hat{f}(\zeta)$ is really the best extrapolation one can conceive and notice that, especially when ε is close to ε_o, it may differ considerably from the best dispersion yield, $\hat{h}(\zeta)$ (for $\varepsilon < \varepsilon_o$, it does not even exist!). Explicit formulae both for $\hat{f}(\zeta)$ and $\gamma(\zeta)$ can be found in ref[40].

4.3. M_o, The Lowest Bound For The Maximum Of The Amplitude Modulus On Γ_2.

From the ε_o section it is already clear that given the error-channel condition (3.1.1), there exists[41] a minimal value M_o of the stabilizing lever M below which there do not exist any analytic functions $f(z)$ satisfying (3.1.1) and (3.1.2) simultaneously. This important number is given by the transcendent equation

$$\varepsilon = \varepsilon_o \left[h, \ M_o/\varepsilon \right] \tag{4.3.1}$$

with fixed ε and h. The solving of (4.3.1) on a computer is much eased by the fact that, M being a stabilizing lever, the sets $\mathcal{F}_1$ and $\mathcal{F}_2$ controlled by M_1, M_2 are $\mathcal{F}_1 \subset \mathcal{F}_2$ if $M_1 < M_2$ and hence[41] ε_o is a strictly decreasing function of M. M_o is a very sensitive device[42] for checking whether h(z) are really boundary values of holomorphic functions, since, if they are not, it is very hard (at least for small ε) to slip a holomorphic function inside the error corridor and thus the corresponding M_o gets very, even extremely high. Indeed, for vanishing errors, h(z) completely determines an analytic function together with all its singularities and so, no finite M_o may be found satisfying (3.1.2).

It is probably worth emphasizing that, if the M_o test fails to reveal any singularity structure, apriorically any other test gets superfluous, since this means that there really <u>are</u> holomorphic functions satisfying (3.1.1) and (3.1.2) with small M's.

5. CUTKOSKY'S PROBABILISTIC APPROACH

5.1. Statistics and L^2-Problems

In the preceeding chapter we looked for amplitudes passing through well defined error channels. However, the latter have at most attached to them certain probabilities and hence all extrapolation results should be assigned a certain confidence level. Similar probabilities should be attached to the different positions (values) of the stabilizing levers. As it has been explained in the preceding section, the χ^2 of the data fit may always be made arbitrarily small if we allow M on Γ_2 to increase indefinitely. Since the estimate of a least squares fit to the data on Γ_1 is given by $\chi^2 = \frac{1}{\pi} \int_{\Gamma_1} |f(z) - h(z)|^2 / \varepsilon^2(z) d\theta$ it is perhaps natural to think that $h(\theta)$ and $\varepsilon(\theta)$ induce on the space of possible amplitudes a probability $\exp(-\chi^2)$; the effect of the stabilizing lever may then be added by multiplying $\exp.(-\chi^2)$ by a factor $\exp(-\theta)$, such that, $(h|_{\Gamma_2} \equiv 0)$,

$$\exp\left\{-(\chi^2 + \theta)\right\} \equiv \exp\left\{-\frac{1}{\pi}\int_{\Gamma_1}\frac{|f(\theta) - h(\theta)|}{\varepsilon(\theta)^2}d\theta - \frac{1}{\pi}\int_{\Gamma_2}\frac{|f(\theta)|^2}{M^2(\theta)}d\theta\right\} \quad (5.1.1)$$

$\chi_c^2 \equiv \chi^2 + \theta$ is usually referred to as the Cutkosky modified χ^2 [43]. Although it can be shown that (5.1.1) is not a probability measure on H^2 since the latter turns out to have zero total measure, it is nevertheless clear that the "most probable" function according to it is that which minimizes χ_c^2 i.e.

$$\check{h}(z) = \frac{1}{C(z)}\left(Ch\right)_+(z) \equiv \frac{1}{2\pi i\, C(z)}\oint_\Gamma \frac{h(z')\,C(z')}{z' - z}\,dz' \quad (5.1.2)$$

where $C(z)$ is a C-outer function (3.1.3b) bringing $\varepsilon(z)$ and $M(z)$ to the same level, 1 . Moreover, using (5.1.1) one can calculate[44] finite averages of relevant functionals R, like the value of the amplitude at a given point and its dispersion. The key formulae are

$$\langle R \rangle = \lim_{N\to\infty} \frac{\int\cdots\int e^{-\Sigma|f_n - h_n|^2}R(\tilde{f}_1\cdots\tilde{f}_m)d\tilde{f}_1\cdots d\tilde{f}_m}{\int\cdots\int e^{-\Sigma|f_n - h_n|^2}d\tilde{f}_1\cdots d\tilde{f}_m} \quad (5.1.3)$$

and the Riesz theorem (for <u>linear</u> $R_L(\tilde{f})$ on the existence of an $R \in H^2$ such that $R_L(\tilde{f}) \equiv (R,\tilde{f}) \equiv \sum R_n^* \cdot \tilde{f}_n$. Here $\tilde{f}_n$ and $\tilde{h}_n$ are the (positive)

Fourier coefficients of the C-weighted amplitude and data function[1].

Ross[46] and then Shepard and Shih[47] modified this simple χ_c^2-test in order to allow for different errors on the real and imaginary parts, which is of great value for the real experimental situation. Another important direction is that of Pietarinen[48] who modifies (5.1.1) so as to give finite mesure to H^2 or to its subspaces of, say, differentiable functions. He achieves this by adding to Cutkosky or Ross χ_R^2 a term

$$\Theta_p = \frac{1}{2\pi} \int_{\Gamma_1 + \Gamma_2} |f'_{(z)} - h'_{(z)}|^2 \, d\theta \equiv \sum k^2 |c_k^2| \ , \quad (5.1.4)$$

or more generally, $\sum \{ c_k^2 / |\Delta_k|^2 \}$, $\Delta_k \sim 1/k^3$. It is interesting, as Mrs. Caprini[49] has noticed, that the Pietarinen χ_p^2 also ensures a finite dispersion for the value of the amplitude, even on the cuts, in contradistinction to the usual Cutkosky χ_c^2.

Finally we would like to send the intereested reader to the paper of G. Nenciu[50] on the frequently encountered problem of combining the information available along the cuts with that existing in some isolated interior points, as in formfactor problems. He achieves this goal by isolating in the Hilbert space the hyperplane of functions containing that information which originated in the interior points.

---

1)

As an example, the mean value of the weighted amplitude $f(z)$ at some $z = z_o$ is, $(x_n = \tilde{f}_n - \tilde{h}_n)$,

$$\langle \tilde{f}(z_o) \rangle = \lim_{N \to \infty} \frac{\int \ldots \int e^{-\sum |x_n|^2} R_n^*(x_n + \tilde{h}_n) \, dx_1 \ldots dx_n}{\int \ldots \int e^{-\sum |x_n|^2} \, dx_1 \ldots dx_n}$$

where the R_n^* are such that $\sum R_n^* f_n \equiv \frac{1}{2\pi i} \int \frac{\tilde{f}(z')}{z'-z_o} \, dz'$. Since the terms linear in x_n do vanish, we get $\langle \tilde{f}(z_o) \rangle \equiv \sum R_n^* h_n$, and hence $\langle \tilde{f}(z_o) \rangle = \frac{1}{2\pi i} \int \frac{h(z')}{z'-z_o} \, dz'$. Further[44], for its dispersion $\Delta \tilde{f}(z)$ one finds $\langle (\Delta f(z_o))^2 \rangle = 1/(1-|z_o|^2)$. For applications, see[45].

5.2. The Complete L^2 Problem And The M_o Bound

The drawback of the simplified L^2 approach is that the error channel condition

$$\| f(\xi) - h(\xi) \|^2_{L^2(\Gamma_1)} \equiv \frac{1}{\pi} \int_{-\pi/2}^{\pi/2} \frac{|f(e^{i\theta}) - h(e^{i\theta})|^2}{(\mathcal{E}(\theta)/\mathcal{E})^2} \, d\theta < \mathcal{E}^2 \qquad (5.2.1)$$

and the stabilizing condition

$$\| f(\xi) \|^2_{L^2(\Gamma_2)} \equiv \frac{1}{\pi} \int_{\pi/2}^{3\pi/2} \frac{|f(e^{i\theta})|^2}{(\mathcal{M}(\theta)/\mathcal{M})^2} \, d\theta < \mathcal{M}^2 \qquad (5.2.2)$$

mix in (5.1.1) in an uncontrollable way. The proper way to handle this problem is to look for all holomorphic functions f (ξ) which satisfy (5.2.1) and (5.2.2) separately. To this end, I. Sabba-Stefanescu[51] expands $\tilde{f}(e^{i\theta}) = f(e^{i\theta}) \, C(e^{i\theta})$ in terms of the polynomials $P_n(e^{i\theta})$ orthogonal on Γ_1, so that for any finite approximant N, one has

$$\sum_{0}^{N} |\tilde{f}_n - \tilde{h}_n|^2 < \mathcal{E}^2 - \mathcal{E}_N^2 \qquad (5.2.3)$$

$$\sum_{0}^{N} g_{mn} \tilde{f}_m \tilde{f}_n \le \mathcal{M}^{(N)2} \qquad (5.2.4)$$

with $g_{mn} = \frac{1}{\pi} \int_{\pi/2}^{3\pi/2} P_m P_m^* \, d\theta$, $\tilde{h}_m = \frac{1}{\pi} \int_{-\pi/2}^{\pi/2} \tilde{h} \, P_m^* \, d\theta$, $\tilde{f}_m = \frac{1}{\pi} \int_{-\pi/2}^{\pi/2} \tilde{f} \, P_m^* \, d\theta$

and where $\mathcal{E}_N$ is the L^2 error $\| \tilde{h} - \sum_{0}^{N} \tilde{h}_n P_n \|_{L^2}$. Equation (5.2.3) represents a N-dimensional sphere around $\{\tilde{h}_n\}_0^N$ and (5.2.4) an ellipsoid arround the origin of the Hilbert space: critical $\mathcal{M}_o^{(N)}$ can be found from the condition of them being tangent. The sequence $\mathcal{M}_o^{(N)}$ is monotonically decreasing to a limit $\mathcal{M}_o$ where only one amplitude, obeying (5.2.1) and (5.2.2) with $\mathcal{M} = \mathcal{M}_o$ is left alive.

For χ^2 addicts, $\mathcal{M}_o$ allows a quick location of poles or zeroes of the amplitudes.

5.3. Cutkosky's Test Function (CTF)

The drawback of the modified χ^2_c test (5.1.1) is that whereas some assumptions about the <u>shape</u> of M (θ) may be invented, it is more difficult to figure out its magnitude. The merit of the CTF device to be described below (for cosine extrapolation to the cuts Γ_2) is that it does not depend on the overall scale factor V, as it works with the ratios[43]

$$r_n = \frac{\alpha^2_{n+1}}{\frac{1}{n}\sum_1^n \alpha^2_i} \qquad n = 1,2,\ldots, k-1 \qquad (5.3.1)$$

where α_n are the coefficients of the amplitude $f(z)$ on the outer (2.1) circle Γ_2 with respect to a basis othogonal with the weight $1/M(\theta)$ so that (5.1.1) $\theta = \frac{1}{V}\sum \alpha^2_n$. It is easy to show that r_n are independent random variables whose distributions P_n (r_n) can be worked out (see Appendix B of [43]) and do not contain V. Clearly[1] one can introduce a new variable $x_n(r_n)$ defined to be Gaussian

$$\int_0^{x_n} e^{-x^2} dx = \int_0^{r_n} P_n(r') \, dr' \qquad (5.3.2)$$

so that $\phi = \sum_n^k \chi^2_n$, known as the Cutkosky's Test Function will have a χ^2 distribution. Then, the best fit is achieved by minimizing (on a computer) the sum : $\chi^2 = \chi^2_{exp} + \phi$. In the literature one can already find many fruitful applications of this method, with excellent results [28-31]. Cutkosky[52] has verified the effect of the CTF device empirically, on some 2500 test functions taken at random, on a computer.

---

1)

The lecturer must confess that, neither to him nor to his close friends, did all these things appear too clear.

5.4. The M_o and $\mathcal{M}_o$ Tests

The special quality of the CTF's is that they do not have to
make assumption on the "position" of the stabilizing lever. Similar
devices are the M_o[42] and $\mathcal{M}_o$[51] tests, since both these quantities
are not related to any specific value of M but are rather the lower
limits of the "spectrum" of all possible (L^∞ and L^2)M's, being thus
characteristic of the whole set of amplitudes passing through the
error channel. Relatively simple computer programs allow the use of
these tests for bias-free determinations of the singularities of the
scattering amplitude. For instance,[42, 51,] if $A(\zeta)$ has a zero at ζ_{oo},
the data function $h(e^{i\theta})$ corresponding to $f(\zeta) = A(\zeta)\,(1-\zeta_o^*\zeta)/(\zeta-\zeta_o)$
being close (especially when ε is small) to the limiting values of a
strongly nonanalytic function, unless ζ_o coincides with ζ_{oo}, the corresponding $M_o(\zeta_o)$ as well as $\mathcal{M}_o(\zeta_o)$ will be very high, except for a
narrow dip around ζ_{oo}.

If moreover some theoretical information is available
(unitarity, Froissart bound, etc) limiting the upper bound of
$M(=M_{true})$, then the only possible values for ζ_o are those for which
$M_o(\zeta_o) \lesssim M_{true}$, the distance between the M_o curve and M_{true} defining
a sort of probability distribution for the location of zeroes (poles)
of the amplitude.

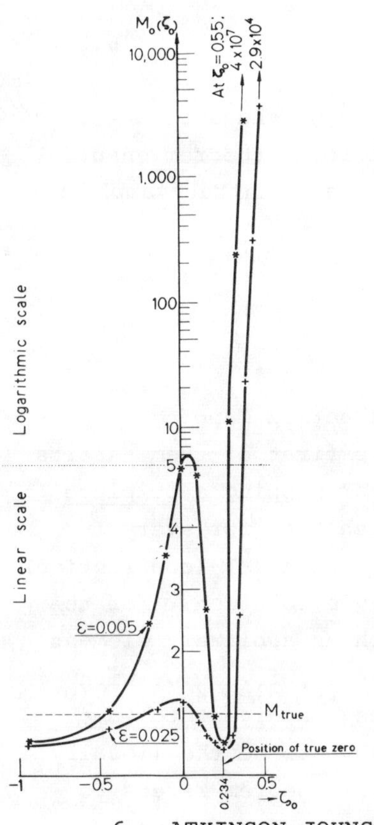

Fig.3. Typical dependence (see[42]) of M_o versus the location of the artificial pole $\zeta_o=\zeta(z_o)$ [here $\zeta(z)=(3o-z+i\sqrt{195z(z-4)})/(14z_o-3o)$] of the function $F_1(z)$ $(1-\zeta_o^*\zeta)/(\zeta-\zeta_o)$ introduced in order to locate the zero of the <u>model</u> amplitude $F_1(z) = (1-\sqrt{4-z})/(1+\sqrt{z-4})$. The dip corresponds to $z_o=3$, where the artificial pole disappear identically. Upper curve corresponds to 1% errors, while lower curve to 5% ones

6. ATKINSON-JOHNSON-WARNOCK BOOTSTRAP

6.1. Mandelstam Representation And Fixed Points Theorems

So far the reader has been left with the impression that all this stuff on ill posed problems is just good to devise sophisticated data fitting. In these last two sections we will try to dispel this feeling by showing how these questions interfere with purely theoretical dynamical schemes.

The original bootstrap[53] program became a really serious mathematical scheme only when fixed point theorems were brought on the stage[54-56] by Atkinson. His basic idea - which is already familiar to most of us - was to think of the sequence of the three principles of the Mandelstam representation (analyticity, unitarity and crossing) as an operator P for which a fixed point theorem[1] can be applied if imbedded into an intelligently devised functional analytic frame. The main difficulty overcome by Atkinson[54-56] and Warnock[57-59] was to invent a norm refined enough to prove that the theorems could be applied for the operator P introduced above. It has been shown that for a given inelastic double spectral function input $v(s,t)$, defined by

$$\rho(s,t) = \rho^{el}(s,t) + \rho^{el}(t,s) + v(s,t) \qquad (6.1.1)$$

where $\rho^{el}(s,t)$ is the elastic spectral function of the s-channel, there exists ranges of v's for which the fixed point theorem ensure the existence of a crossing-symmetric, analytic and unitary amplitude.

----

1)

The theorems commonly used are the Contraction Mapping Principle and the Schauder Theorem. The first of them asserts that if P is some (non-linear) operator mapping a <u>complete</u> subset of a Banach space into itself such that for every φ, ψ one has $\| P_\varphi - P_\psi \| < k\| \varphi - \psi \|$ with $k<1$ (contraction), then there exists in this subset an unique solution ϕ of the equation $P\phi = \phi$. Indeed, starting with an arbitrary element f_o, one can easily prove that the series f_o, f_1, f_2... with $f_{k+1} = Pf_k$ is Cauchy convergent and hence convergent to some element ϕ, which can then easily be shown to be the (locally unique) solution of Pf = f. The Schauder theorem refers to compact convex subset K of a (even infinite dimensional) Banach space asserting that if P is continuous and $P(K) \subseteq K$, then there is at least a fixed point of P, in K.

6.2.   The Froissart-Gribov Scheme

Atkinson, Johnson and Warnock, in their most recent paper[60] work with a nonlinear operator P taking the elastic spectral function $\rho^{el}(s,t)$ and the substraction $\rho(s)$ into $\bar{\rho}^{el}(s,t)$ and $\bar{\rho}(s)$, defined by the following sequence of transformations:

One first starts with a dispersion relation for the t-channel absorptive part (for $s_+ > 4$)

$$(6.2.1)$$

$$A_t(s_+,t) = \rho(t) + \frac{1}{\pi} \int_4^\infty ds' \left(\frac{1}{(s'-s_+)} + \frac{1}{(s'-u_-)} \right) (\rho^{el}(s',t) + \rho^{el}(t,s') + v(s',t))$$

and with the Froissart-Gribov representation for partial waves

$$A_\ell(s_+) = \frac{\delta_{\ell,0}}{\pi} \int_4^\infty \frac{ds' \, \rho(s')}{s' - s_+} + \frac{2}{\pi(s-4)} \int_4^\infty dt \, Q_\ell\left(1 + \frac{2t}{s-4}\right) A_t(s_+,t) \qquad (6.2.2)$$

then one constructs the elastic absorptive part of the s-channel (for physical cosines)

$$A_s^{el}(s,t) = \sum_{\ell=0}^\infty (2\ell+1) P_\ell(z_s) \sqrt{\frac{s-4}{s}} \, |A_\ell(s_+)|^2. \qquad (6.2,3)$$

Considering the l.h.s. of (6.2.3) as a <u>data function along</u> $\Gamma_1 = \{ 4-s < t < 0 \}$, one can find an extrapolating function $\hat{A}_s^{el}(s,t)$ stable up to the t-cuts $\Big($in terms of the optimal (2.1.9) mapping $w(1 + 2t/(s-4))\Big)$

$$\hat{A}_s^{el}(s,t) = \sum_0^{M_c} a_m \left(w^m + w^{-m} \right) \qquad (6.2.4)$$

where the real coefficients a_n are to be found by a best fit in the Tchebycheff meaning $\left(L^\infty \right)$ of the l.h.s. of (6.2.3) on Γ_1 (here $w^m + w^{-m} \equiv \cos m\,\theta$).

One can compute then again back the (iterated) elastic spectral function and substraction constant on the t-cut $\left(W\big|_{\Gamma_2} = R\,e^{i\theta}\right)$ as

$$\bar{\rho}^{el}(\Delta,t) = \operatorname{Im}\hat{A}_s(\Delta,t) = \sum_{0}^{n_c} a_n \sin(n\,\theta(\Delta,t))(R^n - R^{-n})$$

$$\bar{\rho}(\Delta) = \hat{A}_s(\Delta,\infty) = \sum_{0}^{n_c} a_n\, i^n (R^n + (-1)^n R^{-n}) \qquad (6.2.5)$$

(if no noise was introduced in the previous steps, n_c is to be taken equal to infinity). In the original paper of Atkinson, Johnson and Warnock, an (weighted) orthogonal expansion is used, in terms of Tchebycheff polynomials in $z = \frac{1}{2}(W + W^{-1})$

$$b_n = \frac{2}{\pi(1+\delta_{no})} \int_{-1}^{1} \frac{dz}{(1-z^2)^{1/2}}\, T_n(z)\, A_\rho^{el}(\Delta, t(\Delta, z)) \qquad (6.2.4')$$

$$\qquad (6.2.5')$$

$$\bar{\rho}^{el}(\Delta,t) = \sum_{n=e}^{\infty} b_n(\Delta)\, \operatorname{Im} T_n(z(\Delta,t_+))$$

$$\rho(\Delta) = \sum_{n=0}^{\infty} b_n(\Delta)\, T_n(z(\Delta,\infty))$$

A Solution of the "GF-scheme" is provided by a fixed point of the operator P defined by the sequence of operations (6.1.1) (6.1.5) i.e. is a point ρ^{el}, ρ such that $(\rho^{el},\rho) = (\bar{\rho}^{el},\bar{\rho})$, where $(\bar{\rho}^{el},\bar{\rho}) \equiv P(\rho^{el},\rho)$.

The amplitude A (s,t) constructed through the Mandelstam representation using this fixed point ρ^{el} and ρ (as well as the input v), is then the desired crossing-symmetric, unitary, amplitude.

The norm which brings in force the fixed point theorems (Schauder theorem) for P is

$$\|(f(s),\, g(s,t))\| = \sup(\|f\|_1,\, \|g\|_2) \qquad \text{with}$$

$$\|f\|_1 = \sup_{\Delta_1 \neq \Delta_2} \left\{ \left[|f(\bar{\Delta})| + \frac{f(\Delta_1) - f(\Delta_2)}{\left|\frac{\Delta_1 - \Delta_2}{\Delta_1 \Delta_2}\right|^\mu} \right](\ln \bar{\Delta})^{1+\varepsilon}\, \bar{\Delta}^{-1/2} \right\} \qquad (6.2.6)$$

$$\|g\|_2 = \sup_{\substack{\Delta_1 \neq \Delta_2 \\ t_1 \neq t_2}} \left\{ \left(|g(\bar{\Delta},\bar{t})| + \frac{g(\Delta_1,t_1) - g(\Delta_2,t_2)}{\left|\frac{\Delta_1 - \Delta_2}{\Delta_1 \Delta_2 \bar{t}}\right|^\mu + \left|\frac{t_1 - t_2}{t_1 t_2 \bar{\Delta}}\right|^\mu} \right)(\ln \bar{\Delta}\, \ln \bar{t})^{1+\varepsilon} \right\}$$

where $\bar{s} = \min(s_1, s_2)$, $\bar{t} = \min(t_1, t_2)$ and $0 < \mu < \frac{1}{2}$. These rather
horribly looking norms (6.2.6) are found convenient in problems
involving Cauchy integrals because they are especially well suited
for the evaluation of principal value integrals $\mathcal{J}(s)$, and namely
for the evaluation both of their modulus and their Hölder constants
$\left(|\mathcal{J}(s_1) - \mathcal{J}(s_2)| / (s_1-s_2)^{\mu} \right)$ and, hence, of their (6.2.6)-norms
themselves ! . It is then shown in[60] that a fixed point of P can
be found in the Banach space $\mathcal{B}$ (of function pairs) endowed with the
norm (6.2.6), if $\|v\|_2$ exists and is sufficiently small. To this end
one fist shows[60] that if $\|v\|_2$ exists and $(\varrho, \varrho^{el}) \in \mathcal{B}$,then the
various integrals and infinite sums in eq.(6.2.1)-(6.2.5) all converge
so that $\bar{\varrho}$ and $\bar{\varrho}^{el}$ are well defined and bounded functions; then it is
shown that the mapping P is equivalent with the "Mandelstam mapping"
M in which the condition (6.2.3) is replaced by the unitarity condition
relating the elastic spectral function to the t-absorptive parts
$A_t(s,t)$. Then the fixed point of P is the same as that of the mapping
M_t^{54-60}.

6.3.   The Chew-Mandelstam Schemes

New mappings can be conceived with partial waves dispersion
relations

$$A_\ell(s) = \frac{1}{\pi} \int_{-\infty}^{0} \frac{\operatorname{Im} A_\ell(s')}{s' - s}\, ds' + \frac{1}{\pi} \int_{4}^{\infty} \frac{\sqrt{\frac{s'-4}{s'}}\, |A_\ell(s')|^2}{s' - s}\, ds' +$$

$$+ \frac{1}{\pi} \int_{16}^{\infty} \frac{1 - \eta_\ell^2(s')}{4\sqrt{\frac{s'-4}{s'}}\,(s'-s)}\, ds' \qquad\qquad (6.3.1)$$

where [1] $\quad \operatorname{Im} A_\ell(s_+) = - \int_{-1}^{+1} dz\, P_\ell(z)\, \operatorname{Re} A_t\left(s_+, \frac{4-s}{2}(1-z)\right) \qquad (6.3.2)$

and $\quad \dfrac{1 - \eta_\ell^2(s)}{4\sqrt{s-4}/\sqrt{s}} = \dfrac{2}{\pi(s-4)} \int_{4}^{\infty} dt\, Q_\ell\left(1 + \frac{2t}{s-4}\right)\left(\varrho^{el}(t,s) + v(s,t)\right) \qquad (6.3.3)$

---

[1]
     For what follows in Chapter 7, it is important to notice that,
once $\varrho(s,t)$ is constructed (see further) in terms of $|A_\ell|^2$ and
$v(s,t)$, one could get on the left cut not only $\operatorname{Im} A_\ell(s_+)$, but the
complete boundary values $A_\ell(s_+)$, simply using the Mandelstam representation for $s < 0$ (followed by a partial wave projection).

A.J.W. envisage two different "Chew Mandelstam" (CM) mappings corresponding to two different ways of iterating the set of equations (6.3.1), (6.2.4), (6.2.5), (6.2.1) and (6.2.3):

CM_1; suited for small, high partial amplitudes, (great l's): one starts with a set of values for the squares of the moduli $|A_\ell|^2$ and computes first the "inputs" (6.3.2) and (6.3.3) (via eq. (6.2.4), (6.2.5) and (6.2.1)) which, together with $|A_\ell|^2$ themselves determine completely the r.h.s. of (6.3.1) and hence the iterated $\overline{A}_\ell$. (At any finite iteration step, unitarity is violated, as in general $|\overline{A}_\ell|^2 \neq |A_\ell|^2$). One proceeds further with the new $|\overline{A}_\ell|^2$.

CM_2; (N/D scheme): The inputs" (6.3.2) and (6.3.3) computed as in the CM_1 scheme from given $|A_\ell|^2$ are solely introduced in (6.3.1) (not so are the old $|A_\ell|^2$!) and one proceeds further by solving the N/D equations for the interated $\overline{A}_\ell$. This scheme is suited for low partial waves, which may be resonant (small l's). For details we refer the reader to the excellent paper of A.J.W.[60].

7.   OPTIMAL N/D INTEGRAL EQUATIONS

7.1.   Tautologies and Optimization of N/D Equations

The usual input of the N/D equations is the imaginary part of the amplitude on the left hand cut Γ . Nevertheless there is no a priori reason to throw away information on the real part, as the latter can be computed on Γ in essentially the same way as the imaginary one: this can be done either by direct analytic extrapolation from the nearby physical regions or via spectral function as in the AJW bootstrap (see the footnote of section 6.3). Henceforth we take as input the whole amplitude on the left hand cut Γ .

As it has been emphasized in the Introduction, for "exact" input data usually there exist infinitely many tautological theories yielding exactly the same output ; in our case "exact data" means l.h. cut data which are really boundary values of meromorphic and unitary (on the r.h. cut, γ) functions. If however _inexact_ data are used, the symmetry with respect to the tautology group is broken [1]

---

1) i.e. different tautological methods yield different results: there may exist subgroups which remain tautological (e.g. the group of changing the subtraction point in the D equation) but these are irrelevant for our purposes.

and it makes sense to find that tautology for which the departure of
the solution from the exact one is least. For N/D equations this leads
to an extremal L^1 problem which, happily, can be completely solved[61],
one obtaining thus the equation which is most insensitive with respect
to the errors

$$\left| A(z) - \mathcal{A}(z) \right| < \mathcal{E}(z) \qquad z \in \Gamma \tag{7.1.1}$$

of the input data $\mathcal{A}(z)$. (We take $\mathcal{E}(z) = M$ and $\mathcal{A} = 0$ on that
distant Γ region where $A(z)$ is not known at all).

7.2. An Extremal L^1 Problem

We work in the complex plane $z(s)$ in which the l.h. cut Γ is
mapped onto the unit circle $|z| = 1$ while the r.h. one, γ , comes onto

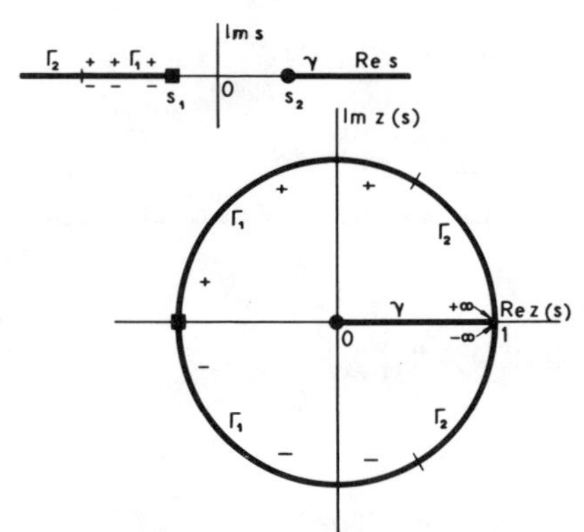

Fig. 4

the radius $[0,1]$. Using appropriate
C-weights, we follow the N/D prescript-
ions directly for $\widetilde{A}=AC$ which satisfies
(7.1.1) (with respect to $\widetilde{\mathcal{A}}=AC$) with
constant $\mathcal{E}$ all along Γ .
We take $\widetilde{A}=N/D$, with N holomorphic in
the unit disk and D in the whole z
plane cut along $\gamma =[0,1]$. Upon substituting the Cauchy integral for N into
the one for D (we assume CDD poles
are absent; see however[62])one gets
the following integral equation for D:

$$D(z) = 1 + \frac{1}{2\pi i} \int_{\Gamma} dz'' \, D(z'') \, \widetilde{A}(z'') \, G_z(z'') \tag{7.2.1}$$

$$\text{with } G_z(z'') = \frac{1}{\pi} \int_0^1 dz' \frac{\widetilde{\rho}(z')}{(z'-z)(z'-z'')} \quad \text{where } \widetilde{\rho}(z') = \sqrt{\frac{s'-4}{s'}} \Big/ C(z')$$

<u>Since</u> $N = D\widetilde{A}$ <u>is holomorphic</u> inside the unit disk one may add to $G_z(z'')$
in (7.2.1) any function $F_z(z'')$ holomorphic with respect to z" and
arbitrary in z, without altering the solution. Hence, an alternative
equation to (7.2.1) is

$$D(z) = 1 + \frac{1}{2\pi i} \int_{\Gamma} dz'' \, D(z'')\widetilde{A}(z'')(G_z(z'') + F_z(z'')) \tag{7.2.2}$$

Furthermore, one can show that the F's exhaust all the relevant tauto-
logies of eq.(7.2.1). If now we use instead of $\widetilde{A}$ the data function $\widetilde{\mathcal{A}}$,

the solutions $\mathcal{D}(z)$ of the approximative equations <u>become F-dependent</u> and one is led to seek   that $F=F^O$ for which $\|\mathcal{D}-D\|$ is least. Now, at least far enough from eigenvalues, the solutions are closest if the kernels $\mathbf{K}$ and $\mathcal{K}$ are closest. Using the Banach space $C(\Gamma)$ of complex functions continuous on Γ, with the norm

$$\|D\|_C = \sup_{z \in \Gamma} |D(z)| \tag{7.2.3}$$

it can be proved[61] that

$$\sup_A \|\mathbf{K} - \mathcal{K}\|_C = \sup_A \sup_{z \in \Gamma} \frac{1}{2\pi} \oint_\Gamma |dz''| \, |\tilde{A}(z'') - \hat{\mathcal{A}}(z'')| \cdot$$

$$\cdot |G_z(z'') + F_z(z'')| \le \frac{\varepsilon}{2\pi} \sup_{z \in \Gamma} \oint |dz''| \cdot |G_z(z'') + F_z(z'')| \tag{7.2.4}$$

so that the optimal $F=F^O$ is that for which

$$\sup_{z \in \Gamma} \|G_z + F_z\|_{L^1} \equiv \sup_{z \in \Gamma} \frac{1}{2\pi} \oint |dz''| \cdot |G_z(z'') + F_z(z'')| \quad \text{is}$$

minimal. The usual way one solves such extremal problems is to turn them into their L^∞ dual ones. For this it is necessary first to show[61] that:

$$m^O \equiv \inf_{F_z} \sup_{z \in \Gamma} \|G_z + F_z\|_{L^1} = \sup_{z \in \Gamma} \inf_{F_z} \|G_z + F_z\|_{L^1} \tag{7.2.5}$$

so that the duality principle

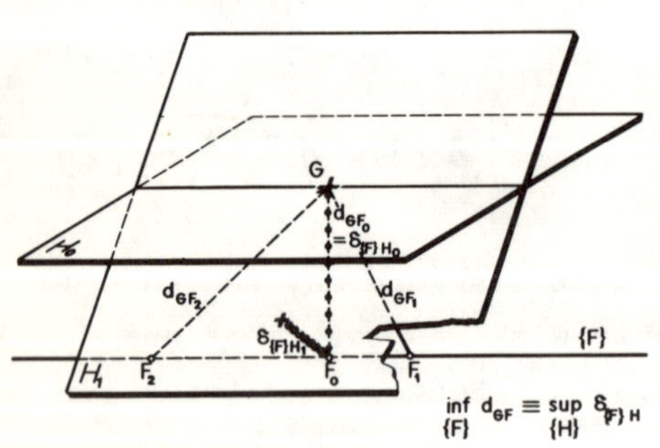

Fig.5 A geometrical view of the duality principle: Points are to be interpreted as elements of a L^1 space, while planes mean level surfaces of linear functionals over L^1, i.e. elements of the dual space L^∞. Then, the smallest (L^1) distance d_{GF} between a given point G and points of a linear subset F coincides with the greatest distance $\delta_{{F}H}$ between the linear subset $\{F\}$ and the planes whose normals are orthogonal to $\{F\}$, passing through G

$$\inf_{F_z \in H^1} \left\| G_z(z'') + F_z(z'') \right\|_{L^1} = \sup_{\substack{h_z \in H^\infty \\ |h_z| \leq 1}} \left| \frac{1}{2\pi} \oint_\Gamma h_z(z'') G_z(z'') \, dz'' \right| \qquad (7.2.6)$$

may be used. Now, since the discontinuity of $G_z(z'')$ across γ is $\tilde{g}(z'')/(z''-z)$, with positive $\tilde{g}$,

$$\sup_{z \in \Gamma} \sup_{\|h\|=1} \left| \frac{1}{2\pi i} \oint_\Gamma dz'' \, h_z(z'') G_z(z'') \right| \equiv \sup_{z \in \Gamma} \sup_{\|h\|=1} \left| \frac{1}{\pi} \int_0^1 dz' \, \frac{\tilde{g}(z') h_z(z')}{z' - z} \right| \leq$$

$$\leq \sup_{z \in \Gamma} \frac{1}{\pi} \int_0^1 dz' \, \frac{\tilde{g}(z')}{|z'-z|} \leq \frac{1}{\pi} \int_0^1 dz' \, \frac{\tilde{g}(z')}{1-z'} \qquad (7.2.7)$$

But the bound $\frac{1}{\pi} \int_0^1 dz' \, \frac{\tilde{g}(z')}{1-z'}$ of (7.2.7) is <u>actually attained</u> in the point $z=1$ by the holomorphic function $h^o_{z=1}(z'') \equiv 1$ and hence is identical with m^o of (7.2.5). Now, since (7.2.6) and since

$$\left| \frac{1}{2\pi} \oint dz'' \, h^o(z'') G_z(z'') \right| \equiv \left| \frac{1}{2\pi} \int_0^{2\pi} d\theta'' \, z'' h(z'') \left(G_z(z'') + F^o_z(z'') \right) \right| \, , \text{ it}$$

is necessary that $z''h^o(z'')(G_z(z'') + F_z(z''))$ should have constant phase all along Γ. This enables us to write $\left(a \equiv (z''G_z h^o_z)(o) \right)$:

$$F_z(z'') = \frac{1}{z'' h^o_z(z'')} \, \frac{a}{2\pi i} \oint d\theta \, \frac{e^{i\theta}+z}{e^{i\theta}-z} \, \mathrm{Im} \left\{ \frac{e^{i\theta} h^o_z(e^{i\theta}) G_z(e^{i\theta})}{a} \right\}, \quad (7.2.8)$$

so that the optimal F_z for $z=1$ (corresponding to $h^o_{z=1}(z'') \equiv 1$) is such that

$$F^o_{z=1}(z'') + G_{z=1}(z'') = \frac{1}{\pi} \int_0^1 dz' \, \frac{\tilde{g}(z')}{1-z'} \, \frac{\mathcal{P}(z'',z')}{z''} \qquad (7.2.9)$$

where $\mathcal{P}(z'', z) \, dz''/z'' \equiv (\partial \tilde{g}/\partial \eta) \, d\theta''$ is the Poisson kernel for the unit circle. As it is further shown in[61], the Poisson kernel observes the bound $m^o \equiv \frac{1}{\pi} \int_0^1 dz' \, \frac{\tilde{g}(z')}{1-z'}$ also for all other points $z \in \Gamma$ and hence, the optimal equation for $\mathcal{D}$ reads:

$$\mathcal{D}(z) = 1 - \frac{1}{2\pi^2 i} \oint_\Gamma \frac{dz''}{z''} \, \mathcal{D}(z'') \mathcal{A}(z'') C(z'') \int_0^1 dz' \, \frac{g(z')}{C(z')(z'-z)} \, \mathcal{P}(z'', z) \qquad (7.2.10)$$

7.3. <u>Best Analytic Extrapolation and Unitarity</u>

A new and completely different way of building unitary and holomorphic amplitudes from (error effected) boundary values on the l.h. cut Γ takes its source in the methods of best analytic extrapolation described in Chapter 4. The crucial fact[63] which made unitarity to fit so good into this scheme is a theorem which shows that if one replaces ε with ε_o (4.1.3) <u>the inequality</u> (3.3.1) $\left| \tilde{f}(\zeta) - \tilde{h}(\zeta) \right|_\Gamma < \varepsilon$ turns into the <u>equality</u>

$$\left| \tilde{f}(\zeta) - \tilde{h}(\zeta) \right|_\Gamma = \varepsilon_o \tag{7.3.1}$$

satisfied by the unique holomorphic function $f_o(\zeta)$ which survives for $\varepsilon = \varepsilon_o$. To find an S-matrix partial wave, S_ℓ satisfying on the right h.semicircle Γ_2 the elastic (on $\Gamma_{2,1} \subset \Gamma$) and inelastic (on $\Gamma_{22} = \Gamma_2 \setminus \zeta_1$) unitarity

$$|s_\ell|_{\Gamma_{21}} = 1 \quad , \quad |s_\ell|_{\Gamma_{22}} = e^{-2\eta(\zeta)} \tag{7.3.2a}$$

and approximating the data function $\mathcal{S}_\ell$ on the left h. cut Γ_1

$$|s - \mathcal{S}_\ell|_{\Gamma_1} < \varepsilon \cdot \varepsilon_1(\zeta) , \quad \varepsilon \to \min \tag{7.3.2b}$$

one first uses the C-weights with $|c_1(\zeta)|_{\Gamma_1} = 1/\varepsilon_1(\zeta), |c_2(\zeta)|_{\Gamma_{22}} = \exp\{2\eta(\zeta)\}$, (both equal to 1 on the remainder of Γ) so that the problem (7.3.2) is reduced to $|\bar{s}_\ell - \bar{\mathcal{S}}_\ell| < \varepsilon \to \min$

$$\text{and} \quad |\bar{s}_\ell|_{\Gamma_2} = 1 \tag{7.3.3}$$

Here $\quad \bar{s}(\zeta) = c_1(\zeta)\, c_2(\zeta)\, s(\zeta) .$

Taking now the equation $\varepsilon = \varepsilon_{oo}$ where ε_{oo} is, in analogy to M_o, the solution of

$$\varepsilon_{oo} = \varepsilon_o\left[h, \ 1/\varepsilon_{oo} \right], \tag{7.3.4}$$

we get for the unique function left, $\bar{s}_\ell^o(\zeta)$,

$$\left| \bar{\mathcal{S}}_\ell - \bar{s}_\ell^o \right| = \varepsilon_{oo}, \quad |\bar{s}^o|_{\Gamma_2} = 1 .$$

Hence $s_\ell^o = \bar{s}_o/c_1 c_2$ is the required best unitary and holomorphic extrapolation from the left cut data $\mathcal{S}_\ell$; $\bar{s}_o$ can be found, via $\tilde{\bar{s}}_o = c_o\, \bar{s}_o$, using the recurrent methods of section 4.2 [1].

---

[1] One can equally well work $f \equiv \wp A_\ell$; in this case $h(\zeta)$ on Γ_2 is no more zero, but $i/2$.

A final remark: since in fact we deal here with a cut to cut
extrapolation where boundedness alone is insufficient for stability,
we have to make sure that the functions $S_\ell^o(\zeta)$ remains, for instance,
Hölder continuous : this can always be achieved restraining to a
finite (not too high, as in section 2.2!) number of steps N in the
recursive procedure 4.2 [1] .

---

[1]

This would enhance a bit ε_{oo}, as $\varepsilon_{oo}^N > \varepsilon_{oo}^\infty$,but, as in
section 2.2, one should keep a balance between the accuracy of the
data fitting on Γ_1 and the growth of the Hölder continuity constant.

REFERENCES

1.   J. Fischer, S. Ciulli, J.E.T.P. $\underline{41}$, 256 (1961) and
     Sov. Phys. J.E.T.P.$\underline{14}$, 185 (1962)
2.   S. Ciulli, J. Fischer, Nucl. Phys. $\underline{24}$, 465 (1961)
3.   W.R. Frazer, Phys.Rev. $\underline{123}$, 2180 (1961)
4.   J.E. Bowcock, J.C. Stoddard, Nucl. Phys. $\underline{42}$, 156 (1963)
5.   M. Islam, Phys. Rev. $\underline{138}$, B226 (1965)
6.   C. Lovelace, Nuovo Cim. $\underline{25}$, 730 (1962)
7.   I. Ciulli, S. Ciulli, J. Fischer, Nuovo Cim. $\underline{23}$, 1129 (1962)
8.   S.D. Drell, Proceedings of the Aix en Provence Conference, $\underline{v.2}$,
     129 (1962)
9.   D. Atkinson, Phys. Rev. $\underline{128}$, 1908 (1962)
10. J. Hamilton, T.D. Spearman, Ann.of Phys.$\underline{12}$, 172 (1961)
11. J. Hamilton, P. Menotti, T.D. Spearman, W.S. Woolcock,
     Nuovo Cim.$\underline{20}$, 519 (1961)
12. J. Hamilton, T.D. Spearman and W.S. Woolcock, Ann.of Phys.$\underline{17}$,
     1 (1962)
13. J. Hamilton, P. Menotti, G.C. Oades, L.L.J.Vick, Phys.Rev.$\underline{128}$,
     1881 (1962)
14. R.E. Cutkosky, B.B. Deo, Phys.Rev.Lett. $\underline{22}$, 1272 (1968)
15. R.E. Cutkosky, B.B. Deo, Phys.Rev. $\underline{174}$, 1859 (1968)

16.   S. Ciulli, Nuovo Cim. 61A, 787 (1969)

17.   S. Ciulli, Nuovo Cim. 62A, 301 (1969)

18.   S.L. Walsh, Interpolation and Approximation by Rational Functions,
       Amer.Math.Soc. Providence RI 1956

19.   B.B. Deo, M.K. Parida, Phys.Rev.Lett. 26, 1609 (1971)
       see also S. Ciulli, Nuovo Cim. 61A, 463 (1969)

20.   H. Nielsen, Lyng Petersen, E. Pietarinen, Nucl.Phys.B22, 525(1970)

21.   F. Elvekjaer, H. Nielsen, Rutherford Preprint (1971)

22.   F. Elvekjaer, H. Nielsen, G.C. Oades, to be published

23.   F. Elvekjaer, Short range πN P-wave Interaction, Rutherford
       Preprint (1972)

24.   H. Nielsen, Nucl.Phys. B30, 317 (1971)

25.   F. Elvekjaer, The Up-down Ambiguity for Pion-Pion T=O S-wave,
       Aarhus, Preprint (1972)

26.   J. Hamilton, New Methods in the Analysis of π-N Scattering,
       Springer Tracts 57, 41 (1971)

27.   J. Hamilton, Pion-Nucleon Scattering Theory, CERN 71-14 (1971)

28.   Y.A. Chao, E.Pietarinen, Phys.Rev.Lett. 26, 1o6o (1971)

29.   Y.A. Chao, Phys.Rev.Lett. 25, 3o9 (1970)

30.   R.C.Miller, T.B. Novey, A. Yokosawa, R.E. Cutkosky, H.R. Hicks,
       R.L. Kelly, C.C. Shih, G. Burleson, Nucl.Phys. B37, 401 (1972)

31.   R.E. Cutkosky, C.C. Shih, Phys.Rev. D4, 2750 (1971)

32.   G. Steinbrecher, Phys.Rev. 174, 1794(1968)

33.   A. Martin, Contribution to Professor N.N.Bogolubov's Festschrift
       (1969)

34.   S. Ciulli, G. Nenciu, Lund Conference (June 1969) and
       Commun.Math.Phys. 26, 237 (1972)

35.   J. Fischer, J. Formánek, P.Kolár, I.Vrkoč, Stable Cut to Cut
       Extrapolation, paper presented at the Int.Symp.of Elem.Part.
       Reinhardsbrunn 1972

36.   J.E. Bowcock, G. John, Nucl. Phys. B11, 695 (1969)

37.   P. Prešnajder, J. Pišút, Nucl.Phys. B22, 365 (1970), see also
       J. Pišút, P. Prešnajder, Nucl.Phys. B12, 110 (1969), J. Pišút,
       P. Prešnajder, J. Fischer, Nucl.Phys. B12, 586 (1969), and
       P. Prešnajder, J. Pišút, Nucl.Phys. B14, 489 (1969)

38.   S. Ciulli, J. Fischer, Nucl.Phys. B24, 537 (1970)

39.   G. Nenciu, Nuovo Cim.Lett. 4, 96 (1970)

40.   S. Ciulli, G. Nenciu, Optimal Analytic Extrapolations, Prague
       Colloquium on Theoretical Physics (Sept.1970), and J.Math.Phys.
       in press.

41.   S. Ciulli, G. Nenciu, Nuovo Cim. $\underline{8}$, 735 (1972)

42.   I. Caprini, S. Ciulli, C. Pomponiu, I. Sabba-Stefanescu, Phys.Rev. $\underline{D5}$, 1658(1972)

43.   R.E. Cutkosky, Ann. of Phys. $\underline{54}$, 350 (1969)

44.   J. Pišút, P. Prešnajder, Nuovo Cim.$\underline{3A}$,603 (1971)

45.   P. Lichard, Pr. Prešnajder, Nucl. Phys. $\underline{B33}$, 605 (1971)

46.   G.C. Ross, Nucl.Phys. $\underline{B31}$, 113 (1971)

47.   H.K. Shepard, C.C. Shih, Nucl.Phys. $\underline{B42}$, 397 (1972)

48.   E. Pietarinen, Dispersion Relations and Experimental Data, Nordita preprint 1972

49.   I. Caprini, private communication

50.   G. Nenciu, Analytic Extrapolation from Boundary and Interior Points, sent to Nuovo Cim.

51.   I. Sabba-Stefanescu, On the Analytic Extrapolation in L^2 Norm, submitted to Nucl.Phys.

52.   R.E. Cutkosky, Convergence Test Functions, an Empirical Study, Carnegie Mellon, internal report CAR-882-26 (1972)

53.   G.F. Chew, S. Mandelstam, Phys.Rev. $\underline{119}$, 467 (1960)

54.   D. Atkinson, Nucl. Phys. $\underline{B7}$, 375 (1968)

55.   D. Atkinson, Nucl. Phys. $\underline{B8}$, 377 (1968)

56.   D. Atkinson, Nucl. Phys. $\underline{B13}$,415 (1968)and $\underline{B23}$, 397(1970)

57.   R.L. Warnock, Phys. Rev. $\underline{170}$, 1323 (1968)

58.   H.Mc.Daniel, R.L. Warnock, Phys.Rev.$\underline{180}$, 1433 (1969)

PHENOMENOLOGICAL STUDY OF TWO BODY HADRON SCATTERING

G.L. Kane

Physics Dept., University of Michigan, Ann Arbor, Michigan

and

Rutherford Laboratory, Chilton, Didcot, Berkshire, England

INTRODUCTION

In looking for the "hydrogen atom" of high energy hadron physics, many workers over the years have turned to high energy two body reactions. At the present time, in view of the apperently increasing complexity of the experimental data, the proliferation of attempts at phenomenological description, and the absence of any compelling theoretical insights, an objective observer would hardly be encouraged about our prospects for success. In these lectures I will

(1) discuss the present structure of the experimental data and recent attempts to describe it;

(2) present in detail one model (a "realistic" absorption model) which now appears to be capable of providing a useful description of almost all two-body reactions (including elastic and quasi-two-body reactions) at energies above the resonance region. Whether the reader accepts the assumptions of the model or not, it at least provides a way of organizing the experimental data and showing that the apparent complexity of experimental data can be understood in rather simple terms.

(3) consider how we might view duality and exchange degeneracy from a phenomenological point of view rather than a theoretical one. The main question is what kind of model can simultaneously describe the high energy data and have the duality properties shown by the experimental data .... especially the property that some amplitudes have the same structure in t at low and high energies. From one point of view this leads to a discussion of selection rules for Regge-Regge cuts, while from another it leads us to consider dual amplitudes with unitarity

cuts, second sheet resonance poles, and absorption.

One of the main reasons for phenomenological study of experimental data is to decide whether we have available the concepts needed to understand the data or whether we are just missing some essential idea. If a theory were available, of course, one would simply calculate and compare with data. I think there is currently no theory where this is possible for hadron interactions. We have certain ideas available: Regge behaviour, unitarity, analyticity, etc. . There are essentially three possibilities. The data might behave in a way that can probably be understood in terms of the available ideas, or it might appear not to be understandable because we do not see clearly enough the implications of the available ideas, or we might really be missing an essential concept. My personal view leans toward the first of these (the data behaves reasonably in terms of our present ideas) and I will try to convey that impression to the reader. That is not to say that we are near to a theory; on the contrary, I will argue that the data can only be described by including very important unitarity effects at all energies, and that means it will be very hard to construct a theory. In essence, I will try to convince the reader that hadrons really are strongly interacting particles.

FORMALISM

We need only a little formalism, of a general nature. In figure 1, we define some notation.

We can carry out our entire discussion in terms of s-channel helicity amplitudes. Formally, of course, any set of amplitudes is as good as any other. But for HE two-body reactions it seems apparent now that if there is any set of amplitudes with simple physical structure, it is the s-channel helicity amplitude.

We give each particle a helicity label λ, and we imagine matrix elements taken between helicity states:

$$M_{\lambda_c \lambda_d ; \lambda_a \lambda_b}(s,t) = \langle P_c P_d \lambda_c \lambda_d | M | P_a P_b \lambda_a \lambda_b \rangle$$

For a given reaction the total number of amplitudes is

$$(2s_a + 1)(2s_b + 1) \quad (2s_c + 1)(2s_d + 1) \qquad \text{since } \lambda_a \text{ can take on}$$

$2s_a + 1$ values, etc.; for a photon $2s + 1 = 2$. For hadron interactions we assume parity conservation. Under parity a helicity goes into minus itself and s, t are unchanged, so

$$M_{\lambda_c \lambda_d; \lambda_a \lambda_b}(s,t) \to M_{-\lambda_c -\lambda_d; -\lambda_a -\lambda_b}(s,t) = \eta_p M_{\lambda_c \lambda_d; \lambda_a \lambda_b}(s,t) \tag{1}$$

The last step says that parity invariance requires that we get back

$$M_{\lambda_c \lambda_d; \lambda_a \lambda_b}(s,t)$$

up to some phase η_p which we can choose to be ± 1.

As a result of this, only about half of the amplitudes for any processs are independent.

It may be worth emphasizing what is needed to make a model. For some given reaction there will be several complex helicity amplitudes that are independent (one for $\pi\pi$ scattering, two for $0^- + \tfrac{1}{2}^+ \to 0^- + \tfrac{1}{2}^+$ such as $\pi N \to \pi N$, four for $\pi N \to \pi \Delta$, six for $\pi N \to \rho N$, five for $np \to pn$, ...). Then we must be able to give the s and t dependences of each amplitude; in general they could all behave differently. In addition we must give the relative strength of each. There are very few places where one amplitude dominates a reaction.

If we specialize to the standard two-body case of scattering in the X-Z plane with incoming beam along the Z axis we get for the partial expansion

$$M_{\lambda_c \lambda_d; \lambda_a \lambda_b}(s,t) = \sum_J (2J+1)\, d^J_{\lambda\mu}(\theta)\, M^J_{\lambda_c \lambda_d; \lambda_a \lambda_b}(s) \tag{2}$$

$$\lambda = \lambda_a - \lambda_b, \quad \mu = \lambda_c - \lambda_d$$

For us the main point of this section is the behaviour of this as $\Theta \to 0$; using the known behaviour of d^J

$$d^J_{\lambda\mu}(\Theta) \xrightarrow[\Theta \to 0]{} (\sin \Theta/2)^{|\lambda-\mu|} = (\sin \Theta/2)^n \qquad (3)$$
$$n = |(\lambda_a-\lambda_b)-(\lambda_c-\lambda_d)|$$

we have, for $\Theta \to 0$, for any s-channel helicity amplitude

$$M_{\lambda_c\lambda_d;\lambda_a\lambda_b} \sim (\sin \Theta/2)^n. \qquad (4)$$

We want to go one step further, though only with a heuristic argument. Looking at Fig. 1, suppose the "exchanged state", whatever it might be, has definite parity [definite normality is sufficient]. Then we know that if we put all helicities at only one vertex into their negatives, leaving the other vertex alone, we again get back what we started with up to a sign. But since

$$(m_a = \lambda_c-\lambda_a, \; m_b = \lambda_d-\lambda_b), \quad n = |m_b-m_a| \qquad (5)$$

by changing the sign of m_b or m_a we cause the amplitude to vanish faster than $(\sin \Theta/2)$ as $\Theta \to 0$. It is obvious from this argument (and also correct) that an amplitude with definite parity exchanged must vanish in fact as

$$(\sin \Theta/2)^{|m_a|+|m_b|} = (\sin \Theta/2)^{n+x}, \; n+x = |m_b|+|m_a|. \qquad (6)$$

When $x > 0$ (x is always even) the amplitude is called "evasive". Amplitudes with $n = 0$ and $x = 2$, which need not vanish at $\Theta = 0$ from a general argument but do for any definite parity exchange (and so vanish at $\Theta = 0$ for any elementary or Reggeised exchange) play an important role in many reactions, particularly π exchange processes.

THE DATA APPEARS COMPLICATED

To get a feeling for the apparent complexity of the data, let us begin by considering various aspects where "simple" ideas have led people to expect a certain behaviour. Because most readers are familiar with them, some material is used here that is only derived later on. Anyone who has not encountered these examples before should put this section near the end of the lectures.

(a) π-exchange reactions. Several years ago it was generally considered obvious that $\gamma p \rightarrow \pi^{+} n$ and $np \rightarrow pn$ would have forward turnovers at high energies, because the π exchange contribution would be big and it flipped the nucleon helicity in the s-channel. In terms of the results derived in the previous section, pion exchange (Reggeized or not) contributions have $n + x = 2$ so the amplitudes vanish linearly in t as $t \rightarrow 0$.

The data, however, showed a sharp peak on a scale of order $m_{\pi}^2 = .02$ Gev^2; the "simple" expectation, that there should be a forward turnover, was wrong.

The explanation is generally thought to be that in one amplitude with $n = 0$, $x = 2$, the pion contribution does not have to vanish at $t = 0$ unless it has definite parity; then it is evasive. Some other slowly varying contribution is present and interferes destructively with the pion pole, so that at $t = 0$ we see only the other contribution, while by $t \approx m_{\pi}^2$ the destructive interference has produced a large decrease in cross section because the pion pole term varies rapidly on that scale.

The other contribution could be anything, of course, but the regularities are such (seen in $\gamma N \rightarrow \pi N$, $pn \rightarrow np$, $\pi N \rightarrow \rho N$ and at different energies) that it is only reasonable to say that it is something closely associated with π exchange. Such an effect is provided by unitarity connections.

At high energies the absorption model is an attempt to approximate the effects of unitarity. If one begins with an amplitude which vanishes at $t = 0$, makes a partial wave expansion, absorbs away essentially all the s-wave, most of the p-wave, some d-wave, etc., it is clear that after resuming the delicate cancellation between partial waves that gave the zero is destroyed and the amplitude no longer vanishes.

We will illustrate this quantitatively in a definite absorption model below. However one thinks of it, the "other contribution" has the same quantum numbers as the π exchange apart from angular momentum and parity, a strength related to the π exchange strength, and an s and t dependence different from the π pole but partially determined by it.

To summarize: the "simple" expectation is wrong; it is probably wrong because of unitarity corrections to the simple exchange contribution.

(b)   Next consider polarization in $\pi^- p \to \pi^0 n$. At HE only ρ exchange is allowed. Many workers expected that because there was only one exchange it would give both helicity amplitudes [the nucleon can flip helicity or not] the same phase. Then the polarization, proportional to Im $M_{++} M_{+-}^*$, would vanish.

Experimentally the polarization was significant, at least 25%, and rather constant from 5 GeV/c to 11 GeV/c.

Again, unitarity corrections in the absorption approximation do the right thing. The amplitude with n = 0 is large in the forward direction so it feels the absorption of the low partial waves rather strongly, while the n = 1 amplitude which vanishes at t = 0 is less affected by the absorption. The real and imaginary parts in each amplitude are affected differently since they contain different partial waves in general. Thus the two amplitudes get different phases and polarization is generated. Whether the results are quantitatively correct is model dependent, but most models start right at small t.

(c)   Next consider nonforward zeros of amplitudes. In the view of most theorists what is "simple" is a picture where all amplitudes for (for example) ρ or ω exchange vanish when the trajectories pass zero, α_ρ or $\alpha_\omega = 0$.   This occurs at $-t \approx \frac{1}{2}$ GeV^2.

This would imply, among other things, that in $\pi^{\pm}$ p elastic scattering where the ρ is the main contribution that changes sign between $\pi^{\pm}$, or in K$^{\pm}$ elastic scattering, where the ω is the main contribution that changes sign, one should observe a zero in the differences of differential cross sections. This zero is observed, of course ..... it is the well known cross over zero. But it occurs at $-t \sim 0.15$ to 0.2 GeV^2 rather than at 0.5 GeV^2. Physically that is a very big difference.

In addition, the zeros cannot just be associated with the exchanges, since they do not occur in non-elastic reactions where the same exchanges dominate. The ω exchange dominates $\gamma p \to \pi^0 p$, for example, in an amplitude with the same coupling at the nuclear vertex, and gives maximum instead of a zero at $-t \approx 0.2$.

Unitarity corrections again do the right thing. Whatever the detailed shape of the definite parity exchange contribution in t, by removing mainly low partial waves the t distribution gets sharper and a zero is introduced or moves toward t = 0 in n = 0 amplitudes where the absorptions is strong. For amplitudes with n > 0 as in $\gamma p \to \pi^{o} p$ the zeros will occur further out in -t.

Finally, it would have been "simple" if in a given amplitude the zero would be at the same t value for both real and imaginary parts. But for the n = 0 ρ exchange in πNscattering this is not so (Ringland and Roy(1971), Halzen and Michael (1971)) and probably it is not so for the vector and tensor K^{*} exchanges (Barger and Martin (1972)). For the vector exchanges the zero in ImM is near $-t = 0.2$ GeV^{2} but the zero in the real part is further out in -t (around 0.4 GeV^{2}) if it occurs at all. Although the absorption model will separate the zeros of the real and imaginary parts because their partial wave structure is different, the traditional versions of the model give less of an effect than is observed, and have the zero in ReM closer to t = 0 than that in ImM.

(d)   The final "simple"view we have to give up is the possibility that one could understand experimental data by only specifying quantum numbers in the s-channel (as proposed by the naive strong absorption model) or in the t-channel (as required by any model with only poles exchanged) but not both.

That the t-channel quantum numbers do not suffice is clear from $\pi^{-} p \to \pi^{o} n$, where there is t-channel ρ exchange but the data shows a completely different zero structure in the amplitudes with s-channel net helicity flip n = 0 or n = 1. That the s-channel quantum numbers are not sufficient is not completely established experimentally but is probably the case, with tensor exchanges f, A_{2}, K^{**} having zeros at different t values than vector exchanges ω, ρ , K^{*} in the same amplitudes (Phillips 1971, Barger & Martin 1972).

Thus any model which will help us understand the data must simultaneously depend on s and t-channel quantum numbers for each reaction. Later on we will discuss how this might come about.

Before we go to construct actual models one more bit of perspective is needed. We are trying to see whether we have available the concepts to understand the data or not. It is then important <u>to keep all the data for all reactions in mind</u>; or, conversely, it is dangerous to use one model for one process, a modified model for another, etc. We would be misled by such a procedure.

The reader can choose to stop thinking about the data at any stage. It should be clear by now that the complexity of techniques needed to describe the data is closely related to the complexity of the data. The author feels that the complications arise because

unitarity modifications are very important. Whether they can be introduced in a "simple" way is not clear, but we will try below.

BUILDING MODELS

In my judgement, it is possible that in the subject we are discussing theoretical derivations have not helped to understand data very
much; perhaps they have even been misleading on occasion. In addition,
there are standard treatments of Regge models in books and reviews.
Consequently I will proceed in a somewhat unconventional way, trying to
add a minimum of theoretical assumptions.

SOMETHING IS EXCHANGED

It is clear from experimental data that we should think in terms
of the exchange of known particles in some sense. When a known particle
can be exchanged in a process the cross section is "big" and has a forward peak, and the converse holds too. We will summarize this here by
defining an exotic state as one with quantum numbers not belonging to
any of the well known particles; then non-exotic exchanges are "big"
and have forward peaks while exotic ones are small.

REGGE POLES

What is exchanged from the point of view of s, t dependence?
If it were an elementary particle of spin J and mass and width m_R
and Γ_R we would have

$$M(s,t) \sim a_J(t) s^J \quad \text{as} \quad s \to \infty$$

which is not allowed by the Froissart bound (which we believe),

$$\sigma_T \leq c \ln^2 s$$

The simplest way to solve this problem must be to have many
partial waves correlated in such a way that when they are all "exchanged"
the power law increase disappears and we are left with a reasonable
behaviour.

In addition, some experimental cross sections fall with energy
approximately like a power, and sometimes a power that changes with t.

But conceivably this could arise from competition with other channels rather than being a property of the exchange, so it is not a compelling argument.

For these reasons, we assume

$$M(s,t) \sim \beta(t)\, s^{\alpha(t)} \tag{7}$$

where β and α are essentially arbitrary functions of t at present. The Froissart bound then requires that $\alpha(t \leq 0) \leq 1$.

Now consider the phase of M . Here we use some conventional theory which can be made rigorous and can probably be believed. Our treatment will be brief. Define $\nu = (s-u)/2m$, where $s + t + u = 4m^2$, and only consider equal mass spinless scattering. We want to show that an amplitude with a power law behaviour s^ν must have a phase given by $e^{-i\pi\nu/2}$.

We assume that our amplitude is a real analytic function in a cut ν complex plane. Then above the unitarity cut on the right we have $M(\nu_R + i\varepsilon)$. At fixed t we interchange s and u and we want to get the same amplitude back (crossing symmetry) so we require that the amplitude below the left hand cut, $M(-\nu_R - i\varepsilon)$, be the same as what we started with

$$M(\nu_R + i\varepsilon) = M(-\nu_R - i\varepsilon). \tag{8}$$

Using real analyticity we can go above the left hand cut so

$$M(\nu_R + i\varepsilon) = M^*(-\nu_R + i\varepsilon) \tag{9}$$

which is the usual form of the crossing condition. We finally write it as

$$M(\nu_R + i\varepsilon) = M^*(-(\nu_R - i\varepsilon)) \tag{10}$$

One way to satisfy this is to put

$$M(\nu) = f(-i\nu) \tag{11}$$

i.e. M can only be a function of the variable $-i\nu$ rather than just ν. To check this, we have

$$\text{LHS} = f(-i\nu_R + \varepsilon)$$

$$\text{RHS} = f(-(i\nu_R - \varepsilon))^* = [f(i\nu_R + \varepsilon)]^* = f(-i\nu_R + \varepsilon)$$

as required. Thus, crossing and analyticity require that our amplitude only be a function of the variable $-i\nu$. For $s \to \infty$ we can equivalently use $-is$.

For crossing-odd amplitudes one has a minus sign in equ.(9) and the equivalent statement is that the amplitude must be i times a function of -is.

We can immediately conclude that our Reggeon exchange amplitude with correct phase must be given by $(-i = e^{-i\pi/2})$

$$M(s,t) = \beta(t)\ s^{\alpha(t)} e^{-i\pi\alpha(t)/2} \left\{ \begin{array}{l} 1 \\ i \end{array} \right. \tag{12}$$

where 1 or i is used to make the amplitude real at particle poles (equivalent to crossing even or odd). Thus for ρ exchange the pole will occur at $\alpha = 1$ so the i is used, while for π or f exchange the pole is at $\alpha = 0$ or 2 so the 1 is used. Along a given trajectory part-icle poles can only occur in $\beta(t)$ at every other integer to maintain the reality properties. In the conventional treatment, the trajectory is called even or odd signature for the 1 and the i respectively.

Next consider how the poles appear. We want $M(s,t)$ to have poles at physical particles at every other point along a trajectory and only at physical values of J. One simple way to do this is to write

$$\beta(t) \sim \Gamma\left(\tfrac{1}{2}(J - \alpha(t))\right) \tag{13}$$

where J is the spin of the lowest allowed physical state; e.g. $J = 0$ for the pion trajectory, $J = 1$ for the ρ trajectory, $J = 2$ for the A_2 trajectory, $J = \tfrac{1}{2}$ for the nucleon trajectory etc.

Thus a form for a Regge pole which embodies everything we know about it is

$$M_{(\lambda)}(s,t) = \beta_{(\lambda)}(t)\, \Gamma\left(\tfrac{1}{2}(J - \alpha(t))\right)\left(\frac{s}{s_0}\right)^{\alpha(t)} e^{-i\pi(\alpha(t) - J)/2}\, (-t)^{\frac{n+x}{2}} \tag{14}$$

where $\alpha(t)$ and $\beta(t)$ are arbitrary apart from some analyticity re-strictions and can be taken real for t < 0. The subscript (λ) stands for $\lambda_c\lambda_d;\ \lambda_a\lambda_b$. In applications we will always assume that β factorises into $\beta_{\lambda_c\lambda_a}(t)\ \beta_{\lambda_d\lambda_b}(t)$.
This is quite different from the usual form

$$\beta(t)\ \frac{\pm 1 - e^{-i\pi\alpha(t)}}{\sin \pi\alpha}\ \left(\frac{s}{s_0}\right)^{\alpha(t)} \tag{15}$$

but is just as reasonable without further physical input restrictions.

One important consequence for equation (14) (which may be some-thing seen in the data) is that the Regge pole actually remembers the range of the force, because of the presence of J in the propagator to avoid ghost states. This is indicated in Fig.2 where $\Gamma(z)$ is plotted vs.z. As t varies from O to -1 GeV^2, putting trajectories $\alpha(t) = J + \alpha'(t-m^2)$ with $\alpha' = 1$GeV^{-2}, the argument of $\Gamma((J-\alpha)/2)$ varies most rapidly for $\tilde{\pi}$ exchange, less for ρ , even less for A$_2$, etc. That is, <u>tensor exchange is significantly less peripheral than vector ex-</u>change--- as would be expected very naively from the exchange of a heavier particle. Here the effect arises because J and m^2 increase together and to avoid ghost states the amplitude must have a knowledge of the lowest allowed J.

EXCHANGE DEGENERATE REGGE POLES

It is a fact that within a few percent $\sigma_T(K^+p)$ and $\sigma_T(K^+n)$ are constant in energy over a wide range. Since $\sigma_T \sim$ ImM(s,o) by the optical theorem, and the Pomeron contribution gives a constant σ_T, we must have to a good approximation, for all energies, at t = 0,

$$\text{Im}\,(\omega+f) = \text{Im}\,(\rho+A_2) = 0.$$

The two contributions must vanish separately by isospin arguments. If we assume σ_T = constant for all exotic channels we can require similar conditions for $\rho+f$ in $\pi^+\pi^+$ scattering, etc.

Since exotic channels have no resonances we could require (to guarantee that) that at any t the exotic amplitude have zero imaginary part. Then we get important relations among the Regge poles. (One could think of doing this for more complicated amplitudes with cuts, etc., but then one could not get simple results for all t.)

We can see this most easily in an example. Consider $\pi^+\pi^+$ scattering. Then

$$M(s,t) = i\beta_\rho(t)\,\Gamma(\tfrac{1}{2}(1-\alpha_\rho(t)))(se^{-i\frac{\pi}{2}})^{\alpha_\rho}+ \beta_f(t)\,\Gamma(\tfrac{1}{2}(2-\alpha_f(t)))(se^{-\frac{i\tilde{\pi}}{2}})^{\alpha_f} \tag{16}$$

and we want

$$\text{Im}\,M(s,t) = 0. \tag{17}$$

To satisfy this at all energies we clearly need $\alpha_\rho = \alpha_f \equiv \alpha$. Then

$$\text{Im}\,M \sim \beta_\rho\,\Gamma(\tfrac{1}{2}(1-\alpha))\cos\frac{\pi\alpha}{2} - \beta_f\,\Gamma(\tfrac{1}{2}(2-\alpha))\sin\frac{\pi\alpha}{2} \tag{18}$$

so, for example

$$\beta_\rho = \frac{\Gamma(\tfrac{1}{2}(2-\alpha))}{\Gamma(\tfrac{1}{2}(1-\alpha))}\,\frac{\sin\frac{\pi\alpha}{2}}{\cos\frac{\pi\alpha}{2}}\,\beta_f \tag{19}$$

$$M(s,t) = \beta_f \, s^\alpha \left[\frac{i \sin \pi\alpha/2}{\cos \pi\alpha/2} e^{-i\pi\alpha/2} + e^{-i\pi\alpha/2} \right] \Gamma\left(\frac{2-\alpha}{2}\right) \tag{20}$$

$$= \beta_f \, s^\alpha \, \Gamma\left(\frac{2-\alpha}{2}\right) / \cos \pi\alpha/2$$

This has poles now when $\alpha = 2,4,6,\ldots$ from the Γ function and at $\alpha = 1,3,5,\ldots$ from the zeros of $\cos \pi\alpha/2$, so it simultaneously contains the ρ and f trajectories and is real as desired. Note also that if β_f is featureless as expected then β_ρ has acquired a zero at $\alpha = 0$, $\alpha = -2$, etc; when only the ρ can contribute as in $\pi^-\pi^+ \to \pi^0\pi^0$ (or $\pi^- P \to \pi^0 n$) this zero should appear.

The general phenomenon is that in an EXD pole model the zeros in an amplitude will appear whenever a trajectory goes through a value that is Nonsense (i.e. not a physically allowed J value in that amplitude) and Wrong Signature (i.e. even integers for an odd signature state and vice versa). In principle the zeros can appear in different amplitudes in various ways, but in practice view the only interesting alternative is a linear zero in every helicity amplitude.

NAIVE ABSORPTION MODELS -- ELASTIC AND ENHANCED

Now we want to begin to try to take into account the effects of unitarity. By using factorising Reggeized pole amplitudes we have included the structure of the exchanged system in our models, but we have not considered the effects due to the strongly interacting nature of the external particles.

No one knows how to do that, of course, but several reasonable approaches have all led to the same answers (eqn.(27) below), so it should at least give us a good idea of the kind of effects that will occur. The various forms of absorption models should be viewed as attempts to approximately take unitarity effects into account. It should probably be emphasized that they are not viewed by their originators as attempts to fit data but as models for the behaviour of strongly interacting particles. Although the approximate models constructed may not yet be good ones, it would seem quite remarkable if two body hadron interactions were not strongly affected by unitarity.

In this section we only discuss non-diffractive reactions. We will return to the Pomeron processes later. We also ignore spin, assuming spinless particles here and stating the general results below.

Assume that in the absence of unitarity corrections for the external particles the HE amplitude for some reaction is R (s,t), with partial wave expansion

$$R(s,t) = \sum_{\ell} (2\ell+1)\, R_{\ell}(s)\, P_{\ell}(\cos\theta)$$

We will refer to R(s,t) as a <u>pole term</u>, and we will assume that it has particle poles all of definite normality (either all of spin J have parity $= (-1)^{J}$ or all of spin J have parity $= (-1)^{J+1}$).

Next we assume that the effect of the initial and final state interactions is to shift the phase in an amount given by the diffractive elastic scattering of the initial or final states (the Sopkovich prescription). We assume that the full amplitude M, including all effects, is given in each partial wave by

$$M_{\ell}(s) = e^{i\delta_{\ell}^{f}(s)}\, R_{\ell}(s)\, e^{i\delta_{\ell}^{i}(s)} \tag{21}$$

where the $\delta_{\ell}^{f,i}$ are the phase shifts of the elastic scattering in the final or initial states respectively. Each δ_{ℓ} is mainly imaginary because the scattering is diffractive; the δ_{ℓ} are related to the elastic amplitude M^{EL} by

$$S_{\ell}^{EL} = e^{2i\delta_{\ell}} = 1 - iq\, M_{\ell}^{EL}/4\pi W \tag{22}$$

In practice we also assume that M_{ℓ}^{EL} is the same for initial and final states, although one could carry out calculations using eqn. 34 if so desired. Then we can substitute $e^{2i\delta_{\ell}}$ from eqn.(22) into eqn. (21) , giving

$$M_{\ell}(s) = R_{\ell}(s) - \frac{iq}{4\pi W}\, R_{\ell}(s)\, M_{\ell}^{EL}(s) \tag{23}$$

It is a little easier to see what happens if we switch to an impact parameter representation. Writing

$$\ell + \tfrac{1}{2} \approx qb \quad\text{and}\quad P_{\ell}(\cos\theta) \approx J_{0}(b\sqrt{-t})$$

we have

$$R(s,t) = 2q^{2}\int_{0}^{\infty} b\,db\, J_{0}(b\sqrt{-t})\, R(s,b) \tag{24}$$

$$R(s,b) = \frac{1}{2q^{2}}\int_{0}^{\infty} \sqrt{-t}\, d(\sqrt{-t})\, J_{0}(b\sqrt{-t})\, R(s,t) \tag{25}$$

and eqn.(23) becomes

$$M(s,b) = R(s,b) - \frac{iq}{4\pi W} R(s,b) M^{EL}(s,b) \qquad (26)$$

with $M(s,t)$ given by a transform as in eqn.(24). Note from (23) or (26) that the absorption correction is additive in the partial waves and is constructed by multiplying amplitudes in impact parameter.

When we think about the structure of eqn.(26) we see that we can think of the cut as a double scattering in an appropriate sense (Fig.3); the amplitudes involved are always physical on-mass-shell ones. This is just the meaning of multiplying amplitudes in impact parameter space.

We are immediately led to ask why we did not include other intermediate states in addition to c and d (Henyey, Kane, Pumplin, Ross 1968).

For example, the Reggeon could excite a state c^* which could make a diffractive transition to c. The size of the contribution of these states is an important question.

If the pole term had zeros to start with, the pole-cut interference will move then in toward zero. One subtlety we have ignored above is that M is complex. Its real and imaginary parts will have separate zeros because they always have a different zero structure in the pole, but the final zeros will be nearby - - it is a complex zero in t.

In a practical case, such as ρ or A_2 or π exchange if we use pole terms of the form given in equation and if we enhance the absorption by about a factor of 1.5 to take account of intermediate states in addition to the elastic one, we find a zero near $-t = 0.2$ in both real and imaginary parts of the amplitude. For the ρ the real part zero is closer to $t = o$ than 0.2 and the imaginary part zero is a little further out.

<u>S P I N</u>

For understanding real data it is essential to take account of spin. Different helicity amplitudes behave in completely different ways as functions of both s and t.

In our approximations we just replace $J_o(b\sqrt{-t})$ by $J_n(b\sqrt{-t})$ in equations (24),(25) for a SCHA with net helicity flip n. The form of M^{EL} is just the same. We are assuming that M^{EL} does not flip

s-channel helicities, which is certainly a good approximation at nucleon
vertices, and presumably everywhere. The Reggeon term will have a factor $(-t)^{\frac{n+x}{2}}$
from equation (6) ,because of angular momentum conservation and definite
t-channel parity.

Thus for an arbitrary SCHA we have, combining equations (26),
(22) and (24)

$$M^{(n)}_{\lambda_c\lambda_d;\lambda_a\lambda_b}(s,t) = 2q^2 \int_0^\infty b\,db\, R^{(n)}_{\lambda_c\lambda_d;\lambda_a\lambda_b}(s,b)\, S^{EFF}(s,b)\, J_n(b\sqrt{-t}) \tag{27}$$

The Reggeon is given by equation (14). Since the absorption
only depends on n as far as helicity dependence goes, <u>all amplitudes
with the same n are absorbed the same way</u>. This is one of the main
predictions of the absorption model.

<u>QUALITATIVE SYSTEMATICS - BESSEL FUNCTIONS, RANGE OF FORCES</u>

Consider all amplitudes with x = 0. Suppose the Reggeon term,
which decreases like a Gaussian in b, is negligible beyond some radius
R. If S^{EFF} is very small at small b, absorbing away essentially all
of the s-wave, and grows to one at large b, then bRS^{EFF} is only large
near R $\approx$ b so we can write (Ross, Henyey, Kane 1970; Dar 1963)
$bRS^{EFF} \sim \delta(b - R_o)$. Then

$$M_{\lambda_c\lambda_d;\lambda_a\lambda_b}(s,t) \sim \gamma_{\lambda_c\lambda_a}\gamma_{\lambda_d\lambda_b}\, J_n(R_o\sqrt{-t}) \tag{28}$$

The γ's are constants associated with the sizes of the pole
vertices. In this naive form the systematics are very clear. We expect zeros in the SCHA at the zeros of J_n. If R = 1 fermi for n = 0,
the first zero is at -t = 0.23 GeV^2; for n = 1 it is at 0.6, and
for n = 2 it is at 1.06. Even in the naive absorption model the real
and imaginary parts of the amplitude will not have exactly coincident
zeros because the integration over the phase separates them. An edge
which is spread out instead of a delta function will multiply J_n by

a smooth function such as e^{bt}; for a reasonable interpretation we must have $b \lesssim 2$ GeV^{-2}.

It should be emphasized that these systematics are only meaningful in a model which includes significant contributions from intermediate states in addition to the elastic ones --- an enhanced absorption model.

If the range R were the same for all Reggeons we would then have a universal s-channel model with all amplitudes of a given n having the same s and t dependence regardless of the exchanged Reggeon. We would only need to know the nature of the exchange to get the relative size of the contributions to real and imaginary parts (from the phase at t = 0) and to different amplitudes.

Although it is not yet certain, it appears that one of the two main things that have been learned from experiment in the past year is that in the real world not all Reggeons have the same range R--- rather, it appears that the range of the force decreases as the mass of the lightest particle that can be exchanged on the trajectory increases. That may not surprise readers familiar with other areas of physics.

Alternatively, we could say that because J and M^2 increase together the range of the force decreases with increasing J. Indeed, that is just the result we get from our naive approach above, where we saw that we wanted a propagator

$$\Gamma\left(\frac{J-\alpha}{2}\right).$$

We will see later how to get a complete overview of the data --- in addition to taking into account the range of forces even for Reggeon exchange, we must be careful about phases. The naive systematics from eqn.(28) are still rather useful to get a qualitative view of how the amplitudes behave and to have a simple qualitative view of a large body of experimental data.

If we go back one step to see how the systematics arise we can get a feeling for how changes will occur. The four main amplitudes needed to understand two body reactions are shown in Fig. 4.

The n = x = 0 amplitude behaves as in the spinless case above. For n = 1, x = 0 both pole and cut vanish as $\sqrt{-t}$ in the forward direction; then the cut slope is smaller so they intersect at a larger -t than for the n = x = 0 case, near 0.6. For n = 2 the intersection is even further out. Even with the enhanced absorption it can be seen that for small t only n = 0 amplitudes are significantly affected;

this is because the other amplitudes all vanish for forward scattering and thus feel much less absorption.

For amplitudes with $n > 0$ and for $-t \lesssim 0.4$ GeV^2 it is a detailed quantitative question to distinguish between models. For example, the $n = 1$ amplitudes from the naive absorption model and from an exchange degenerate pole model are hardly different there. Since all models will have approximate exchange degeneracy near $t = 0$ because the data does, it seems safe to assume that amplitudes with $n > 0$ are approximately degenerate at small t, even in an enhanced absorption model.

Finally, consider the $n = 0$, $x = 2$ amplitude (Fig.4 d) which is responsible for the sharp π exchange peaks. The pole has the extra evasive factor of t but the cut does not have to vanish since $n = 0$. They interfere destructively. Thus the amplitude is given by the cut at $t = 0$, while it is essentially zero when the pole reaches its peak. For π exchange the pole is of the form $te^{at}/(m_\pi^2 - t)$ and its peak is at a t value smaller than $-t = 0.05$ GeV^2; the sharp forward peak is just the one seen in $np \to pn$, $\gamma p \to \pi N$, $\pi N \to \rho N$. For other exchanges, the pole peaks at larger $-t \sim 0.15$ GeV^2 and the cut is smaller in magnitude because of the missing forward peak of the pole so one sees a zero at about $-t = 0.1$ and $-t \sim 0.3$--- the numbers are rather model dependent. An important point for actually understanding data is that π and A_2 exchange always occur together in practic so the pole term peaks very early and is very broad.

By giving these four amplitudes their proper weights in a cross section or polarization, one can get a good qualitative picture of most data. Although $n = 0$ and $n = 1$ cross sections both have dips, an incoherent sum of both with about equal weights produces a monotonic cross section as in Fig.5. With a bit more $n = 0$ amplitude it can be seen from the figure that one could have essentially an exponential cross section. Bearing in mind the identity

$$1 = J_0^2(z) + 2\sum_{k=1}^{\infty} J_k^2(z)$$

one can see the general pattern.

WHEN SHOULD THE NAIVE ABSORPTION MODEL WORK?

Harari has suggested that the systematics of eqn.(28) with

$M_\lambda \sim J_n(R\sqrt{-t})$, should only apply to the imaginary part of the amplitude (Dual Absorption Model). His argument that it should apply to the
imaginary part is very good: for duality to apply to experimental data
and give fixed t zeros as a function of s the amplitude must be dominated by peripheral contributions at all s ($J \sim qR$ with $R \sim 1$ fermi).
At low s we get dual results by local averaging in πN reactions ---
then the imaginary part is big but the real part averages to a small
contribution. Thus the peripheral partial waves only need to dominate
in the imaginary part, and Harari suggests they only do dominate there
--- the real parts will be obtained from some analyticity assumptions.
For vector and tensor meson exchanges this view appears to give reasonable systematics; for pion and baryon exchange processes it is not useful, and perhaps not correct. If tensor exchange $n = 0$ imaginary amplitudes do not have the $-t = 0.2$ GeV^2 zero, then it is essentially wrong;
this may show up in hypercharge exchange reactions.

A different point of view suggests itself if one takes seriously the idea that absorption is an approximate attempt to take unitarity
effects into account. Then we expect that the cut we find will be
approximately valid. Suppose we decompose our amplitude at each t into
a piece pointing in the direction of the pole and a piece normal to that,
in the complex plane.

We would expect that <u>the piece of the amplitude pointing approximately in the direction of the pole, with a cut approximately 180°
out of phase, is correctly given by the naive absorption model.</u>

Thus for vector exchange (ρ) the pole starts at 45° and rotates
toward 90°; it is largely imaginary where there is data. So we expect
the imaginary part to be well given by the absorption model but perhaps
not the real parts. (I am only speaking in terms of real and imaginary
parts to simplify comparison with what is known ... it should still be
the parts parallel and perpendicular to the pole that are best and worst
determined by the naive absorption model). For π exchange the pole is
mainly real so the real part should be well described by the NAM, as
is the case. For tensor meson exchange the amplitude starts at -45°
and rotates toward 0°; it is mainly real. If we allow for the apparent
shorter range it is indeed the real part of the $n = 0$ amplitude that is
as expected in the analysis of Barger and Martin; it really has a zero,
while the imaginary part does not have an absorptive model shape. At
$u = 0$ the N (nucleon) exchange is mainly imaginary and the Δ exchange
is at about -45°; the Δ should behave largely like the A_2.

In all cases, we expect the part normal to the main pole direction to be very sensitive to finite energy corrections to the high energy absorption ideas and to any aspect of phases that we have gotten wrong. Probably the main thing we have learned from the data (mainly from the $\pi^- p \to \pi^0 n$ polarizations of Guisan et al) in the past year or so is that such phase corrections are indeed important ... and we have learned how we might account for them (see below).

HOW THE AMPLITUDES LOOK

Here I will discuss how the $n = 0$ and $n = 1$, $x = 0$, amplitudes are currently thought to behave. They are the flip and non-flip amplitudes in spin zero, spin one-half scattering, for example. The reader can combine them into cross sections in any desired combination. There is little direct experimental data for any other amplitudes, but we will see in some detail how the others will look in a realistic absorption model below. Since a good deal of the relevant analyses are reviewed in detail in Phillips' Amsterdam rapporteurs talk and Michael's Oxford conference talk, I will only give a very selective discussion.

Fig. 6 shows πN SHCA at 6 GeV/c. They are labelled $F^{I_t}_{\lambda_b \lambda_a}$ in terms of the t-channel isospin and the s-channel nucleon helicities. For the moment, only notice the "data"... these are from the (essentially) model independent analysis of Halzen and Michael (HM 1971). Since one does not measure the overall phase they are defined as parallel ($\parallel$) to F^0_{++} (the dominant amplitude) or normal ($\perp$) to F^0_{++} . We can speak in terms of real and imaginary parts since the rotation is not great (the solid and dashed lines respectively show the $\parallel$, $\perp$ or Re, Im amplitudes from an absorption model description of the data (Hartley, Kane 1972)).

From the "data" for $F^1_{\lambda'\lambda}$ we learn that

(i) $\mathrm{Re}F^1_{++}$ has a zero further out than $\mathrm{Im}F^1_{++}$, contrary to the NAM expectation.

(ii) $\mathrm{Re}F^1_{+-}$ does not have a zero by the time $\mathrm{Im}F^1_{+-}$ does, contrary to the NAM expectation, but expected in an EXD pole model where it would have a double zero.

(iii) $\mathrm{Im}F^1_{++}$ has a zero around $-t \sim 0.2$, the crossover zero ... in fact, this was well known long ago and is contrary to what is expected in an EXD pole model, but expected in the NAM.

(iv) For both amplitudes at small t, ReF/ImF > 1, contrary to what is expected in an EXD pole model where the amplitude should rotate toward 90°, from about 45° at t = 0.

(v) The shape of ImF^{1}_{+-} is consistent with both models. For F° the main result is that F°_{+-} is comparable in size to F^{1}_{+-}.

Barger and Martin (1972) have found (Fig.7) similar results for K^{*} around 4 GeV/c, in an analysis with assumptions. They find K^{*} (tensor) amplitudes of a very different sort, with a zero in ReF_{++} near -t = 0.7 GeV^{2} and a double zero centred around -t = 0.65 GeV^{2} for ImF_{++}. Very crudely, the tensor amplitudes, which are rotated by 90° from the vector by signature, behave like the vector with Re $\leftrightarrow$ Im but with zeros at a larger t value as if they were more central. Whether they are more central because of the pole term structure or the effects of absorption is an open question.

Most of the evidence is consistent with tensor amplitudes such as these for f and A_{2} also, but there is no clean evidence. The results are remarkably non-exchange degenerate ... a glance at Fig.7 shows that $ImK^{*} - ImK^{**}$ can never be very small compared to $ReK^{*} - ReK^{*}$,for example.

Of the many approaches to understanding the data we have room for a discussion of two.

1. Take as a basic starting point an EXD pole model, add cuts from naive elastic absorption, and look toward the dual theories both for guidance, and for other contributions. One would expect the other contributions to be lower lying ones in the J-plane such as Regge-Regge (R*R) cuts, and it is widely known that in $\pi^{-} p \rightarrow \pi^{\circ} n$ the $\varrho^{*} f$ cut does the right sort of thing. [A R*R cut is one constructed from two Reggeons via the second form in equation (26) rather than from a Reggeon and an elastic scattering.] But Finkelstein (1971) showed that in EXD theories R*R cuts obey some surprising selection rules, and then Worden (1972) and a Saclay group (Girardi et al, 1972) showed independently that in πN scattering in EXD theories the $\varrho^{*}f$ cut will not appear --- it is cancelled by the $\omega^{*}A_{2}$ cut. At that stage it appears rather difficult for such theories to produce any description of the data. However, the Saclay group argues that in the dual theories the cancellationg does not occur for the appropriate absorbed R*R cut --- the R*P*R cut. On the contrary, for these the two add,giving a result large enough to account for the πN data. So far they have only looked

at $\pi N \rightarrow \pi N$ scattering but the authors are rather optimistic about dealing with the data. Their biggest problem will presumably be with hypercharge exchange reactions where various authors (Michael 1969, Irving et al. 1971, Kwiecinski 1971) have argued that one cannot simultaneously describe cross sections and polarizations in EXD theories with R*R cuts, and with π exchange processes, especially np $\rightarrow$ pn and related processes.

We will discuss their arguments in some detail below.

2. One could assume the naive enhanced absorption model is basically right but needs to be understood better, particularly in its phases. This was pursued by Ross(1972), Hartley and Kane (1972) and Martin and Stevens (1972), all of whom showed in different but related ways that by using an elastic amplitude with an appropriate t-dependent phase (which might be the actual elastic phase) in equation (27) one could reproduce the observed πN amplitudes and data.

Very recently there has been more progress in this area and it now appears to me that along these lines it is possible to make a unified and simple treatment of almost all two body reactions, including s and t dependence of elastic, quasi-elastic, and normal two body reactions.

The main modifications needed to construct this "realistic" absorption model (RAM) are to allow the Reggeons to use the simple propagator (in equation (27)) which remembers the range of the force, and to have a model for the energy dependence and phase of the Pomeron. I will describe the RAM in detail below.

REGGE - REGGE CUTS IN EXCHANGE DEGENERATE THEORY

Here we describe in some detail the arguments of Finkelstein, Worden, and the Saclay group concerning the implications of an exchange degenerate theory.

It is "well known" that an amplitude pictured as in Fig.8a will not have a J-plane cut unless the lines n and n' (representing intermediate states) represent states with signature--- alternatively put, the amplitudes $a + R_1 \rightarrow c + R_2$ and $\bar{R}_1 + b \rightarrow \bar{R}_2 + d$ must both have third double spectral functions. In a dual theory this can only occur if neither of the channels $a + R_1$ and $\bar{c} + R_1$ are exotic --- otherwise there are no resonances in an exotic channel so no Veneziano term is present and so no double spectral function. This is essential Finkelstein's rule.

Finkelstein gives some applications of his selection rule; e.g. in KN scattering no R^*R cuts can occur. Worden (1972) has shown that the practical implications are even stronger --- no $\rho * f$ or $\omega * A_2$ cuts are allowed in N scattering.
On the surface of it, this leaves very little in the way of contributions to provide agreement with data in an EXD theory---it seems to be accepted that the ρ pole plus elastic or enhanced absorption, even allowing for some phase dependence in M^{EL}, does not allow one to describe the data in an EXD model.

Worden's argument is illustrated as follows (his paper should be consulted for the general argument). Consider the sum in Fig.8b; it is split into two terms by summing over K_i^* with even c-parity, and K_j^* with odd c-parity. By Finkelstein's rule it is zero. In the processes $K^o p \rightarrow K_j^* N \rightarrow K^+ n$ the ρ and A_2 imaginary parts exactly cancel, so $M_\rho / M_{A_2} = i \tan \pi \alpha /2$. Near $t = 0$, $\alpha \approx \frac{1}{2}$ so $\tan \pi \alpha /2 \approx 1$. Then one can replace the charged A_2 line in the first term and the neutral A_2 line in the second by ρ^+, ρ^o lines. If the tensor and vector multiplets have the same d/f ratios, the two terms then have the same SU(3) properties and one can rotate them into the terms in Fig. 8c, the desired result.

The point of Girardi et al. is essentially as follows. First, the $\rho * f$ cut one might have expected is exactly cancelled, as pointed out by Worden, in the n = o amplitude. In the n = 1 case, however, the two poles do not enter symmetrically because one has a $\sqrt{-t}$ with it, and there is still a $R * R$ contribution of order $i\pi/2(\ell ns-i\pi/2)^2$ relative to the pole. The same cancellation occurs for $R*R*P$ cuts, where the Pomeron is attached at the end.

When the Pomeron is in the middle, however, they propose that there is a sign change and the results add. This is the $R*P*R$ cut. The reason is the crossing requirement for a meson baryon amplitude. For a $R*R$ cut to contribute in $\pi^- p \rightarrow \pi^o n$, if one exchange has a real phase the other has a rotating one, as in Fig.9a. Then one finds in an EXD theory that $\rho * f$ cansels $\omega * A_2$. But for a $R*P*R$ diagram crossing requires both exchanges to be real or both rotating, as in Fig.9b. Then $\rho * P * f$ and $\omega * P * A_2$ odd. The reader can check which diagrams contribute by labeling quark lines and tracing them through; this can only be done for the allowed diagrams.

Having identified a contribution it is still necessary to calculate it and the theory does not help here. The Saclay group chooses its sign to give constructive interference with the imaginary part of the $\rho * P$ cut in $\bar{\pi}N$, as is needed. They choose its magnitude by using a factor $i/8\bar{\pi}$ with each $*$ as one would do in an elastic absorption model. They calculate the s and t dependence by using an eikonal prescription.

They then obtain good results for $\bar{\pi}N$. Since the $R * P * R$ cut has an intercept $\alpha(o) \lesssim o$, the extra contribution goes away fairly rapidly, and the old absorption model results are obtained at a rather low energy.

If this approach can also provide a reasonable description of hypercharge exchange processes and of a pair such as $np \rightarrow pn$ and $\bar{p}p \rightarrow \bar{n}n$ it could be very important. On the other hand, the various cut contributions are increasingly important at lower energies and things may look very complicated there.

POMERON

Before we construct our Realistic Absorption Model we must consider seriously the Pomeron. It is normally assumed that an adequate description of the Pomeron for use in an absorptive model is the purely imaginary diffractive form $P \sim i\sigma_T S e^{At}$.

Sometimes shringage is assumed, with $A = A_o + \alpha' \ln s$, and sometimes a real part is provided at $t = 0$. These forms give an adequate description of elastic data for $-t \lesssim 0.75$ GeV^2 at accellerator energies (not at the ISR). There are three pieces of information which suggest that this simple view will not suffice. Most dramatic is the increase in slope in elastic scattering at small t which was first noticed by Carrigan(1970) and then clearly observed at the ISR (Barbiellini et al, 1972). The data appear to show that the smallest t part of $d\sigma/dt$ shrinks significantly with increasing energy. A contribution with these properties to the Pomeron used in absorption could have a big effect, rather different from the standard exponential term.

Secondly, the determination of the isoscalar amplitude in $\bar{\pi}$N scattering discussed above showed a significant real part varying with t. If one separates a plausible f exchange out one obtains a Pomeron real part like the one shown in Fig.10 or Fig.6.

Again, its effect on absorption is significant.

Third, for some time there have been increasingly strong theoretical arguments, beginning with some Russian work (Verdiev et al,1963) clarified somewhat by Finkelstein and Kajantie in 1968, and made considerably more rigorous recently in inclusive theory (see the review talk by F. Low in the proceedings of the III Multiparticle Conference, Zakopane, 1972), that a factorizable Pomeron pole with $\alpha(0) = 1$ must decouple at $t = 0$ from all particles. Although it is not at all clear what direction the theory will take, we must be careful about detailed conclusions concerning absorption models drawn using a naive Pomeron pole (fixed or moving).

As it turns out we can proceed fairly far along these lines. It is now possible to take a form for the Pomeron amplitude which does show the effects observed at the ISR, the small t slope change and the shrinkage characteristics, and which does have the real part suggested by the phenomenological analysis. Although its theoretical status is not particularly respectable, it certainly is not a factorizable pole, so it may be safe from that side too. Using this Pomeron we can construct a "Realistic Absorption Model" where the same Pomeron used in the absorption is used to describe the elastic scattering data from πN at 6 GeV/c to pp at $s = 2800$ GeV^2, including polarization effects.

It should be emphasized that this result is really for the _full elastic amplitude_ arising from Pomeron exchange and not for a simple t-channel singularity; we can continue to use the exchange language and call it Pomeron, but it may actually arise mainly from s-channel unitarity effects. The original t-channel singularity (if there is one) could be completely buried.

Indeed, if we take seriously the Freund-Harari separation of Pomeron from Regge contributions and the duality identification Pomeron-background, Regge-resonances, then the following extension of the Freund-Harari hypothesis to the dynamical origin of the particles and the Pomeron seems plausible. Suppose we assume that the existence of the resonances can be understood in a bootstrap sense. What about the Pomeron? Since the elastic scattering is in fact a small part of the total cross section (about 15%) it is possible that in fact _the Pomeron is simply to be calculated from the s-channel unitarity sum rather than generate itself in a self-consistent way_. Thus Pomeron and background are calculated from coherent unitarity effects, while Regge and resonances will be understood as arising in a bootstrap.

With this background we can obtain a form for the Pomeron as follows. We imagine that in the unitarity sum some fraction of the contribution behaves mainly in a peripheral way, as if its central partial waves were absorbed away, while the rest behaves as a central piece with a smooth fall-off in the contribution of partial waves of increasing impact parameter.

Then the central part will contribute something to the imaginary part of the Pomeron which we can approximate in the old way as e^{bt}, while the peripheral part may contribute something like

$$e^{b_0 t}\, J_0(R\sqrt{-t})$$

to ImP (the exponential provides a spread out edge in impact parameter rather than a delta function one).

In addition, we expect that the edge piece will contribute a real part

$$R\sqrt{-t}\, e^{b_1 t}\, J_1(R\sqrt{-t}).$$

For example, if the imaginary part arose from something like a Breit-Wigner shape in impact parameter it would give a real part which was essentially the derivative, as above. More generally, analyticity arguments will do this---putting a peaked imaginary part into a dispersion relation will produce a real part like that.

If one felt that the outer peak in the real part in impact parameter space (there must be two peaks to give zero total area if the real part vanishes at $t = o$) should be at the same impact parameter as that of the imaginary part then the R in the real part will be somewhat smaller than that of the imaginary part.

Finally, we expect that edge part to remember its peripheral or exchange origins so we assume $R^2 = R_0^2\, \ln s$ as would arise from underlying Regge singularities.

Thus we put

$$P(s,t) = si\left(a e^{bt} + a_0 e^{b_0 t}\, J_0(R_0\sqrt{-t\ln s})\right)$$
$$+ sa_1 e^{b_1 t}\, J_1(R_0'\sqrt{-t\ln s})\sqrt{-t} \tag{29}$$

The quantities a, a_0,...R are not exactly known but none of them are free. At some reasonable energy R must be about 1 Fermi. R' must be a little smaller than or equal to R. The thickness of the edge is measured by b_0 and b_1 so they must be less than about 2 GeV^{-2}

while b measures the size of the central region and could be of order
3 GeV^{-2}. The relative size of a_1 and a_0 should be typical of the
areas of the real and imaginary parts of a Breit-Wigner. Since absorptive
effects appear to be strong the total area arising from the edge should
be comparable to that from the centre. Remarkably enough, with a Pomeron of the form given above (and these constraints) we can obtain a quantitative description of all the experimental elastic data and simultaneously remove the two main problems of the NAM-- phase and insufficient shrinkage. We will describe the absorption model effects below;
here we note the elastic scattering results and how they can be understood. In Fig. 10 we show the real and imaginary parts of the Pomeron
separately, for pp scattering at two separated energies and for N from
6 GeV/c, just scaled by the ratio of total cross sections. The real
part is observable in two ways --- it provides the phase variation and
it prevents a dip from occurring in $d\sigma/dt$.

Fig. 11 shows a description of this data with the above Pomeron.
The piece with the larger slope at small t is due to the J_0 and thus
to the edge of the proton. The small t shrinkage is just the increase
of the radius in J_0 like $(\ln s)^{\frac{1}{2}}$. The data does not show any need for
shrinkage in the central term.

It is important to note that with this form σ_T = constant, the
energy dependence of J_0 decoupling at t=o. Also, equation (29) is written in a simple, transparent form, ignoring questions of analyticity
and signature. One can write a much more complicated form with good
analyticity and signature and essentially the same numerical behaviour.

REALISTIC ABSORPTION MODEL

Our goal is to get a comprehensive overview of all relevant
two body data. To keep the model simple both conceptually and for
practial work some approximations and oversimplifications have been
made. It is to be hoped that occasionally some detail or prediction
goes wrong so that we can learn about the behaviour of contributions
we have ignored such as lower lying areas in the J-plane, R$*$R cuts,
etc., and especially Regge residues, which (appropriately defined) are
assumed constant here.

To specify a model we must describe (equation 49)

    (i)   The Regge pole input;
    (ii) Which contributions are important;

(iii) The effective S-matrix for the absorption, S^{EFF}.

The main theoretical input is needed in (iii).

(i)   Regge pole input. We assume all poles are described by eqn.(14). The trajectories are treated as linear.

(ii)   We will try to see whether all reactions can be described by elastic scattering plus exchanging the vector, tensor, and pseudoscalar nonets.

(iii) S^{EFF} is constructed as follows. First write

$$S^{EFF} = 1 - \frac{iq}{4\pi W} M^{EFF} \tag{30}$$

$$M^{EFF} = M^{DIFF} + P \tag{31}$$

That is, we assume that the elastic intermediate state contributes only through the Pomeron exchange P, and that the sum over inelastic states can be lumped together in a term getting contributions from diffractively produced intermediate states. The Pomeron is given by eqn.(29). The diffractive term is expected to be more peripheral (RHK). By arguments similar to the qualitative picture used above for the Pomeron we expect

$$M^{DIFF} = i\left(c\,e^{dt} + C_o J_o(R\sqrt{-t})e^{d_o t}\right) + C_1 e^{d_1 t} J_o(R\sqrt{-t}) \tag{32}$$

where the central part $c\,e^{dt}$ is significantly smaller than it was for the Pomeron, but the edge parts can be comparable in size to the Pomeron ones. The real part here is allowed not to vanish as $t \to o$ since we do not know of any reasons to make the diffractive production amplitude purely imaginary at $t = o$. (For more discussion of the form of M^{DIFF} see my 1972 Zakopane lectures). It is expected that the integrated diffractive cross section will be qualitively consistent with data, i.e. we expect

$$\int |M^{DIFF}|^2 dt \sim \tfrac{1}{2} \int |M^{EL}|^2 dt$$

Note that we use P in S^{EFF}, not $M^{EL} = P + f + \omega + \ldots$

This is done mainly for simplicity. In addition, if these are included and convoluted with exchanges they generate R$*$R cuts in the
amplitude. The argument of Worden that these cancel out in some reactions with other cuts that would have to be inserted by hand en-
courages us to leave them out. On the other hand, in the real world
one might be rather sceptical and feel that such cancellations would
at best be incomplete -- then these terms might be needed. In the absence of a theory one can only tell by systematic phenomonology so we
assume they are not present and later on their role can be studied.
Since M^{DIFF} is not determined separately it could mock the effect of
such terms. When amplitudes are available at two fairly separated
energies it will be easy to tell.

The qualitative effect of this procedure (RHK) is illustrated
in Fig.12. Part a shows a sketch of the impact parameter distribution
of the Pomeron, part b the impact parameter distribution for the diffraction, more peripheral, and part c the sum. Comparing this with the
approximation of scaling the Pomeron by a factor shows clearly the more
physical nature of the present procedure. Essentially, the enhanced absorption fills in the disk rather than overabsorbing a lot at the centre.
This also provides us with the clue we need to see how to go from one
kind of particles to another. Suppose we have fixed on the absorption
for πN scattering. Then we go to KN reactions with a decrease in σ_T,
we can simply scale down M^{EFF} by the ratio of $\sigma_T(KN)/\sigma_T(\pi N)$, because
for small σ_T the result will be largely linear in σ_T. But for reactions
with larger σ_T things will not scale linearly; rather, the disc in
impact parameter will fill up as σ_T increases, approaching a black disc
right out to its edge instead of just near the centre. Note that for
a given reaction all exchanges are absorbed with exactly the same S^{EFF}.
For hyperexchange meson-baryon reactions we assume that the appropriate
scaling is as the average of $\sigma_T(KN)$ and $\sigma_T(\pi N)$. Although it is not
clear that at a finite energy we should use the same absorption for
both reactions in a line reversed pair we do so in order to see the
effect of the absorption.

There are no parameters that are free to vary widely to fit
data. Every parameter has a simple physical interpretation and is only
a parameter because of our ignorance of its value. Hopefully they can
be eliminated as parameters soon. Once we have determined S^{EFF} for a
class of reactions any further experiment can be predicted if the
coupling strengths are known.

In Fig. 13a - f we show some available results. At the time

of writing we have treated a few reactions at a time; soon we will have combined them all to be sure no surprises arise. It is clear that we can handle current data; what happens with predictions has yet to be seen. Note the good results for $np \rightarrow pn$ and $\bar{p}p \rightarrow \bar{n}n$ with all exchanges ($\bar{\pi}$, A_2, P) absorbed in exactly the same way, with the same s^{EFF}. Other approaches may have trouble here. For a given <u>amplitude</u> the results are not very sensitive to small changes in quantities we only know approximately, but cross sections and polarizations are generally very sensitive. The proper way to go to a new reaction is to calculate with the best available s^{EFF} couplings, trajectories, and then vary them within the limits of our ignorance. Observables which are stable can be trusted; those which change a lot can only be fixed by further input.

We should remark on the energy dependence of s^{EFF} --- by including in it no low lying contributions we may get effects which go away too slowly with s, but it should not be very wrong if we have only erred in not including the f and ω from elastic scattering. On the other hand, our results will be different from those of the Saclay group as the energy changes because their corrections are very low lying contributions directly to the amplitude. The J_o contributions to our s^{EFF} shrink with increasing s and reduce the size of the cut, leading to much more shrinkage at low and medium energies in the RAM than in the NAM. Thus even this long standing problem of absorption model seems to be solved with our Pomeron and s^{EFF}.

<u>DUAL ABSORBED AMPLITUDES</u>

It seems very likely that some form of absorption effects will have to be included in any model that can describe data. So far we have looked at high energies. In recent years, however, we have learned that sometimes scattering amplitudes have the unexpected property that they have the same structure at HE and LE; this is the phenomenologists' definition of duality. It is observed in the data.

If the absorption is necessary at HE, and we want phenomenological duality , what is the absorption dual to? What is the effect of the absorption at LE? Can it be fitted in? Here I will briefly describe a treatment by a Saclay-Michigan collaboration (G.Cohen-Tannoudji, R. Lacaze, F.S. Henyey, D. Richards ,W.J. Zakrzewski, G.L.Kane,1972). We find, and show explicitly in a model, that one can not only solve these problems but that the solution is very natural---essentially the absorption continued to LE can be <u>interpreted</u> as arising from the opening of inelastic channels, as one might hope. A number of HE and LE properties of the amplitude get tied together in interesting ways.

We want to construct a model with as much unitarity (U) as possible. At HE we include U in the absorptive model sort of approximation. At LE we want to have resonances that are second sheet poles, unitarity bounds on partial waves, and the absorptive effects due to the opening of inelastic channels.

The essential point is to note that duality in π N works because the peripheral resonances dominate ($J \sim qR$). This comes about at HE because of strong absorption removing many lower partial waves, due to competition with many channels. At LE the central resonances (of the Regge pole or Veneziano amplitude) are absorbed out because competing channels open up in lower partial waves (from angular momentum barrier effects), and because the $2\ell +1$ multiplies something bounded independent of ℓ .

The implications of this approach for constructing amplitudes with absorption at all energies can be studied in the paper by G.Cohen-Tannoudji et al (1972).

136

<u>REFERENCES</u>

1. G. Barbiellini et al, contributed paper to the IVth International Conference on High Energy Collisions, Oxford 1972.

2. V. Barger, A.D. Martin, Phys.Lett.39B, 379 (1972).

3. V. Barger, R.J.N. Phillips, Phys.Rev. <u>187</u>, 2210 (1969).

4. G. Cohen-Tann-Tannoudji, G.L.Kane, C. Quigg, Nucl.Phys. <u>B37</u>, 77 (1972).

5. J. Finkelstein, Nuovo Cimento <u>5A</u>, 413 (1971).

6. J. Finkelstein, K.Kajantie , Phys.Lett. 26B,305 (1968).

7. G. Girardi, R. Lacaze, R. Pechanski, G. Cohen-Tannoudji, F. Hayet, H. Navalet, Sacley Preprint DPh-T/72/36.

8. F. Halzen, C. Michael, Phys. Lett. <u>36B</u>, 367(1971).

9. H. Harari, Phys. Rev. Lett. <u>26</u>, 1400 (1971).

10. H. Harari, A. Schwimmer, Phys. Rev. <u>D5</u>, 2780 (1972).

11. F. Henyey, G.L. Kane, J. Pumplin, M. Roos, Phys.Rev. <u>182</u>, 1579(1969); Phys.Rev.Lett. <u>21</u>, 946 (1968).

12. B.J. Hartley, G.L. Kane, Phys.Lett. <u>B39</u>, 531 (1972).

13. A. Irving, H.D. Martin, C. Michael, Nucl.Phys. <u>B32</u>, 1(1971).

14. G.L. Kane, Phys. Lett. <u>40B</u>, 363 (1972).

15. A. Krzywicki, J.Tran Thanh Van, Phys. Lett. <u>30B</u>,185 (1969).

16. S. Kwiecinski, Proceedings of the XIth Cracow School of Theoretical Physics 1971.

17. A. Martin, P. Stevens, Phys.Rev. <u>D5</u>, 147 (1972).

18. C. Michael, Nucl. Phys. <u>B 13</u>, 644 (1969).

19. R.J.N. Phillips, Amsterdam Conference Rapporteur's talk, 1971.

20. G.A. Ringland, R. Roberts, D.P. Roy, J. Tran Thanh Van, Rutherford preprint RPP/c/39/1972.

21.   G.A. Ringland, D.P. Roy, Phys. Lett. $\underline{36B}$,110 (1971).

22.   M. Ross, Phys. Lett. $\underline{38B}$, 321 (1972).

23.   M. Ross, F. Henyey, G.L. Kane, Nucl. Phys. $\underline{B23}$,269 (1970).

24.   I. Verdiev, O.V. Kanchelli, S.G. Matinyen, A.M. Popova,
        K.A. Ter-Martinrosyan, J.E.T.P. $\underline{44}$, 341 (1963)
        (Sov.Phys. J.E.T.P. $\underline{19}$, 1148 (1964)).

25.   R. Worden, Phys. Lett. $\underline{40B}$, 260 (1972).

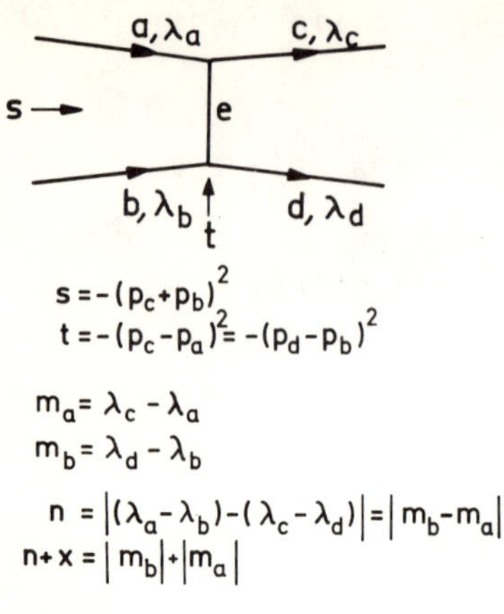

Fig. 1

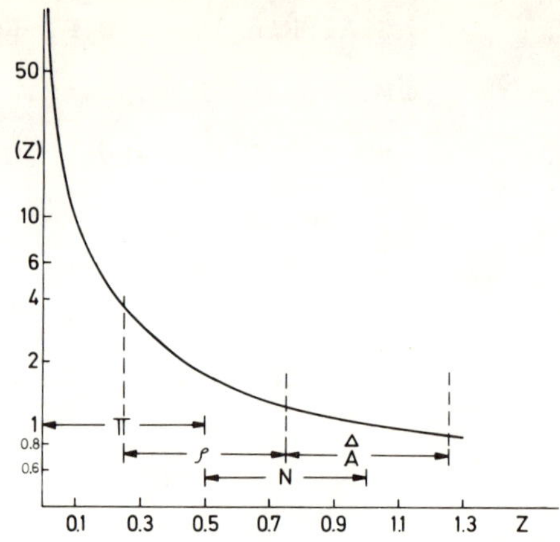

Fig. 2

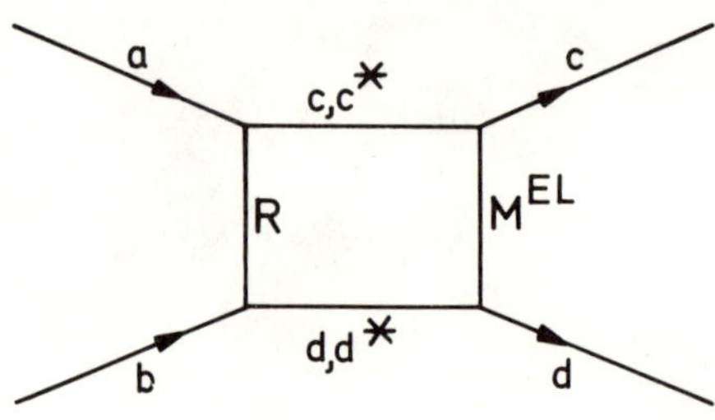

Fig. 3

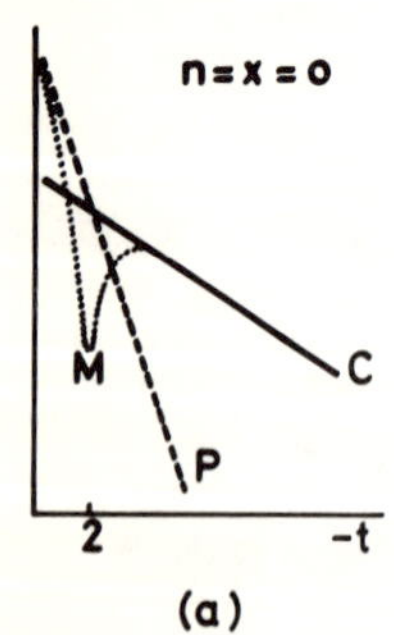

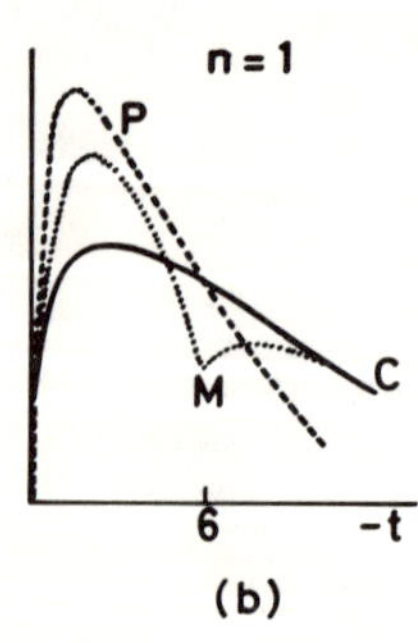

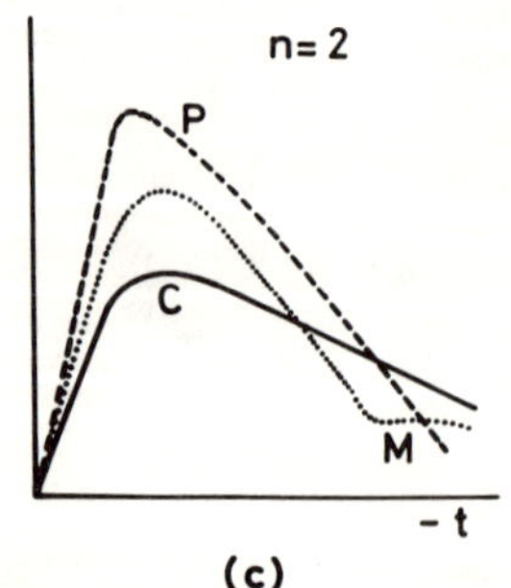

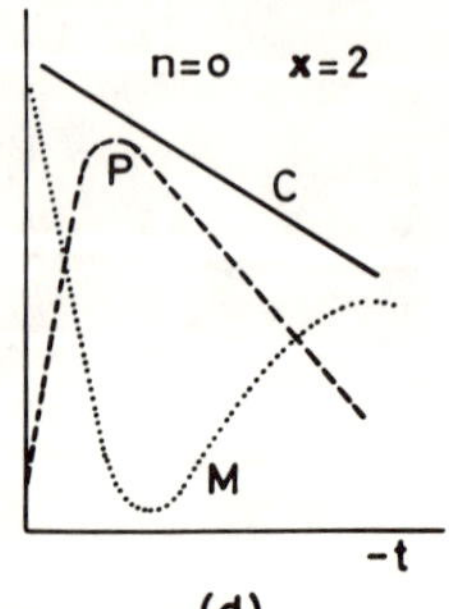

Fig. 4

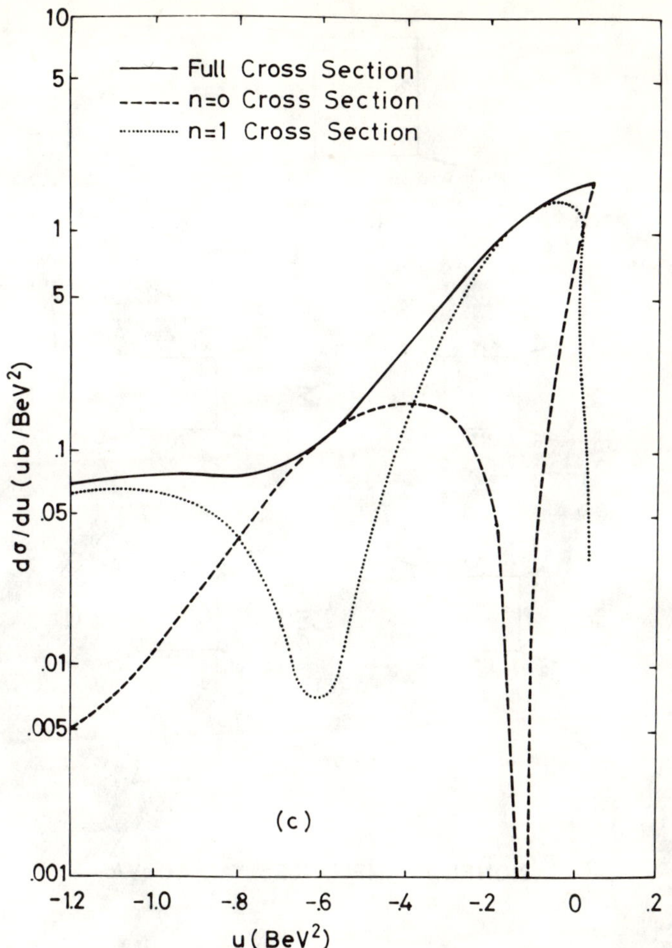

Fig. 5

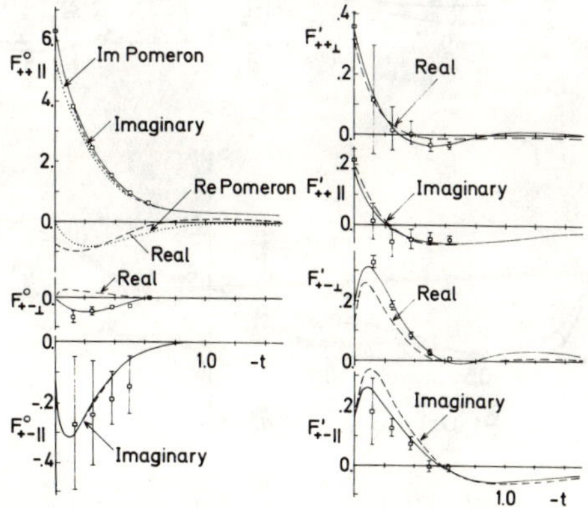

Fig. 6

a
n
c
R₁
R₁
b
n'
d
(a)

K⁰
Kᵢ*
K⁺
A₂⁺
A₂⁺
p
N*
n

K⁰
Kⱼ*
K⁺
A₂⁺
A₂⁰
p
N*
n

(b)

π⁻
πᵢ*
π⁰
ρ
f
p
N*
n

π⁻
Bⱼ*
π⁰
A₂
ω
p
N*
n

(c)

Fig. 7

NONFLIP AMPLITUDES AT 4 GeV/c
K*
K**
(ImK*)_M
(ImK*)_Λ
(ImK**)_Λ
(ImK**)_M
(ReK*)_M
(ReK*)_Λ
(ReK**)_M
(ReK**)_M

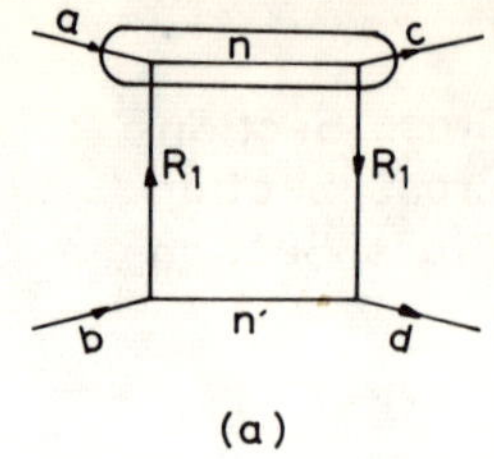

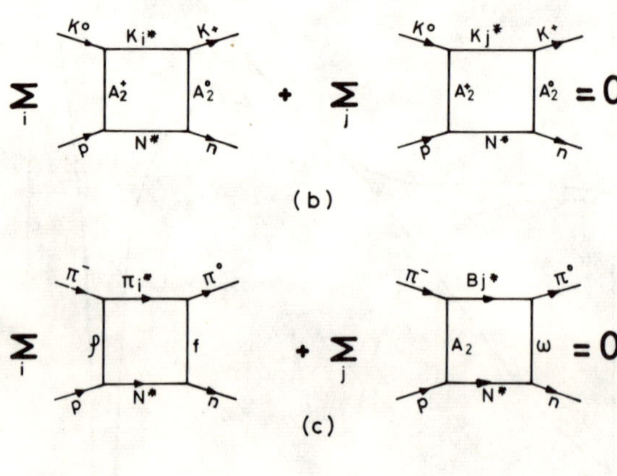

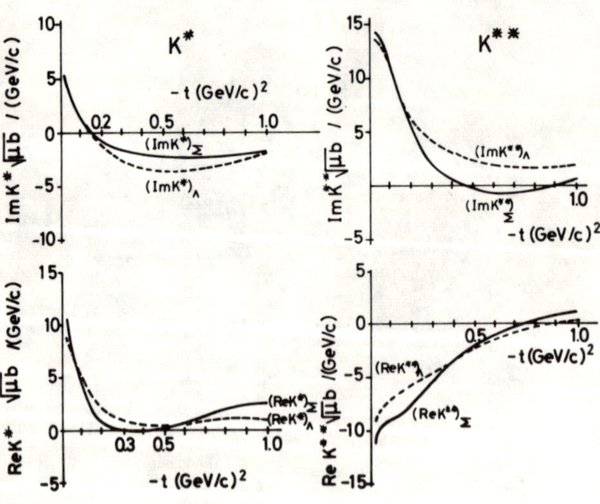

Fig. 8

(a)

n
p̄
π̄
p
P
p
n
e^{-iπα}
p̄
p
n
π°
p
N
n

ρ ∗ f + ω ∗ A₂ ≃ o
n
p̄
π̄
p
P
p
n
Pomeron
n
n̄
π°
n
N
p
(b)
ρ∗P∗f + ω∗P∗A₂ ≃ 2 ρ∗P∗f

Fig. 9

Pomeron Amplitude
ImM
1.0
0.8
0.6
0.4
0.2
0.1
0
-0.1
10^{-2}M/s
S = 2808 GeV²
S = 60
S = 12 πN
(Χσ_T(pp)/σ_T(πp))
ReM
0.2 0.4 0.6 0.8 1.0 1.2 1.4 1.6
-t (GeV/c)²

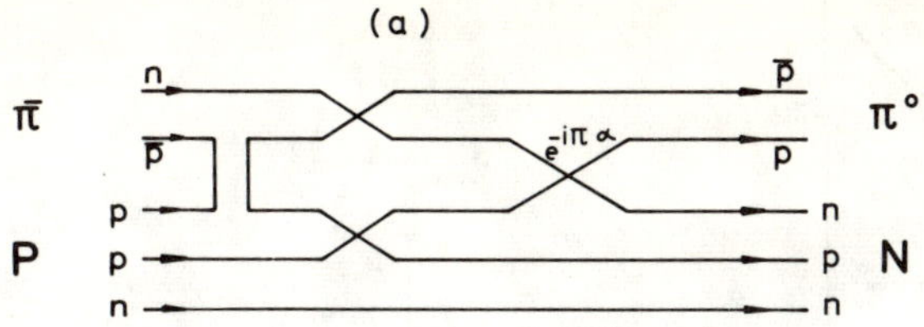

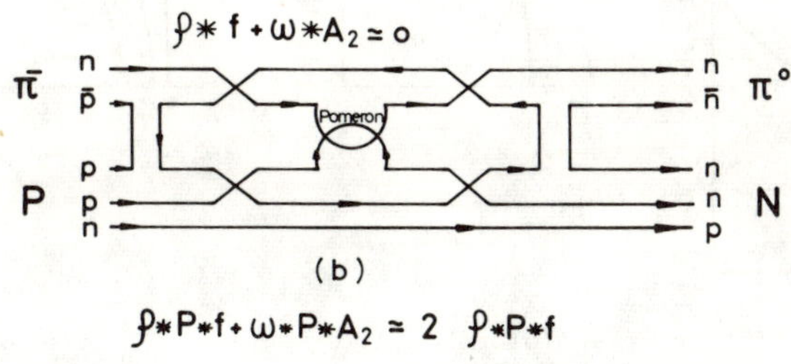

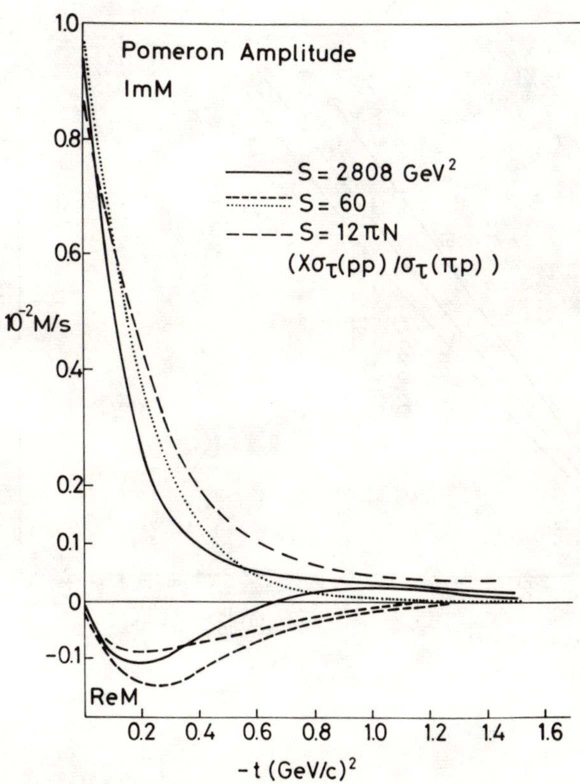

Fig. 10

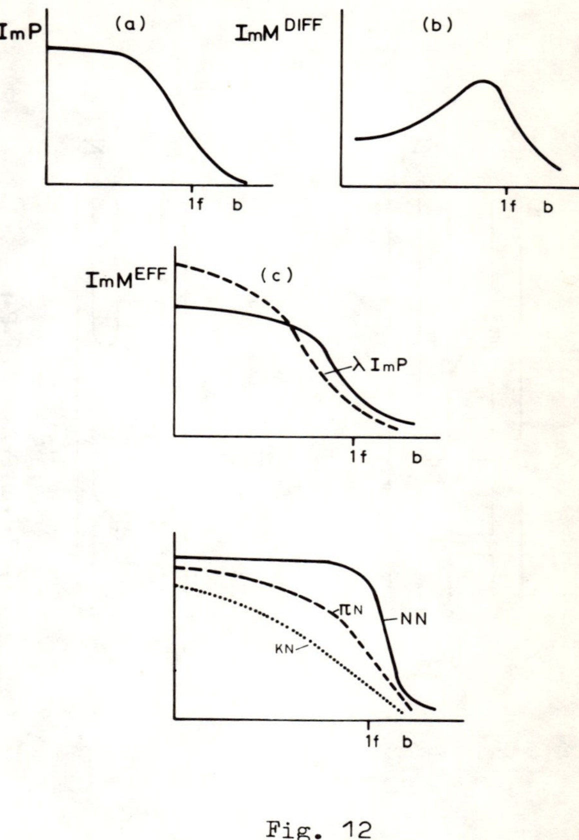

Fig. 11

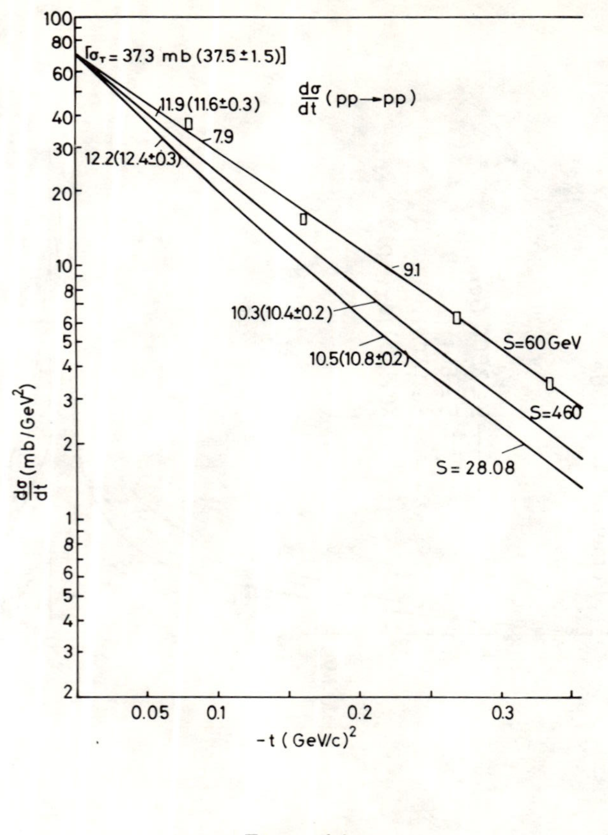

Fig. 12

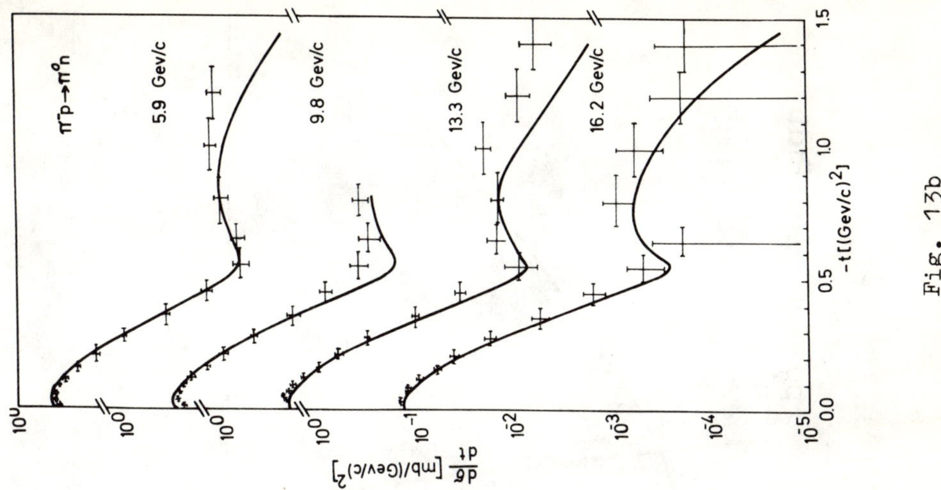

Fig. 13b

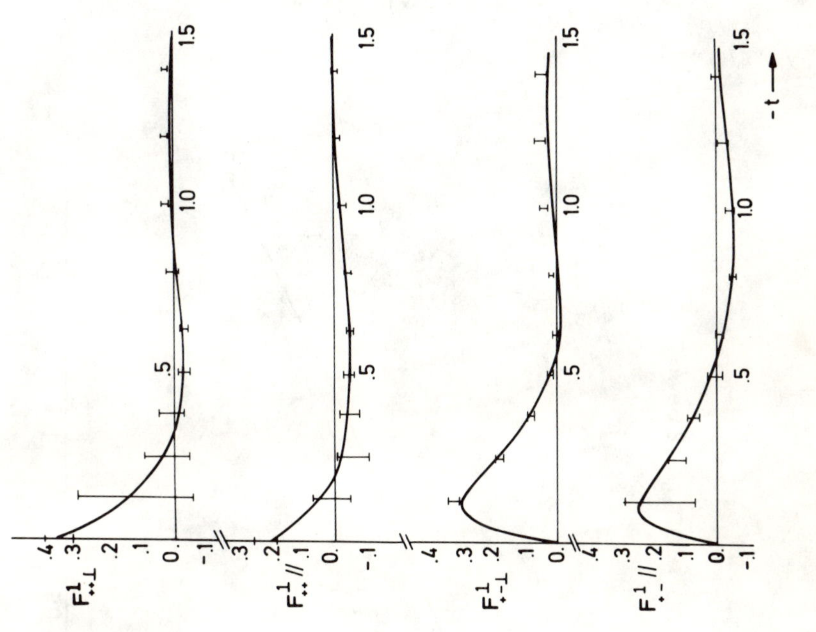

Fig. 13a

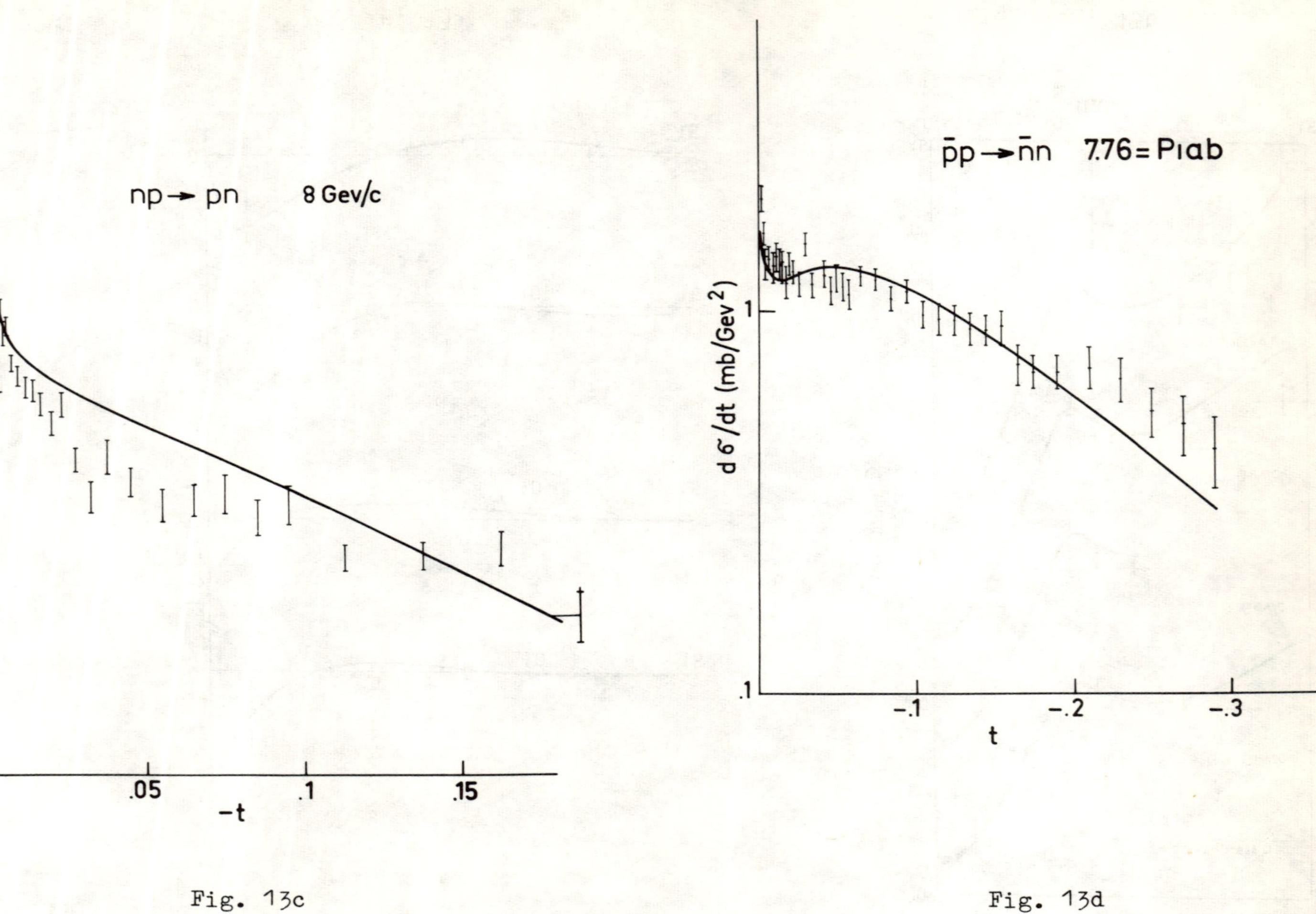

Fig. 13c

Fig. 13d

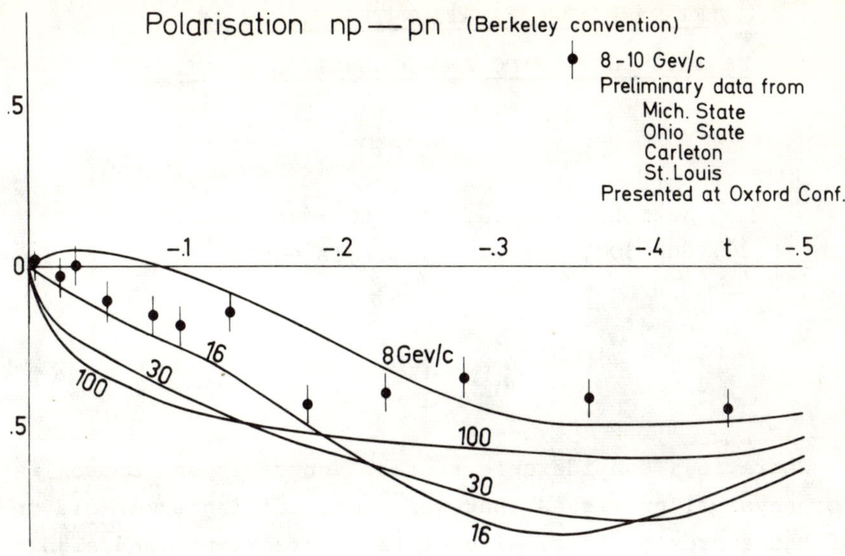

Fig. 13e

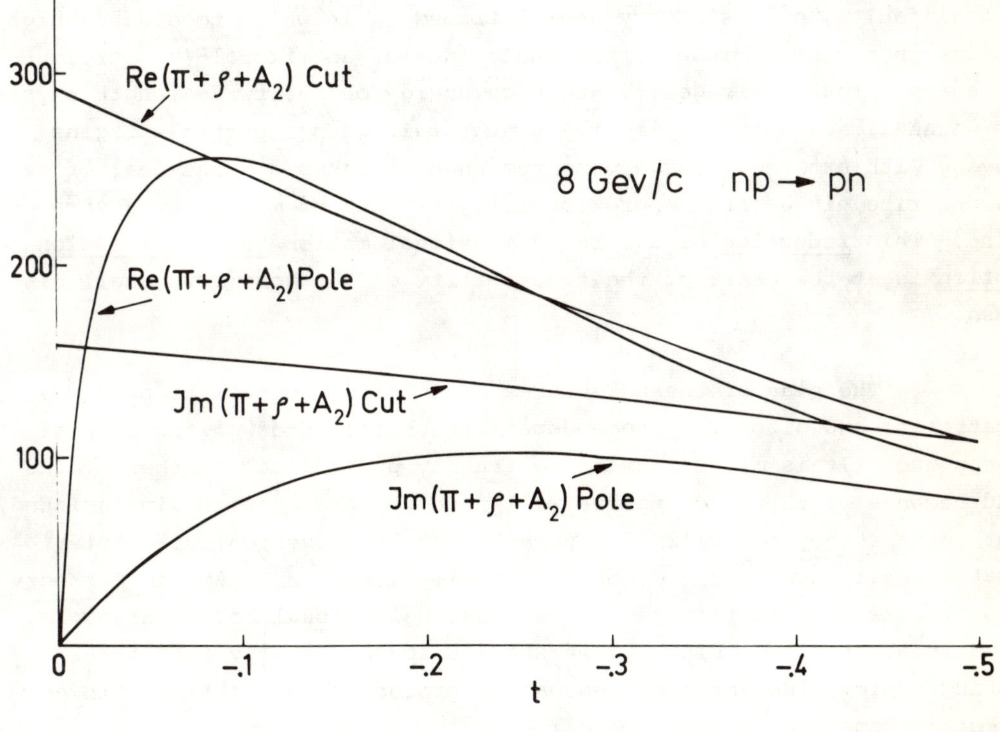

Fig. 13f

<u>EIKONAL APPROXIMATION TECHNIQUES IN</u>

<u>ELASTIC SCATTERING AND PRODUCTION PROCESSES</u>

Henry D.I. Abarbanel

National Accelerator Laboratory
P. O. Box 500, Batavia, Illinois 60510, U.S.A.

There is a considerable attractiveness in any approximation scheme for physical processes that while maintaining some hold on the nature of the approximations, size of the corrections, and significance to real physical phenomena also yields tractable, often analytic expressions for important measurable quantities. The subject of these lectures, the eikonal approximation in elastic scattering and production processes, has precisely all these virtues and, despite its occasional faults, offers a very useful framework in which to discuss high energy phenomena.   Indeed, the whole idea of an eikonal- or straight-line-approximation is generic to high energy or shortwave length physics. As we shall see quite explicitly herein it is when a particle begins moving with extremely large momentum that it makes a great deal of sense to describe, in a first approximation, its path of motion by a straight line. This <u>reduction</u> of a three dimensional motion <u>to one dimensional motion</u> is at the heart of the tractability of the scheme we shall discuss.

The plan of these lectures is to first go back to potential scattering and discuss in some detail the basic ideas of the eikonal technique. (It is my purpose to be frankly pedagogical in these talks and because of that the experts will find the start quite simpleminded, but perhaps they too will find something useful eventually.) After that I shall turn to the eikonal approximation in quantum field theory and discuss how the familiar two dimensional eikonal form emerges for elastic scattering. Finally, we shall discuss a really most interesting topic: inelastic processes and production of particles in an eikonal framework.

A short bibliographical note before we begin. There are innumerable references to the eikonal method in physics and we will refer, without elaborate apology, to a selected subset of them. The standard source of ideas on the eikonal approximation in non-relativistic quantum theory is the set of lectures by Glauber at the 1958 Boulder Summer School[1]. Further development of those ideas was given by Blankenbecler and Sugar[2] and their followers[3] and in a more modern context by Bjorken and his collaborators[4]. For a review of these older ideas I can recommend without hesitation the lectures I gave at the 1971 Boulder Summer School[5] at which time a rather different set of topics was emphasized. The subject of production processes in eikonal approximation has been most attractively pursued by Sugar[6] and his collaborators and we shall follow his lead in our own discussion.

1.   <u>POTENTIAL SCATTERING</u>

As for so many ideas we have about modern physics, non-relativistic quantum mechanics provided the ground in which the eikonal approximation was introduced into modern physics[1]. There are at least two ways to state the basic idea and since both are illuminating we will consider both.

The basic problem that we want to address is the evaluation of the scattering matrix element

$$f(\vec{k}_f, \vec{k}_i)$$

(1)

for a particle of mass m to go from initial momentum $\vec{k}_i$ to final momentum $\vec{k}_f$ in the presence of a potential $V(\vec{r})$ which we take, for the present, to be energy independent. This matrix element is related to the differential cross section into a solid angle Ω about $\hat{k}_f$[7] by

$$\frac{d\sigma}{d\Omega_{\hat{k}_f}} = \left| f(\vec{k}_f, \vec{k}_i) \right|^2$$

(2)

and is related to the usual T-matrix by

$$T(\vec{k}_f, \vec{k}_i) = -\frac{2\pi}{m} f(\vec{k}_f, \vec{k}_i)$$

(3)

This T-matrix element is given in terms of the full solution $\psi_{k_i}^{(+)}(\vec{r})$ to the Schrödinger equation with incoming plane waves, $\exp(i\,\vec{k}_i \cdot \vec{r})$, and outgoing spherical waves by the familiar

$$T(\vec{k}_f, \vec{k}_i) = \int d^3r \; e^{-i\vec{k}_f \cdot \vec{r}} \, V(\vec{r}) \, \psi_{k_i}^{(+)}(\vec{r}) . \tag{4}$$

It satisfies the standard integral equation

$$T(\vec{k}_f, \vec{k}_i) = \tilde{V}(\vec{k}_f - \vec{k}_i) + \tag{5}$$

$$+ \int \frac{d^3q}{(2\pi)^3} \, \tilde{V}(\vec{k}_f - \vec{q}) \; \frac{1}{\frac{k^2 - q^2}{2m} + i\varepsilon} \, T(\vec{q}, \vec{k}_i),$$

where $k = |\vec{k}_i| = |\vec{k}_f|$, $q = |\vec{q}|$, and

$$\tilde{V}(\vec{\Delta}) = \int d^3r \, e^{-i\vec{\Delta} \cdot \vec{r}} \, V(\vec{r}) . \tag{6}$$

Our notation established, we are ready to consider large incident momenta k_i and think how we are to approximate (4) or (5). Of course, we need some criterion of "large" k_i to begin with. One will certainly agree that if k_i is large compared to the primary fourier components in the potential $\tilde{V}(q)$, then the potential will not severely disturb the motion of the projectile and we have the basis for an approximation which is not dependent on the details of $\tilde{V}$. The size of a typical q in $\tilde{V}(q)$ is the inverse of the range, a, of the potential. So let us agree that $k_i a \gg 1$. Furthermore let us imagine that the potential is "smooth" in momentum space (this essentially says $k_i a \gg 1$ for then many particle wave lengths will fit into any variation of the potential), and that its strength V_o is small compared to the initial energy $E = k^2/2m$. Then, turning to (4) we may imagine approximating $\psi^{(+)}$ by an incident plane wave modulated by some function which takes into account the small disturbance due to V:

$$\psi_{k_i}^{(+)}(\vec{r}) = \left(\exp i\vec{k}_i \cdot \vec{r}\right) B(\vec{r}) . \tag{7}$$

The Schrödinger equation provides an exact equation for $B(\vec{r})$, of course, but we wish to imagine that because V is "smooth" the major variation in $\psi^{(+)}$ comes from the exponential and that spatial derivatives of B are small. One arrives in this way at an approximate equation for B

$$\frac{-2i\,\vec{k}_i \cdot \nabla_r B(\vec{r})}{2m} + V(\vec{r}) \, B(\vec{r}) = 0, \tag{8}$$

whose solution is direct

$$B(\underset{\sim}{b} + \lambda \hat{k}_i) = \exp \frac{im}{k} \int_{-\infty}^{\lambda} d\tau \, V(\underset{\sim}{b} + \tau \hat{k}_i) \, , \qquad (9)$$

choosing $B(-\infty) = 0$ and decomposing $\vec{r}$ into a piece along $\vec{k}_i$ and a piece, $\underset{\sim}{b}$, orthogonal to it. When (7) and (9) are placed in (4) we have an expression for T.

What is the key to our arriving at (9)? Clearly once we dropped higher derivatives of $B(\vec{r})$, because of smoothness in V and thus in its spatial variation, we arrived at a <u>one dimensional Schrö-dinger equation along</u> $\hat{k}_i$. This is saying quite directly that propagation along the initial directions is essentially undisturbed by V, only the amplitude of the wave is modulated. This in its simplest form is the <u>eikonal</u> or <u>straight-line</u> or <u>semi-classical approximation</u>. It is valid when the projectile indeed propagates in essentially a straight line which will be true for large k and small scattering angles, $\cos\theta = \hat{k}_i \cdot \hat{k}_f$, in the smooth potentials we have described. Corrections to the basic approximation of a modulated plane wave are elaborated upon in Ref.5.

Suppose, in fact, that the scattering angle is small so we may write $\vec{r} \cdot (\vec{k}_f - \vec{k}_i)$ as

$$\vec{r} \cdot (\vec{k}_f - \vec{k}_i) \approx \left[\underset{\sim}{b} + \frac{\lambda(\vec{k}_f + \vec{k}_i)}{|\vec{k}_f + \vec{k}_i|} \right] \cdot (\vec{k}_f - \vec{k}_i) \qquad (10)$$

$$\approx \underset{\sim}{b} \cdot (\vec{k}_f - \vec{k}_i) \equiv \underset{\sim}{b} \cdot \underset{\sim}{\Delta} \, , \qquad (11)$$

where $\vec{\Delta} = \vec{k}_f - \vec{k}_i$ is the momentum transfer to V in the scattering. One may now cast the expression for T into

$$T(k,\Delta) = \int d^2b \int_{-\infty}^{\lambda} d\lambda \left[V(\underset{\sim}{b} + \hat{k}\lambda) \exp \frac{im}{k} \int_{-\infty}^{\lambda} d\tau \, V(\underset{\sim}{b} + \tau\hat{k}) \right] e^{-i\underset{\sim}{\Delta} \cdot \underset{\sim}{b}} \qquad (12)$$

$$= -\frac{ik}{m} \int d^2b \, e^{-i\,\underset{\sim}{\Delta}\cdot\underset{\sim}{b}} \left[\exp \frac{im}{k} \int_{-\infty}^{+\infty} d\tau \, V(\underset{\sim}{b} + \tau\hat{k}) - 1 \right] , \qquad (13)$$

recognizing the total derivative in the brackets of (12). This last form is what is generally referred to as the eikonal approximation, and the phase $\chi(k,\underset{\sim}{b}) = \frac{m}{k} \int_{-\infty}^{\infty} d\tau \, V(\underset{\sim}{b} + \tau\hat{k})$ is called the eikonal phase. It is the phase up by the projectile in traversing the straight

150

line path along $\hat{k}$ during the scattering process[8].

This seems as good a time as later to remark on the result
of our modulated wave approximation. First, suppose we hold Δ fixed
and let $k \to \infty$, then the only term in (13) which survives is

$$\int d^2b \; d\lambda \; e^{-i \Delta \cdot b} \; V(b + \lambda \hat{k}) \tag{14}$$

which is the Born approximation. This is the correct result for potentials like $V(\vec{r})$ which do not depend on k. However, there is a
value in (13) which is not possessed by the Born approximation, namely
the eikonal approximation satisfies unitarity in the direct (k) channel.
To see this, write

$$T(k, \Delta) = - \frac{ik}{m} \int d^2b \; e^{-i\Delta \cdot b} \left[S(b, k) - 1 \right] \tag{15}$$

and note that since $S = \exp(i\chi)$, unitarity will follow in the high
energy limit. To be more precise, let the potential depend only on
$|\vec{r}|$, then χ and S depend only on $|b|$ = b. We may do the angular
integration in (15) to find

$$T(k, \Delta) = - \frac{2\pi ik}{m} \int_0^\infty b \; db \; J_0(\Delta b) \left[S(b, k) - 1 \right] \tag{16}$$

The unitarity relation in the high energy limit becomes diagonal, to
order $1/k$, when transformed with $J_0(b\Delta)$, and reads

$$\text{Im} \; T(b, k) \propto | T(b, k) |^2 + O(1/k) , \tag{17}$$

which means

$$S(b, k) \; S^\dagger(b, k) = 1 + O(1/k) , \tag{18}$$

which is true for our solution.

This would seem to provide a very general framework for approximations to elastic scattering which satisfy direct channel unitarity. Indeed, it does. Any choice of V(even energy dependent) leads
to a unitary S in the elastic scattering hilbert space. And there
is the rub. For a large variety of scattering processes: $\pi^\pm p$, $K^\pm p$, pp,

and $\bar{p}p$, the ratio of $\sigma_{elastic}/\sigma_{total}$ at even moderate laboratory momenta is of the order of 1/7 or so[9]. This means there is a lot of inelasticity and we <u>do not</u> want to satisfy unitarity in the elastic sector only. We shall suggest a solution to this later on.

Next suppose the potential V is proportional to the momentum k. Then χ is independent of k and

$$\lim_{\substack{k \to \infty \\ \Delta\ fixed}} T_{k^1}(k,\Delta) = i\,k\,F(\Delta)$$

(19)

from (13). That is amusing since $\sigma_{total} \propto \frac{1}{k}\mathrm{Im}\,T(k,0)$ is then constant. And what if $V \propto k^{1+\varepsilon}$, $\varepsilon > 0$? Then the exponential in the eikonal form (13) oscillates like crazy unless $b \approx \log k$, and one finds

$$\lim_{\substack{k \to \infty \\ \Delta\ fixed}} T_{k^{1+\varepsilon}}(k,\Delta) \sim k\,(\log k)^2 ,$$

(20)

which is not accidentally reminiscent of the Froissart bound[10]. This shows us how unitarity plays its important role.

Next let us turn to a more formal deduction of the eikonal approximation (13) - one which stresses the straight line approximation. We begin with the Schrödinger integral equation (5) in operator form

$$T = V + V\,G_0(E)\,T,$$

(21)

where $G_0(E)^{-1} = E - \dfrac{p^2}{2m} + i\varepsilon$, $\vec{p} = -i\nabla_r$ and $E = \dfrac{k^2}{2m}$.

The solution to this is

$$T = V + V G V = V G G_0^{-1} = G_0^{-1} G V,$$

(22)

where

$$G(E)^{-1} = E - \frac{p^2}{2m} - V + i\varepsilon .$$

(23)

We want to approximate G, the full Green function, and evaluate T as best we can using (22).

We imagine, therefore, that the projectile is incident upon the potential with large momentum k_i. Under the smoothness assumptions we made earlier it behoves us to guess that during the scattering process the values of momentum encountered do not appreciably deviate from k_i. Thus we write $G(E)^{-1}$ in a form which emphasizes the closeness of p (albeit an operator) to k_i (a c-number)

$$G(E)^{-1} = E - \frac{p^2}{2m} - V + \frac{(\vec{p} - \vec{k_i})^2}{2m} - \frac{(\vec{p} - \vec{k_i})^2}{2m} \tag{24}$$

$$= \frac{k^2 + k_i^2 - 2\vec{p}\vec{k_i}}{2m} - V - \frac{(\vec{p} - \vec{k_i})^2}{2m} \tag{25}$$

$$= G_i(E)^{-1} - H_i \tag{26}$$

taking
$$H_i = \frac{(\vec{p} - \vec{k_i})^2}{2m} \tag{27}$$

and for the moment staying off the energy shell $k_i^2 = k^2$. The idea is to treat deviations from k_i, as embodied in H_i, as perturbations, although H_i is a fairly singular operator, and to perform a perturbation expansion of G beginning with

$$G = G_i + G_i H_i G = G_i + G H_i G_i \tag{28}$$

The first term is clearly G_i which upon examination of its structure as given in (25) represents propagation of the particle in one dimension along k_i. This should be no surprise, since we constructed it to do this, but the connection with the equation (8) for the amplitude modulating functions should be noted.

We could just as well have made our expansion about k_f since the smoothness of the potential assures us that p will not significantly deviate from it either. So we would write

$$G(E)^{-1} = G_f^{-1} - H_f \tag{29}$$

where

$$G_f^{-1} = \frac{(k^2 + k_f^2 - 2\vec{p}\cdot\vec{k_f})}{2m} - V(\vec{r}) \tag{30}$$

and

$$H_f = \frac{(\vec{p} - \vec{k_f})^2}{2m} \ . \tag{31}$$

Of course,

$$G = G_f + G_f H_f G = G_f + G H_f G_f \ , \tag{32}$$

which together with (28) yields

$$G = \tfrac{1}{2}\left(G_i + G_f\right) + \tfrac{1}{2} G_f \left(H_i + H_f\right) G_i + G_f H_f G H_i G_i \ , \tag{33}$$

suggesting as a first approximation to G the eikonal Green function

$$G_{Eikonal} = \tfrac{1}{2}\left(G_i + G_f\right), \tag{34}$$

and corresponding T-matrix

$$T_{Eikonal} = \tfrac{1}{2}\left(T_{Ei} + T_{Ef}\right)$$

and

$$T_{E(i\ or\ f)} = V + V G_{(i\ or\ f)} V \ . \tag{35}$$

We will evaluate the incoming eikonal T-matrix, leaving T_{Ef} for the diligent student. First write

$$T_{Ei} = V \left(1 + \frac{1}{\frac{k^2 + k_i^2 - 2\vec{p}\cdot\vec{k_i}}{2m} - V + i\varepsilon} V\right) \tag{36}$$

$$= V G_i \ \frac{k_i^2 + k^2 - 2\vec{p}\cdot\vec{k_i}}{2m} \ . \tag{37}$$

We desire the matrix element

$$T_{Ei}(\vec{k_f}, \vec{k_i}) = \langle \vec{k_f} | T_{Ei} | \vec{k_i} \rangle \tag{38}$$

$$= \lim_{k_i^2 \to k^2} \langle \vec{k}_f \mid V G_i \mid \vec{k}_i \rangle \left(\frac{k^2 - k_i^2}{2m} \right) \tag{39}$$

$$= \lim_{k_i^2 \to k^2} \int d^3r \; e^{-i\vec{k}_f \cdot \vec{r}} \; V(\vec{r}) \langle \vec{r} \mid G_i \mid \vec{k}_i \rangle \left(\frac{k^2 - k_i^2}{2m} \right), \tag{40}$$

which should be very familiar in its last form. Indeed we expect that

$$\lim_{k_i^2 \to k^2} \frac{(k^2 - k_i^2)}{2m} \langle \vec{r} \mid G_i \mid \vec{k}_i \rangle = (\exp i \, \vec{k}_i \cdot \vec{r}) \; B(\vec{r}),$$ and we won't be

disappointed. There are many ways to see this. Perhaps the most straightforward is to note that the differential equation for $\langle \vec{r} \mid G_i \mid \vec{k}_i \rangle$ is

$$\left[\frac{k_i^2 + k^2 - 2\vec{p} \cdot \vec{k}_i}{2m} - V(\vec{r}) \right] \langle \vec{r} \mid G_i \mid \vec{k}_i \rangle = e^{i\vec{k}_i \cdot \vec{r}}, \tag{41}$$

which has the solution

$$\langle \vec{r} \mid G_i \mid \vec{k}_i \rangle = \exp i \, \vec{k}_i \cdot \vec{r} \; i \int_0^\infty dt \; \times$$

$$\times \exp\left\{ \frac{i(k^2 - k_i^2)t}{2m} - i \int_0^t d\tau \; V\left(\vec{r} - \vec{k}_i \frac{\tau}{m}\right) \right\} \tag{42}$$

in convenient parametric form. The on energy shell T-matrix is

$$T_{E_i}(\vec{k}_f, \vec{k}_i) = \int d^3r \; e^{-i\vec{\Delta} \cdot \vec{r}} \; V(\vec{r}) \exp\left[-i \int_0^\infty d\tau \; V(\vec{r} - \vec{k}_i \, \tau/m) \right], \tag{43}$$

which is just what we expect.

The eikonal T-matrix referring to k_f is similarly found to be

$$T_{E_f}(\vec{k}_f, \vec{k}_i) = \int d^3r \; e^{-i\vec{\Delta} \cdot \vec{r}} \; V(\vec{r}) \exp\left[-i \int_0^\infty d\tau \; V(\vec{r} + \vec{k}_f \, \tau/m) \right]. \tag{44}$$

As one can discover in a straight forward calculation, each of T_{Ei} and T_{Ef} reduce to the expression (13) we have called the eikonal form, when the scattering angle between $\hat{k}_i$ and $\hat{k}_f$ is small.

Having commented on some of the implications of the eikonal approximations before, let me close this section with a few remarks.

(1) One can clearly perform the expansion of G in power of a perturbation H about any linear combination $ak_i + bk_f$ of directions which are close to the initial or final path. The literature contains several discussions of the virtues of various choices[8] especially the average momentum[3]

$$\vec{k} = \tfrac{1}{2}(\vec{k_i} + \vec{k_f})$$ (45)

whose main claim _a priori_ seem to be connected with being able to satisfy time reversal invariance, but which appears (for reasons mysterious to me) to possess some superiority on numerical and other grounds.

(2) Despite the fact that (13) satisfies unitarity in the elastic sector it may prove useful in nuclear physics problems where the elastic channel is important or the _form_ may be appropriate for a few channel problem by replacing the real potential by some sort of optical potential

$$V(\vec{r}) = V_1 + i V_2$$

using the V_2 term both to represent absorption out of the elastic channel and to define $_{inelastic}$.

(3) The correction term indicated in (33) to the basic eikonal approximation may be evaluated term by term. The first term

$$\tfrac{1}{2} G_f (H_i + H_f) G_i$$ (46)

leads to the so-called Saxon-Schiff correction[2,3,5] to the eikonal approximation. If the fourier transform of $V(r)$ falls off as a _power_ in q momentum space, then simple estimates[2] indicate that the eikonal plus first correction is an excellent representation of $T(k_f, k_i)$ over all angles. This says physically that for such potentials the projectile chooses to make its momentum transfer $\vec{\Delta}$ in one or two steps (no H or one H) rather than multiple scattering. For a gaussian potential; this is not the case. Then a separate analysis[11] reveals that the basic eikonal is most important.

2.   <u>FIELD THEORY</u>

As we well know field theory is much richer than potential scattering because of the ability to produce particles. In this section we will concentrate on the extraction of the high energy elastic scattering amplitude in a relativistic framework and in the next section we will turn to consideration of how particle production may be treated.

The basic thing to realize is that the field theory may be reduced to a discussion of potential scattering and that our experience with the latter, now enhanced by our eikonal knowledge, will serve as a guide to physical approximations.

So, in what way is field theory just potential scattering? The answer[12] is that if we know the amplitude for a particle of momentum p_1 [four momentum = (p_t, p_x, p_y, p_z)] to go to momentum p_1' in the presence of a c-number external potential $A(x)$, then we may answer any question involving the interaction of those same particles and the quanta associated with the potential. Essentially the amplitude $T_A(p_1',p_1)$ acts as a generating function[13] for emission and absorption of such quanta.

Let us illuminate these remarks by considering the quantum electrodynamics of a dirac particle moving in an external c-number electromagnetic potential $A_\alpha(x)$. Suppose we have calculated $T_A(p_1',p_1)$ to second order in A. This is given by

$$\bar{u}(p_1') \, e\gamma \cdot \tilde{A}(q_1') \, \frac{1}{m - \gamma\cdot(p_1'+q_1')} \, e\gamma\cdot\tilde{A}(q_1) \, u(p_1) \qquad (47)$$

or in configuration space

$$T_A(p_1',p_1) = \int d^4z \, d^4y \, \bar{u}(p_1') e^{-ip_1'\cdot z} \, e\gamma\cdot A(z) S_+(z-y) \times$$
$$\times \, e\gamma\cdot A(y) e^{+ip_1\cdot y} \, u(p_1), \qquad (48)$$

where u and $\bar{u}$ are the standard dirac spinors, γ_α the usual 4x4 matrices, $\tilde{A}(q)$ the fourier transform of $A(x)$, and $S_+(z)$ the standard causal propagator for a dirac particle.

Now knowing this suppose we ask: what is the probability amplitude for the particle to scatter in the potential $A_\alpha(x)$, to first

order, and in the same act radiate a photon of momentum k with spin
wave function $\epsilon_\alpha(k)$? The answer, as we well know,

$$T_A\left(p_1 \to p_1' + k\right) = e^2\,\epsilon_\alpha^*(k)\int d^4z\,d^4y\,\Big\{\ \bar{u}(p_1')\,e^{-i(p_1'+k)\cdot z}\,\gamma_\alpha\ \times$$

$$\times\,S_+(z-y)\,\gamma\cdot A(y)\,e^{ip_1\cdot y}\,u(p_1)\ +\ \bar{u}(p_1')\,e^{-ip_1'\cdot z}\ \times$$

$$\times\,\gamma\cdot A(z)\,S_+(z-y)\,\gamma_\alpha\,e^{i(p_1-k)y}\,u(p_1)\ \Big\},$$

(49)

representing emission after and before the scattering on A. It is the
relation between these two answers that we seek. If we replace in (48)
$A_\alpha(z)$ by $\epsilon_\alpha^*(k)\,e^{-ik\cdot z}$, we obviously arrive at the first term of (49),
and similarly the second term comes from replacing $A(y)$ by $\epsilon_\alpha^*(k)e^{-ik\cdot y}$.

Both of these things can be done at a stroke by taking a
derivative of (48) with respect to $A_\alpha(x)$ and in each of the two terms
inserting the appropriate photon wave function $\alpha_\alpha^*(k,x) = \epsilon_\alpha^*(k)e^{-ik\cdot x}$,
then integrating over all possible space-time points x where the inser-
tion might have occurred. That is

$$T_A\left(p_1 \to p_1' + k\right) = \int d^4x\,\frac{\delta\,T_A(p_1 \to p_1')}{\delta A_\alpha(x)}\,\alpha_\alpha^*(k,x),$$

(50)

which it is easy enough to see is generally true. The funny δ means
literally: replace $A_\alpha(z)$ under an integral by $\delta^4(x-z)$; that is

$$\delta A_\alpha(z)\,/\,\delta A_\beta(x) = g_{\alpha\beta}\,\delta^4(z-x).$$

(51)

Another question one may ask of (48) is what is the amplitude,
to second order in e, for the particle to emit a photon and then reabsorb it? We must open up the potential at $A_\alpha(z)$, say, propagate a
photon from z to y via the causal propagator $D_+(z-y)$ and then allow
the particle to reabsorb the photon at y. This yields

(52)

$$T_{\substack{No\\Emission}}\left(p_1 \to p_1'\right) = \tfrac{1}{2}\int d^4w\,d^4z\,\frac{\delta}{\delta A_\alpha(w)}\,D_+(w-z)\,\frac{\delta}{\delta A_\alpha(z)}\,T_A(p_1 \to p_1'),$$

the 1/2 comes from the possibility of the photon's beginning at either y or z in $T_A(p_1 \to p_1')$. The general answer for $T(p_1 \to p_1')$ no emission is gotten by some simple counting

$$T_{\substack{No \\ Emission}}(p_1 \to p_1') =$$

$$\left[exp\; -\tfrac{1}{2} \int d^4z\, d^4w\, \frac{\delta}{\delta A_\alpha(z)}\, D_+(z-w)\, \frac{\delta}{\delta A_\alpha(w)} \right] T_A(p_1 \to p_1'), \qquad (53)$$

and after taking all derivatives, set $A_\alpha = 0$.

Interaction between particles which is mediated by the quanta of A_α , namely photons, comes from two particles moving in two potentials A_1 and A_2 and from evaluating the probability amplitude to open up an A_1 spot and propagate a photon, via D_+, over to an opened A_2 spot. Thus the amplitude for electron scattering $T(p_1 + p_2 \to p_1' + p_2')$ in the absence of corrections to the photon propagator D_+ coming from internal electrons is

$$T(p_1 + p_2 \to p_1' + p_2') =$$

$$\left[exp\; -\int d^4w\, d^4z\, \frac{\delta}{\delta A_{1\alpha}(w)}\, D_+(w-z)\, \frac{\delta}{\delta A_{2\alpha}(z)} \right] \times \qquad (54)$$

$$\times \left[T_{A_1}(p_1 \to p_1')\, T_{A_2}(p_2 \to p_2') - T_{A_1}(p_1 \to p_2')\, T_{A_2}(p_2 \to p_1') \right]_{A_1 = A_2 = 0} .$$

As promised we now see that a large variety of field theoretic phenomena follow from potential scattering probability amplitudes. Indeed, if one takes (54) and performs exactly the same kind of eikonal approximation on T_{A_1} and T_{A_2} as we have done in the previous section, then it follows[3,5] after some straightforward computation that exactly the form (13) emerges. The "potential" is just that of the photon exchange which yields the Born approximation. Rather than present this derivation here, we will follow a somewhat different route which will lead more smoothly into the discussion of production. All the principles of the direct calculation from (54)[5] are illuminated.

Since we are interested in the scattering of a particle in an external potential let us consider some particle currents $J_\alpha(x)$ interacting with the c-number potential $A_\alpha(x)$. The interaction Lagrangian

$$\mathscr{L}_I(x) = J_\alpha(x) A_\alpha(x) \tag{55}$$

leads via the standard rules of field theory to the S-matrix element to go from some state i to another f

$$S_{fi} = \langle f | T(\exp - i\int d^4x\, J_\alpha(x) A_\alpha(x)) | i \rangle \tag{56}$$

We want to evaluate S_{fi} when the states $|i\rangle$ and $|f\rangle$ are moving very fast in, say, the three direction [14]. So we take a standard state, say a rest state for a single particle, and boost it along the 3-axis very fast, so

$$|i\rangle = \exp - i K_3 \Theta \; |i_o\rangle,$$

where Θ is a boost angle and K_3, the generator of 3-boosts. A particle at rest is taken by this operation from $p_o = (m, 0, 0, 0)$ to

$$e^{-i K_3 \Theta} \; p_o = (m\cosh\Theta, 0, 0, m\sinh\Theta). \tag{57}$$

Under such a transformation it is convenient not to consider separately p_t and p_z, but the combinations

$$p_\pm = p_t \pm p_z \tag{58}$$

for

$$e^{-i K_3 \Theta} \; p_\pm = e^{\pm\Theta} p_\pm \tag{59}$$

since p_+ gets large and p_- small when Θ is large. How does $J_\alpha(x)$ transform? Of course, its components in the α or 1 direction are untouched but

$$e^{i K_3 \Theta} J_\pm(x_+, x_-, \underset{\sim}{x}) e^{-i K_3 \Theta} = e^{\pm\Theta} J_\pm(e^{-\Theta} x_+, e^{\Theta} x_-, \underset{\sim}{x}), \tag{60}$$

where $J_\pm = J_t \pm J_3$ and $x_\pm = x^o \pm x^3$ while $\underset{\sim}{x}$ is the two vector (x^1, x^2).

The matrix element S_{fi} becomes

$$S_{fi} = \langle f_0 | e^{i\Theta K_3} T\left(\exp -i \int \frac{dx_+ dx_- d^2x}{2}\left[\frac{J_+A_- + J_-A_+}{2} - \underset{\sim}{J}\cdot\underset{\sim}{A}\right]\right) \times$$
$$\times\, e^{-iK_3\Theta} | i_0 \rangle \tag{61}$$

$$= \langle f_0 | T\left(\exp\left[-\frac{i}{4}\int dx_+ dx_- d^2x\, e^\Theta J_+(e^{-\Theta}x_+, e^\Theta x_-, \underset{\sim}{x})A_-(x_+, x_-, \underset{\sim}{x})\right]\right) | i_0 \rangle \tag{62}$$
$$+\, O(e^{-\Theta}),$$

or on changing integration variables to $\tilde{x}_- = e^{+\Theta}x_-$,

$$S_{fi} = \langle f_0 | T\left(\exp -\frac{i}{4}\int dx_+ d\tilde{x}_- d^2x\, J_+(0, \tilde{x}_-, \underset{\sim}{x})A_-(x_+, 0, \underset{\sim}{x})\right) | i_0 \rangle$$
$$+\, O(e^{-\Theta}) \tag{63}$$

which if we define

$$\mathcal{J}(\underset{\sim}{x}) = \frac{1}{2}\int dx_-\, J_+(0, x_-, \underset{\sim}{x}) \tag{64}$$

and

$$a(\underset{\sim}{x}) = \frac{1}{2}\int dx_+\, A_-(x_+, 0, \underset{\sim}{x}) \tag{65}$$

takes on the two dimensional form

$$S_{fi} = \langle f_0 | T\left(\exp -i\int d^2x\, \mathcal{J}(\underset{\sim}{x})\, a(\underset{\sim}{x})\right) | i_0 \rangle + O(e^{-\Theta}), \tag{66}$$

however, the time ordering which remains must be discussed before we
are so excited about (66).

To get rid of the time ordering, which we must do in order
for S_{fi} to have the true exponential form exhibited by the eikonal
approximation, it is necessary to make some _ansatz_ about the component
J_+ of the particle source operator. The only "time" left to be ordered
is the dependence of J_+ on x_-, so if we make J_+ a c-number with respect
to its dependence on x_- we may remove the T operation and have essen-
tionally an eikonal or more precisely an exponential form for S_{fi}.
This requires that $J_+(x_+, x_-, \underset{\sim}{x})$ and $J_+(y_+, y_-, \underset{\sim}{y})$ commute at $x_- = y_-$

$$\left[J_+(x_+, x_-, \underline{x}), \; J_+(y_+, y_-, \underline{y}) \right] \Big|_{x_- = y_-} = 0. \tag{67}$$

Now $J_\alpha(x)$ commutes with $J_\beta(y)$ for $(x-y)^2 < 0$, already if it is a local operator. So since

$$(x-y)^2 = (x-y)_+ (x-y)_- - (\underline{x} - \underline{y})^2, \tag{68}$$

we see that automatically (67) is satisfied if $(\underline{x}-\underline{y})^2$ is not zero. The additional assumption[15] is that if x-y is lightlike $[J_+(x), J_+(y)] = 0$.

If this is true, then

$$S_{fi} = \langle f \,|\, \exp -i \int d^2x \, J(\underline{x}) \, a(\underline{x}) \,|\, i \rangle, \tag{69}$$

which is essentially the eikonal answer.

It is not obvious that all currents $J_\alpha(x)$ are such that their plus components commute on the light cone. Indeed, as Weinberg[15] has shown, this places strong dynamical constraints on the matrix elements of J which are tantamount to generalized sum rules of the Drell-Hearn-Gerasimov variety. One implication of this is that if we return to our problem of a dirac particle scattering in an external potential and do not add radiative corrections of a self-energy or vertex correction variety, then the particle must have no anomalous magnetic moment and the eikonal technique "works" only if g = 2. This is consistent with more direct approaches to this question.

In the early work on eikonal approximations the c-number nature of J_+ was arranged in a very simple fashion: it was taken to be a c-number. The physical meaning of this is straightforward. If the particle which is being scattered neither produces other particles nor any quanta of A_α (that is, there are no radiative corrections on the particle line) nor does it change its co-ordinates such as a spin, helicity or, if it is taken composite, its internal wave function, then the transition operator causing the scattering is acting in a one dimensional space of particle variables and it is essentially a c-number. The assumption that all these conditions be met is generic to older treatments of the eikonal approximation. Our exercise above shows us that a more general class of scattering processes may take the exponential form of the eikonal. Clearly if J_α (x) is an operator still, the particle scattering in the potential can make many intermediate states of varying character before it emerges in state $|f\rangle$.

The appearance of the commutator of the particle source, J, on the light-cone might at first seem strange. However, just this kind of quantity is to be expected in relativistic theories on the following heuristic grounds. When we deal with very fast particles, say $p_3 \to \infty$, on the mass shell, $p^2 = m^2$, then we are specifying two out of four components of p; namely p_3 and $p_t = p_3 + (\underset{\sim}{p}^2 + m^2) / 2p_3$ (to leading order), where $\underset{\sim}{p} = (p_x, p_y)$. In co-ordinate space we are thus providing information on the variables conjugate to p_t and p_3; namely t and x_3.

Indeed, we are constraining $x_3 \approx \pm t$ for forward or backward going particles respectively. This places us very near the light cone in a space time diagram and tells us that for $p_3 \to \infty$, the dynamics of scattering is gathering information from the whole light cone and not just the canonical space-like surface of equal t. To put it another way, the "natural" variables for $p_3 \to \infty$ processes would appear to be $x_t \pm x_3$ and $\underset{\sim}{x}$, not separately x_t and x_3. Of course, any description must be equivalent; it is just that one's good sense suggests it will be simpler in these light cone variables.

One more word before going on to production. This same heuristic argument tells us why we keep finding two dimensional integrals. It's simple; we have given p_3 <u>and</u> p_t, thus the dynamics lies in the two dimensional subspace we have said nothing about.

3.   <u>PRODUCTION PROCESSES</u>

In view of some of the fairly drastic approximations which are going to be made in this section, I feel it may be worthwhile to recount a few of the salient features of data on production processes which, thanks to the wide attention directed in that direction for the past several years, is now available in useful form. First, as we have noted before, the study of production or inelastic reactions is bound to be important for high energy physics because experimentally even at AGS or CERN - PS or Serpukhov energies, $p_{lab} \simeq$ 30-70 GeV/c, the total cross section is on the order of 7 to 10 times

$\sigma_{elastic}$. It is unlikely that we will understand the latter or quasi-two body processes taken alone. Second, σ_{total} appears to be extraordinarily constant over enormous ranges of incident energy. Strictly speaking, this is known for pp scattering from $p_{lab} \approx 25$ GeV/c to $p_{lab} \approx 1500$ GeV/c as measured at the CERN-ISR. For Kp and πp scatterings the detailed situation is more fluid. Third, the average transverse momentum, which we have called p , seems to be quite small even at the highest available energies. The distribution in $|p| = p_T$ of detected particles tends to be

$$\frac{dN}{dp_T^2} \sim exp\left(-a\,p_T^2\right) \qquad or \tag{70}$$

$$\sim exp\left(-b\,p_T\right) , \tag{71}$$

with a $\approx$ 3 or 4 $(GeV/c)^{-2}$ or b $\approx$ 6 $(GeV/c)^{-1}$. So most hadron physics is within $0 \leq p_T \leq 0.5$ GeV/c. Fourth, and this is most important for what follows, the distribution in momentum of particles detected at high energies is quite different for particles which are definitely produced (e.g., π or K or $\bar{p}$ in pp collisions) from the distribution of particles which can "come through", e.g. protons in pp collisions. The effect seen is that for "through going" particles there is a pronounced maximum for very large or very small longitudinal momenta when measured in the center of mass. That is, the beam or the target particles go through. The distribution of produced particles, on the other hand, is largest at small center of mass p_{long}. This phenomenon, which is very distinct in the ISR data at $p_{lab} \approx 1500$ GeV/c is popularly known as the <u>leading particle effect</u>.

The rest of this section will be devoted to a model which is eikonalistic and attempts to incorporate many of the features just described[16]. The particular type of process which I have in mind will be Nucleon + Nucleon $\rightarrow$ Nucleon + Nucleon + some produced things (π or K or what you like). Taking a strong hint from the data we treat the nucleons as leading particles and treat their co-ordinates as c-numbers. That is, the nucleons are treated as through going objects whose variables, if they are altered at all, are not altered appreciably. This c-number nature of the nucleons is, of course, precisely the trick that allows us to eikonalize this process.

Since there are always 2 nucleons in both the initial and final states let us label the amplitude for n pions + 2N $\rightarrow$ m pions + 2N as T_{mn}, and let us agree to call all produced particles pions[17]. We know from earlier work the T_{oo} (elastic scattering) takes the form

$$T_{oo}(s,t) = is \int d^2 B \, e^{i Q \cdot B} \, T_{oo}(s, B),$$

(72)

where

$$T_{oo}(s, B) = e^{i X(s, B)} - 1 ,$$

(73)

and s is the usual square of the incoming c.m. energy, and $t = -|Q|^2$ is the four momentum transferred. This suggests that we operate in a space where the nucleon co-ordinates s and B are specified, for there we will have a good chance to construct a set of T_{mn} which are unitary. (Recall that unitarity was a nice feature of eikonal approximations.)

The nucleons, then, will be treated as a source, a <u>c-number source</u>, for pions which will be parameterized by the co-ordinates s and B. The leading particles carry off a large fraction of the initial energy, and we will imagine that the pions can move freely in phase space with no special constraints on them due to energy momentum conservation. That is to say, since the pions come out with small p_{long}, in the c.m., and if their number is not large, as we will shortly impose, then energy momentum constraints are essentially negligible on them. Our problem then is to find the S-matrix which comes from a c-number source $\rho(s,B;x)$ which can emit and absorb pions.

A digression on convenient notation before we solve this problem. It is useful to use instead of s and p_{long}, the dimensionless variable <u>rapidity</u>, commonly called y. The rapidity of a particle of mass m, momentum (p_T, p_{long}) is defined to be

$$y = \frac{1}{2} \log \left[\frac{E + p_{long}}{E - p_{long}} \right] ,$$

(74)

and is essentially

$$\tanh y = v_{long} / c ,$$

(75)

or the angle of "rotation" in the time-z plane to produce a particle of momentum p_{long}.

The incident nucleons have center of mass rapidity $\pm\dfrac{y_o}{2}$ for the beam and target respectively, where $y_o = \log s$ to an excellent approximation. The values of y_o in the real world are not overwhelming: $y \approx 4$ at the AGS, $y_o \approx 6$ at NAL, $y_o \approx 8$ at the ISR, and at the "planned" ISABELLE 200 GeV/c colliding beam facility $y_o \approx 11.5$. The real advantage of rapidity is that invariant phase space is simple

$$\frac{d^3p}{E} = dy\, d^2p_T \quad . \tag{76}$$

Usually, energy momentum conservation is complicated in terms of y, but we have just agreed not to worry about this, so we are spared that misery.

Now we are ready to calculate T_{mn}. We want to determine the pion field $\phi(x)$ in the presence of the c-number source $\rho(y_o,\underset{\sim}{B};x)$ [18]. To do this we must solve the equation

$$[m^2 + \partial^2]\, \phi(x) = \rho(y_o,\, \underset{\sim}{B}\,;\, x). \tag{77}$$

The solution, of course, is

$$\phi(x) = \int d^4y\, D(x-y)\, \rho(y), \tag{78}$$

where

$$(m^2 + \partial_x^2)\, D(x) = \delta^4(x), \tag{79}$$

and we have temporarily dropped the y_o, $\underset{\sim}{B}$ parameters.

Which function $D(x)$ we want is dictated as usual by the boundary conditions of the problem.

We wish to express the out field $\phi_{out}(x)$ for departing pions in terms of $\phi_{in}(x)$

$$\phi_{out}(x) = \phi_{in}(x) + \int d^4z\, \hat{D}(x-z)\, \rho(z), \tag{80}$$

$$(\partial^2 + m^2)\,\phi_{in}(x) = (\partial^2 + m^2)\,\phi_{out}(x) = 0, \tag{81}$$

so

$$(\partial^2 + m^2)\,\hat{D} = 0. \tag{82}$$

Since

$$\phi(x) = \phi_{in}(x) + \int d^4z\, D_R(x-z)\,\rho(z), \tag{83}$$

and

$$\phi(x) = \phi_{out}(x) + \int d^4z\, D_A(x-z)\,\rho(z), \tag{84}$$

where D_R and D_A are the retarded and advanced Green functions satisfying (79), we have

$$\hat{D}(z) = D_R(z) - D_A(z). \tag{85}$$

The integral representation of this difference Green function is

$$\hat{D}(z) = \int \frac{d^4q}{2(2\pi)^3}\, e^{iq\cdot z}\, \delta(q^2 - m^2). \tag{86}$$

It is useful, then, to decompose ϕ_{out} and ϕ_{in} into creation and annihilation operators in momentum space

$$\phi_{in}(x) = \int \frac{d^4q}{2(2\pi)^3}\, \delta(m^2 - q^2)\Big[a_{in}(y,q_T)\, e^{iq\cdot x} + a_{in}^+(y,q_T)\, e^{-iq\cdot x}\Big], \tag{87}$$

and

$$\phi_{out}(x) = \int \frac{d^4q}{2(2\pi)^3}\, \delta(m^2 - q^2)\Big[a_{out}(y,q_T)\, e^{iq\cdot x} + a_{out}^+(y,q_T)\, e^{-iq\cdot x}\Big], \tag{88}$$

where we remember that

$$\int \frac{d^4q}{2(2\pi)^3}\, \delta(m^2 - q^2) = \int \frac{d^3q}{2q_0\, 2(2\pi)^3} \tag{89}$$

$$= \int \frac{d^2 q_T \, dy}{4 \, (2\pi)^3} \; . \tag{90}$$

To guarantee the proper equal time commutation relations for $\phi_{in}(x)$ and $\phi_{out}(x)$ we take

$$[a_{in}(y, q_T), \, a_{in}^+(z, k_T)] = 4 \, (2\pi)^3 \, \delta(y-z) \, \delta^2(q_T - k_T) , \tag{91}$$

and

$$[a_{out}(y, q_T), \, a_{out}^+(z, k_T)] = 4 \, (2\pi)^3 \, \delta(y-z) \, \delta^2(q_T - k_T) . \tag{92}$$

From the solution to our problem, Eq.(85), we now learn

$$a_{out}(y, q_T) = a_{in}(y, q_T) + \tfrac{1}{2} \, \rho(y, q_T) \tag{93}$$

where

$$\rho(y, q_T) = \int d^4 x \, \rho(x) \, e^{-iq \cdot x} \tag{94}$$

with

$$q = (q_0, q_T, q_3) , \qquad q_0 = +\left[|q_T|^2 + q_3^2 + m^2 \right]^{1/2} , \tag{95}$$

and

$$y = \tfrac{1}{2} \log \left[\frac{q_0 + q_3}{q_0 - q_3} \right] \tag{96}$$

as before.

The S matrix is defined to be the operator which takes a_{in} to a_{out} so we have

$$S \, a_{in}(y, q_T) \, S^+ = a_{out}(y, q_T) \tag{97}$$

or

$$[S, \, a_{in}(y, q_T)] = \tfrac{1}{2} \, \rho(y, q_T) \, S . \tag{98}$$

This form suggests we seek a solution of (98) as

$$S = C \, \exp \int \frac{dy \, d^2 q_T}{4(2\pi)^3} \, \lambda(y, q_T) \, a_{in}^\dagger(y, q_T) \times$$

$$\times \exp \int \frac{dy \, d^2 q_T}{4(2\pi)^3} \, \lambda^*(y, q_T) \, a_{in}(y, q_T) , \tag{99}$$

where C is a c-number normalization constant. Then using the operator identity

$$[e^A, B] = [A, B] \, e^A \tag{100}$$

which is true when $[A,B]$ commutes with A and B, we find

$$[S, a_{in}(y, q_T)] = \lambda(y, q_T) = \tfrac{1}{2} \rho(y, q_T) , \tag{101}$$

which solves our problem. The normalization C is found by requiring

$$S S^\dagger = 1 , \tag{102}$$

that is, unitarity. This makes C take the value

$$C = \exp - \tfrac{1}{8} \int \frac{dy \, d^2 q_T}{4(2\pi)^3} \, |\rho(y, q_T)|^2 , \tag{103}$$

using (101). So we have constructed an explicity unitary S-matrix operator which acts in pion space (the nucleons are always present so the "vacuum" is the two nucleon state). Of course if we want T_{mn} we must evaluate

$$T_{mn} = \langle in, m \, | \, S - 1 \, | \, in, n \rangle \tag{104}$$

$$= \langle in, m \, | \, S \, | \, in, n \rangle - \delta_{mn} , \tag{105}$$

where the n pion state $|n, in\rangle$ is

$$\frac{1}{\sqrt{n!}} \, a_{in}^\dagger(y_1, q_{T_1}) \cdots a_{in}^\dagger(y_n, q_{T_n}) \, |0\rangle \tag{106}$$

in a standard manner.

It is amusing to ask for the elastic scattering matrix which is the vacuum to vacuum transition; i.e., no pions in, no pions out:

$$T_{oo}(y_0, \underset{\sim}{B}) = \langle 0|S|0 \rangle - 1 \tag{107}$$

$$= C - 1, \tag{108}$$

and reference to (103) shows that indeed we no longer have a unitary
S matrix in the elastic sub-space. In fact, since C = exp-(real number),
the eikonal phase is pure imaginary and thus the "potential" has become completely absorbing!

How are we to regard this solution of the eikonal production
problem? Since the source function $\wp(y_0, \underset{\sim}{B}; y, q_T)$ is unspecified
there is an enormous freedom in possible unitary answers. One must
think of this as a class of solutions which for any $\wp$ yields a unitary
S-matrix and thereby provides an attractive <u>framework</u> into which one
may put his best guess or theory for $\wp$. There are certain constraints
on $\wp$ which come from experimental knowledge on cross sections, multiplicity, etc., and we shall now look at some of these quantities.

The total cross section is given by the optical theorem as

$$\sigma_T(s) = \frac{1}{s} \, \text{Im} \, T_{oo}(y_0, 0) \tag{109}$$

for s large, and in the present solution is

$$\sigma_T(s) = \int d^2B \left[C(y_0, \underset{\sim}{B}) - 1 \right], \tag{110}$$

and this only suggests that for large s or $y_0 = \frac{1}{2}\log s$, $C(y_0, \underset{\sim}{B})$ is independent of y_0 in order to reproduce the constant total cross sections
that are observed.

The distribution of produced pions, that is, the single
particle inclusive spectrum, is defined to be

$$\frac{d\sigma \left[NN \rightarrow NN + \text{one pion} + \text{anything} \right]}{dy \, d^2q_T \, / (4[2\pi]^3)} = \tag{111}$$

$$\int d^2B \, e^{i\underset{\sim}{Q} \cdot \underset{\sim}{B}} \sum_n \int \frac{dy_1 d^2q_{T1} \cdots dy_n d^2q_{Tn}}{[4(2\pi)^3]^n} \left| \langle in, n | a_{in}(y, q_T) S | 0 \rangle \right|^2.$$

Now

$$\left| \langle in, n \mid a_{in}(y, q_T) S \mid o \rangle \right|^2 =$$

(112)

$$\langle o \mid [a_{in}^+, S^+] \mid n, in \rangle \langle in, n \mid [a_{in}, S] \mid o \rangle$$

and summing on n in (111) we have

$$\frac{d\sigma}{dy\, d^2q_T / (4(2\pi)^3)} = \int d^2 B\, e^{i Q \cdot B}\, \frac{1}{4} \left| \varrho(y_0, B; y, q_T) \right|^2 .$$

(113)

So knowledge about the inclusive distribution is direct knowledge on the source function ϱ .

Another interesting quantity is the n pion cross section $\sigma_n(y_0)$ which is

$$\frac{\sigma_n(y_0)}{\sigma_T(y_0)} = \int \frac{d^2 B\, dy_1\, d^2 q_{T1} \cdots dy_n\, dq_{Tn}}{[4(2\pi)^3]^n} \left| \langle in, n \mid S \mid o \rangle \right|^2$$

(114)

$$= \int d^2 B\, C(y_0, B)^2 \left[\log \frac{1}{C(y_0, B)^2} \right]^n / n! ,$$

(115)

and shows that in B, y_0 space one has a Poisson distribution in n with mean

$$\langle n(y_0, B) \rangle = \log \frac{1}{C(y_0, B)^2}$$

(116)

$$= \frac{1}{4} \int \frac{dy\, d^2 q_T}{4[2\pi]^3} \left| \varrho(y_0, B; y, q_T) \right|^2 ,$$

(117)

which should really come as very little surprise. That is to say, once we ignored the constraint of energy momentum conservation on pion production in y_0, B space, the pions were emitted independently and this is precisely the condition under which a Poisson distribution follows.

The mean multiplicity of particles is given by

$$\langle n(y_0) \rangle = \sum_n n \sigma_n / \sigma_T$$

(118)

$$= \int d^2B \, \langle n(y_o, \underset{\sim}{B}) \rangle \tag{119}$$

$$= \int d^2B \, \frac{dy \, d^2q_T}{4(2\pi)^3} \, \frac{1}{4} \, |\, \wp(y_o, \underset{\sim}{B}; y, q_T)\,|^2 \tag{120}$$

This quantity is known to be approximately log s = y_o, from both cosmic ray and CERN-ISR experiments. By the way, this fact of "small" average multiplicity (relative to the $\sqrt{s}$ which could occur if all the incoming energy went into particle production) provides justification for the input of the present model that a only small number of particles with low center of mass momentum are produced, and so one may approximately ignore the conservation of energy momentum for <u>produced particles</u>, once the leading particles are accounted for.

This really completes our discussion of eikonalized production processes. One might proceed further in two directions; one formal, one phenomenological. The first would consist of generalizing the c-number source approximation to some kind of light cone commutator statement, as was done for elastic processes. The other would be to find attractive phenomenological forms for $\wp(y_o, \underset{\sim}{B}; y, \underset{\sim}{q}_T)$ and predict the results of correlation phenomena to be seen at NAL in, say, two particle inclusive processes. At this time I shall refrain from answering these interesting questions both to give the student something to work on and to give the lecturers at the next summer school something to talk about.

ACKNOWLEDGEMENTS

I am very grateful to the high energy theory group of the A. F. Ioffe Physico-Technical Institute, Leningrad for their hospitality while these lectures were constructed. Also I wish to thank Bob Sugar and Dick Blankenbecler for many useful hours of discussion on eikonalistic questions.

REFERENCES AND FOOTNOTES

1. R.J. Glauber in <u>Lectures in Theoretical Physics</u>, ed. by
   W.E. Brittin and L.G. Dunham [Interscience Publishers, Inc.
   New York, (1959)], Volume I, page 315.

2. R. Blankenbecler and R.L. Sugar, Phys. Rev. <u>183</u>, 1387 (1969).

3. H.D.I. Abarbanel and C. Itzykson, Phys. Rev. Letters <u>23</u>,
   53 (1959).

4. J.D. Bjorken, J.B. Kogut, and D.E. Soper, Phys. Rev. <u>D3</u>,
   1382 (1971).

5. H.D.I. Abarbanel, Lectures at the Summer School in Theoretical
   Physics, University of Colorado, 1971, to be published.

6. R.L. Sugar, Lectures at the Summer School in Theoretical
   Physics, University of Colorado, 1971, to be published.
   R. Aviv, R.B. Blankenbecler, and R.L. Sugar, UCSB preprint
   1972, to be published.

   Essentially an identical construction has been made by
   R. Jengo, C. Calluci, and C. Rebbi in various contributions
   to be published in Nuovo Cimento during 1972 and 1972.
   H. Leutwyler also informed me that much work along these
   lines has been carried out by H. Kastrup during the past
   five years.

7. Three vectors are denoted $\vec{k}$ and a unit vector in the
   direction of $\vec{k}$ by $\hat{k}$.

8. Because $\hat{k}_i \approx \hat{k}_f$ when Θ is small we have ceased distinguish-
   ing between them and set either equal to $\hat{k}$. The actual choice
   of $\hat{k}$ may make some difference in actual uses of the result
   (13). This has been considered in the recent literature by
   S.J. Wallace, Phys. Rev. Letters <u>27</u>, 622 (1971) and
   E. Kujawski, Phys. Rev. <u>D4</u>, 2573 (1971).

   I would like to thank A. Gal for a valuable discussion on
   this matter.

9.   Compare the elastic total cross sections $\sigma_{elastic}$ taken from
G. Giacomelli, CERN-HERA 69-3 (December 1969) to σ_{total} as
reported by G. Fox and C. Quigg in their compilation of
Elastic Scattering Data, UCRL-20001 (January 1970).

10.   In fact this "derivation" was essentially given by Froissart
in his original paper, M. Froissart, Phys. Rev. <u>123</u>,
1053 (1961).

11.   R.L. Sugar, private communication.

12.   See the reprints of R.P. Feynman and J.Schwinger in
<u>Quantum Electrodynamics</u> (Dover Publications Inc., N. Y.,
(1958), J. Schwinger, editor.

13.   Actually a generation <u>functional</u> since it depends on a
function: $A(x)$.

14.   We are more or less following the development in Ref.4

15.   This observation seems to have been made first by
B.W. Lee, Phys. Rev. <u>D1</u>, 2361 (1970) and S.J. Chang,
Phys. Rev. <u>D2</u>, 2886 (1970).

        It was made more explicit by S. Weinberg, MIT-CTP preprint
231, September, 1971 and by E. Eichten, MIT-CTP preprint
237, October, 1971.

16.   Although the presentation I will give is rather different
from Ref. 6, the ideas are the same. Indeed, it was the
work of Sugar and his collaborators which suggested the
present chapter.

17.   This is also not such a dumb idea. At the CERN-ISR the ratio
of produced π's to K's is 8/1 and the π/p ratio is 17/1
and the $\pi/\bar{p}$ ratio is 34/1. This is the result of the
British-Scandinavian collaboration at p_{lab} = 1500 GeV/c as
reported by H. Bøggild at the 3rd International Collogquium
on Multiparticle Reactions at Zakopane, Poland; June, 1972.

18.   This is an ancient and honorable problem which is treated with
great physical sense in the book by E. Henley and W.Thirring,
<u>Elementary Quantum Field Theory,</u> (McGraw Hill Book Co.,1962).

FIELD THEORY AT INFINITE MOMENTUM

Ralph Roskies

University of Pittsburgh, Pittsburgh,Pa. 15213

I. Introduction

Since time is limited, I propose to give only a sketchy out-
line of the history and motivation for studying field theories at
infinite momentum. The $P \rightarrow \infty$ limit was first considered in
connection with the derivation of fixed q^2 current algebra sum
rules[1]. The field-theoretic aspects were first studied by Weinberg[2]
in spinless theories. Interest in the matter was heightened by the
development of the parton model[3]. The subject became of even greater
interest when it was realized that the $P \rightarrow \infty$ limit was equivalent
to quantizing the theory on the light cone and that it led to simplified calculations at high energies[4]. In a parallel development, it
was shown[5] that it was sometimes easier to perform quantum electrodynamics (QED) calculations using the $P \rightarrow \infty$ techniques rather
than those of Feynman.

Our motivation for studying this topic is then:
1. to understand the $P \rightarrow \infty$ limit so as to put parton model
   calculations on a firmer basis,
2. to develop an alternate calculational tool from the usual
   Feynman approach,
3. to understand the connection between ordinary field theory
   and light-cone-formulated field theory.

In the long run, it is also hoped that this study will lead
to a deeper understanding of field theories, and to a new set of
approximations for both QED and hadron physics.

II.   <u>The Rules</u>

The S matrix is related to the invariant matrix element M by

$$S = 1 - (2\pi)^4 i\, \delta^4(P_{final} - P_{initial}) M \prod_{ext.part.} N_i \tag{2.1}$$

where N_i is the normalization factor $1/[(2\pi)^{3/2}\sqrt{2E_i}]$ and E_i is the energy of the ith external particle. We now write the rules for calculating the contributions to M in Old Fashioned Perturbation Theory (OFPT). For the moment we restrict ourselves to spinless particles with a ϕ^3 interaction.

The rules are as follows:

1.   For each Faynman graph of order n, assign a time t_i to the i-th vertex. Then draw $n!$ graphs, corresponding to all permutations of the times t_i with the same topology as the Feynman graph. As an example, to the simple Feynman vertex graph

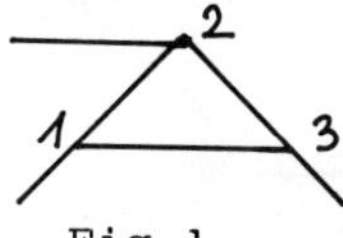

Fig.1

there correspond 6 time-ordered graphs

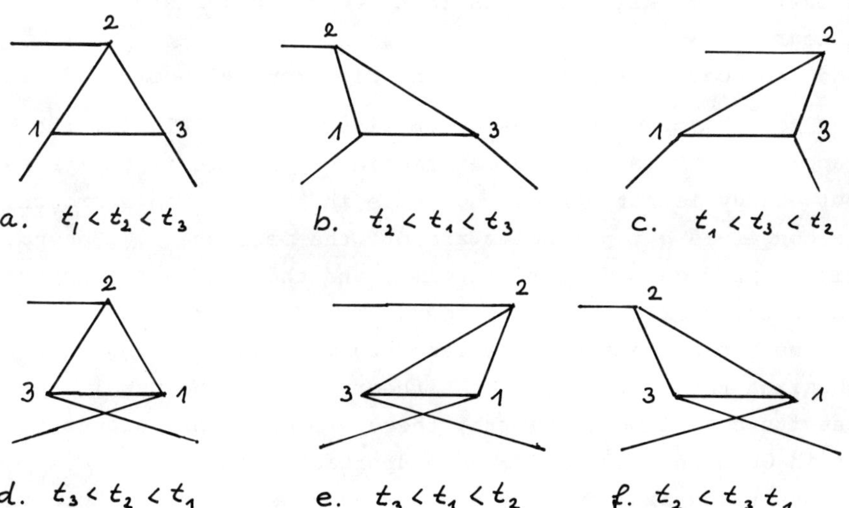

(By convention, time flows from left to right).

Fig. 2

2.   With each line of each time ordered graph, associate a three-momentum.

3.   At each vertex except the last, write a factor $(2\pi)^3 g\,\delta\left(\sum \vec{p_i}\right)$ where g is the coupling constant, and the delta function expresses three-momentum conservation at that vertex. At the last vertex insert only a factor g, since the factor $(2\pi)^3\,\delta\left(\sum \vec{p_i}\right)$ has already been taken out of M in (2.1).

4.   For each internal line write a factor $1/\left[(2\pi)^3 2E_i\right]$ where E_i is the energy of the line in question, calculated on the mass shell, i.e.

$$E_i = \sqrt{p_i^2 + m_i^2} \qquad (2.2)$$

5.   For each intermediate state, i.e. for each state between interaction times, write a factor

$$\frac{1}{E_{inc} - E_{int} + i\varepsilon}$$

where E_{inc} is the total energy of the incoming particles, and E_{int} is the energy of the intermediate state, obtained by adding the single particle energies of all particles in that particular state.

6.   Integrate $d^3 p_i$.

7.   Add the contributions of all the different time ordered graphs.

One sees several features which distinguish OFPT from the Feynman approach. First, every intermediate particle is on its mass shell, but energy is not conserved, while in the Feynman approach, energy is conserved but particles are off the mass shell. Second, all particles propagate forward in time, and the number of particles in a given intermediate state is clear. Third, manifest covariance is gone. One can summarize these last two points by saying that OFPT emphasizes the unitarity of the theory, while the usual approach emphasizes its covariance. Fourth, there are many more graphs to calculate in OFPT than in the Feynman approach.

Points three and four are usually considered serious practical shortcomings of OFPT. However, Weinberg[2] realized that the lack of manifest covariance could be used to good advantage. He argued that since each of the time ordered graphs by itself was not covariant, but their sum was, perhaps it was possible to find a frame of reference in which it was particularly simple to calculate each of these graphs. In particular, there might be a frame in which one could recognize immediately that most of the graphs gave a vanishing contribution. As you may have guessed from the title of these lectures, that frame is the infinite momentum frame.

Let us review his argument. We view the scattering process from a frame moving rapidly in the negative z direction, so that the total incident momentum $\vec{P}$ is large and along the positive z direction. We will show that as $P \longrightarrow \infty$, each of the time ordered graphs tends to a finite limit, often to 0. That each graph tends to a finite limit is not a trivial result, since from covariance only the sum of all the graphs need be independent of P. There might have been cancellations between infinities of specific time ordered graphs.

It should be stressed that letting $P \longrightarrow \infty$ is just a choice of reference frame, and no invariant quantity is getting large.

We parametrize the momentum of the ith line by

$$\vec{p_i} = x_i \vec{P} + \vec{k_i} \tag{2.3}$$

where x_i is a number and $\vec{k_i}$ is a two-dimensional vector in the x-y plane.

Since by definition the total incident momentum is $\vec{P}$ we have

$$\sum_{inc} \vec{p_i} \equiv \left(\sum_{inc} x_i \right) \vec{P} + \sum_{inc} \vec{k_i} = \vec{P}$$

so that

$$\sum_{inc} x_i = 1 \quad , \quad \sum_{inc} \vec{k_i} = 0 \tag{2.4}$$

Because of three-momentum conservation at each vertex, we also have for each intermediate state

$$\sum_{int} x_i = 1 \quad , \quad \sum_{int} \vec{k_i} = 0 \tag{2.5}$$

We can always choose the velocity of the observing frame large enough so that all external particles have their z component of momentum positive, i.e.

$$x_i > 0 \qquad \text{for all external particles.} \tag{2.6}$$

But for internal particles, the x integrations extend over negative x as well.

In the limit $P \to \infty$, we have, from (2.2) and (2.3)

$$E_i = \sqrt{p_i^2 + m_i^2} = |x_i| P + \frac{S_i}{2P} + O\left(\frac{1}{P^3}\right) \qquad (2.7)$$

where

$$S_i = \frac{k_i^2 + m_i^2}{|x_i|} \qquad (2.8)$$

The incident energy is, by (2.6) and (2.4),

$$E_{inc} = \sum_{inc} E_i = P + \sum_{inc} \frac{S_i}{2P} + O\left(\frac{1}{P^3}\right) \qquad (2.9)$$

The energy of an intermediate state is

$$E_{int} = \sum_{int} \left(|x_i| P + \frac{S_i}{2P} \right) + O\left(\frac{1}{P^3}\right) \qquad (2.10)$$

If all the x_i in the intermediate state are positive, then using (2.5) we find

$$E_{int} = P + \sum \frac{S_i}{2P} + O\left(\frac{1}{P^3}\right) . \qquad (2.11)$$

If, however, some x_i are negative, we have

$$E_{int} = \left(1 - 2 \sum_{x_i < 0} x_i \right) P + O\left(\frac{1}{P}\right) \qquad (2.12)$$

Counting powers of P in a graph with n vertices, we obtain:

a) From rule 3

$$\delta\left(\sum \vec{p}_i \right) = \delta^2\left(\sum \vec{k}_i \right) \delta\left[\left(\sum x_i \right) P \right] = \frac{1}{P} \delta^2\left(\sum \vec{k}_i \right) \delta\left(\sum x_i \right)$$

and since there are (n-1) delta functions we obtain a factor $P^{-(n-1)}$.

b) from rules 4 and 6

$$\frac{d^3 p_i}{(2\pi)^3 2 E_i} = \frac{d^2 k_i \, dx_i}{2 |x_i| (2\pi)^3}$$

independent of P.

c) From rule 5 for each intermediate state with all $x_i > 0$, we obtain from (2.9) and (2.11)

$$\frac{2P}{\sum_{inc} s_i - \sum_{int} s_i + i\varepsilon}$$

whereas if some $x_i < 0$, we obtain from (2.12),

$$\frac{1}{\left(2\sum_{x_i<0} x_i\right) P}$$

There are altogether (n-1) intermediate states, and so to obtain a non-vanishing limit as $P \longrightarrow \infty$, each term from rule 5 must contribute a factor P. Thus in all cases we have a finite limit and a non-zero limit only if each intermediate state has all its $x_i > 0$. But since $\sum x_i$ is conserved at each vertex, this is only possible if each vertex has at least one line coming from the past and one line proceeding to the future. Thus of the 6 graphs of Figure 2, only 2a and 2b have non-vanishing limits as $P \rightarrow \infty$. So the passage to infinite momentum has reduced the number of graphs to be calculated.

We have been rather cavalier in counting powers of P, for although P gets large, it is possible that xP is not large, and our expansion (2.7.) in terms of P may not be valid. This is discussed in greater detail in the Appendix. Roughly our analysis is correct for the calculation of renormalized quantities, but must be modified for calculating divergent quantities.

We can now rewrite the rules of calculation. Denoting them by primes, we obtain

1'. For each Feynman graph of order n, draw all time ordered graphs in which each vertex has at least one line from the past and one to the future.

2'. With each line associate an x and k.

3'. At each vertex except the last write a factor $(2\pi)^3 g\, \delta(\Sigma x_i)\, \delta^2(\Sigma \vec{k}_i)$ inserting only g at the last vertex.

5'. For each intermediate state write a factor

$$\frac{2}{\sum_{inc} s_i - \sum_{int} s_i + i\varepsilon}$$

4'. and 6'. Integrate

$$\frac{d^2k_i\, dx_i\, \Theta(x_i)}{(2\pi)^3\, 2\, x_i}$$

7'. Sum over all time ordered graphs.

So much for the ϕ^3 theory.

The presence of spin complicates the situation. In the case of QED for each internal electron line of momentum p we have an extra factor $(\not{p}+m)$, while for a positron we have $(-\not{p}+m)$. (Recall here that p is on the mass shell). For each internal photon there is an extra factor $-g_{\mu\nu}$, and the vertex has a factor γ^μ. We can re-write

$$(\not{p}+m) = 2m \sum_{spin} u(p,s)\,\bar{u}(p,s)$$

and we could associate u and $\bar{u}$ respectively with the two vertices of the given line. This effectively makes the vertex

$$\bar{u}(p)\,\gamma^\mu\,u(p')$$

or a similar expression with u replaced by v.

Our previous counting of powers of P is now upset, since the vertices can also contribute powers of P.A straightforward calculation[6] shows that as $P \to \infty$

$$\bar{u}(p)\,\gamma^\mu\,u(p')$$

is of order P only if

a) $\mu = $ 0 or 3 $\qquad x\,x' > 0$

b) $\mu = $ 1 or 2 $\qquad x\,x' < 0$

$\qquad\qquad\qquad\qquad\qquad\qquad\qquad\qquad\qquad$ (2.13)

Otherwise it is of order 1. Moreover in case a) the coefficient of P is the same whether $\mu = 0$ or $\mu = 3$. (Here x and x' are the x values associated with p and p').

We now show that the contribution to M is of order $P^{N_0+N_3}$, where N_i is the number of external photons of polarization i.

Assume first that all $x_i > 0$, so that case b) never arises. If a vertex is connected to an external photon, it contributes a factor P if and only if the polarization is 0 or 3. If the vertex is connected to an internal photon, because the coefficient of P is the same whether $\mu = 0$ or $\mu = 3$, and because $g_{00} = -g_{33}$ in the photon propagator, the terms of order P and P^2 from these two vertices cancel identiacally, and give an effective vertex of order 1. (The terms of order 1 do not cancel, so that one cannot say that the $\mu = 0$ piece cancels the $\mu = 3$ piece. In fact this longitudinal piece is responsible for the Coulomb force).

Suppose now that some $x_i < 0$. We saw that the matrix element is suppressed by $1/P^2$ for every intermediate state containing a particle with $x < 0$. But such a particle can contribute a factor P^2 to the numerator (a factor P for each of its two vertices). Thus a fermion of negative x can contribute to M in leading order but only if : a) it extends over one time interval only, so that it contributes to only one intermediate state, and b) the fermions at each of its vertices have $x > 0$. Since a photon of negative x contributes no powers of P to the numerator, it can contribute only if every intermediate state containing it also contains a fermion of negative x. This is only possible in the simplest self energy diagram

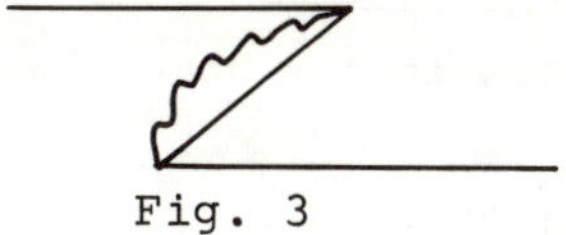

Fig. 3

which will be discussed later. These rules for incorporating fermions of negative x were first derived by Drell, Levy and Yan[6].

Because fermions of negative x can contribute to leading order, our previous criterion of discarding graphs with vertices in which particles were created out of the vacuum or annihilated into it, is no longer valid. But it can be salvaged by modifying the fermion propagator as we now show.

Suppose there is a graph with a positron line $x < 0$. Denote the vertices of the line by V1 and V2 as shown in Fig.4, and denote the momentum and x value by p and x respectively. Since

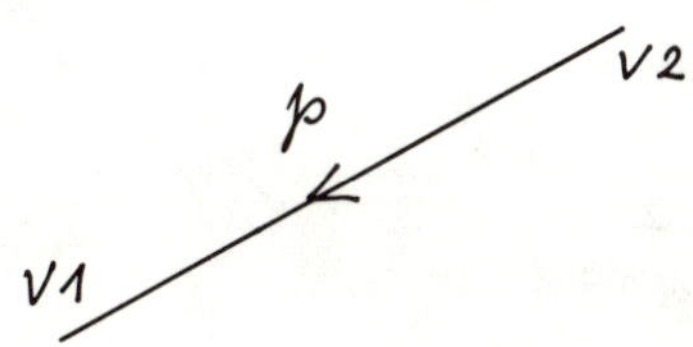

Fig. 4

V1 and V2 are separated by a single time interval, no other particle occurring in the intermediate state between V1 and V2 can have a negative x, since it would extend over more than one time interval. (We will deal with the self energy graph of Fig. 3 separately). The energy denominator associated with the intermediate state between V1 and V2 is

$$\frac{1}{P\left[1 - \sum_{int} |x_i|\right]} = \frac{1}{P\left(1 - \sum_{int} x_i + 2\hat{x}\right)} = \frac{1}{2\hat{x}\,P} \tag{2.14}$$

It is straightforward to see that there must also be a graph in which all other momenta are unchanged except that the time order of V1 and V2 is interchanged, so that it represents an electron going from V2 to V1. Denoting the momentum of the electron by $\hat{p}$, the relation between p and $\hat{p}$ is

$$\vec{p} = -\vec{\hat{p}} \quad , \qquad E_p = E_{\hat{p}} \tag{2.15}$$

The contribution of the line between V1 and V2 and the associated energy denominators to the two graphs is, from (2.14) and (2.15)

$$\frac{(-\hat{\not{p}} + m)}{2\,\hat{x}\,P} + \frac{2\,P\,(\not{p} + m)}{S_{inc} - S_{int} + i\varepsilon} =$$

$$= 2P\left[\frac{\not{p} + m}{S_{inc} - S_{int} + i\varepsilon} + \frac{-2\,E_p\,\gamma^0}{4\,\hat{x}\,P^2} + \frac{\not{p} + m}{4\,\hat{x}\,P^2}\right] \tag{2.16}$$

The third term of (2.16) is negligible compared to the first. Also we have

$$E_p = |\hat{x}|\,P = -\hat{x}\,P \qquad \text{since} \qquad \hat{x} < 0 \tag{2.17}$$

and so (2.16) becomes

$$2P\left[\frac{\not{p}' + m}{S_{inc} - S_{int} + i\varepsilon} + O\left(\frac{1}{P^2}\right)\right] \tag{2.18}$$

where

$$\vec{p}' = \vec{p} \quad , \qquad p'^0 = p^0 + \frac{S_{inc} - S_{int}}{2\,P} \tag{2.19}$$

By changing the propagator of an electron with positive x from $(\not{p}+m)$ to $(\not{p}'+m)$ whenever the electron line extends over a single time interval, we automatically take into account the contribution of all positrons with $x < 0$. Similarly we modify the propagator of positrons with positive x, which extend over one time interval, from $(-\not{p}+m)$ to $(-\not{p}'+m)$. We do not change the propagators of fermion lines extending over more than one time interval.

This replacement of $\not{p}$ by $\not{p}'$ is very reminiscent of the Feynman approach. One takes the fermion off the mass shell ($\not{p}'^2 \neq m^2$), but reduces the number of diagrams. Moreover $\not{p}'$ is the four-vector which enforces four-momentum conservation between the given intermediate state and the external state. To see this recall that

$$E_{inc} = P + \frac{S_{inc}}{2P} \quad , \quad E_{int} = P + \frac{S_{int}}{2P}$$

The energy for all lines in the given intermediate state except for the given fermion is

$$\tilde{E} = P + \frac{S_{int}}{2P} - p^0$$

so that, from (2.18)

$$\tilde{E} + p_0' = E_{inc}$$

However since not all fermion lines extend over a single time interval we do not yet have complete four-momentum conservation.

From (2.13) we see that it is not necessary to modify the propagator of a fermion extending over a single time interval if either of its vertices is connected to an external photon of polarization 0 or 3, although modifying it will not change the answer. This follows because the associated fermion of negative x cannot contribute the necessary P^2 to the numerator.

Our final rules for QED at infinite momentum are then obtained by modifying the infinite momentum rules for the ϕ^3 theory as follows: 2'' - With each internal line associate an x and k. For $\left\{\begin{array}{l}\text{electrons}\\\text{positrons}\end{array}\right\}$ extending over more than one time interval insert a factor $(\pm \not{p} + m)$ where p is the four momentum associated with the line. For fermions extending over one time interval insert the

184

factor ($\pm\ \not{p}' + m$) where p' is related to p by (2.19). For each
internal photon line insert the factor $-\ g_{\mu\nu}$.
3''- Replace g by e γ^μ in 3'.

<u>Renormalization</u>

With these rules one should now be able to calculate any desired S matrix element. But infinities will occur and to get around
them we must understand the renormalization theory. This has been
a notoriously treacherous problem in OFPT. But with the hindsight
provided by a knowledge of renormalization theory in the usual approach,
one can implement the renormalization program in OFPT at infinite
momentum.

Briefly the idea is the following. One isolates the self
energy and vertex parts as in the usual approach, and subtracts counter
terms with formally divergent constants. The problem is to choose an
integral representation for these constants so that subtracting the
integrand of the counter term from the integrand of the unrenormalized
graph yields a well-behaved integrand whose integral is finite. If the
divergent subgraph in question occurs in only one time order, this is
straight-forward as we illustrate in the following example.

Consider the contributions of the graph

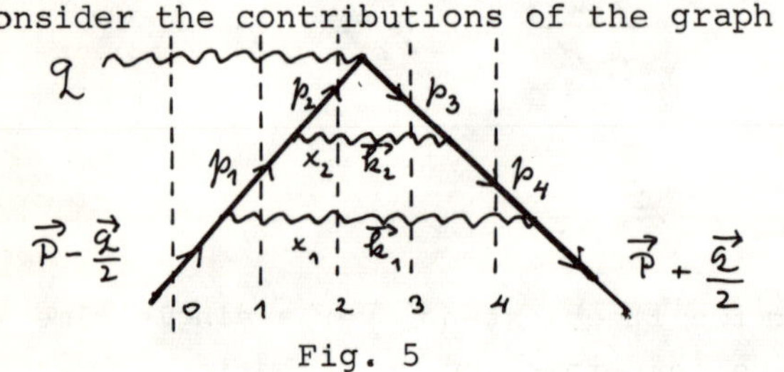

Fig. 5

to the magnetic moment of the electron. Here q is the vector ($0, \vec{q}, 0$)
so that the external photon is spacelike and has x = 0. The vectors
p_1, p_2, p_3, p_4 are determined by three momentum conservations and
the mass shell condition. The intermediate states are denoted by
0,1,2,3,4 as shown. This graph has a reducible vertex piece, so one
must subtract the counter term

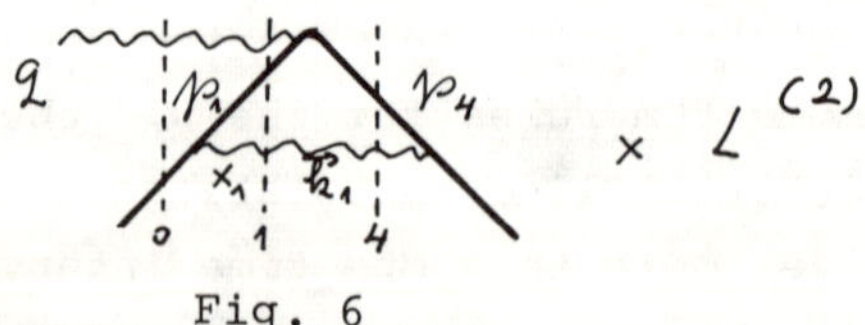

Fig. 6

where $L^{(2)}$ is the coefficient of γ^μ in the divergent subgraph

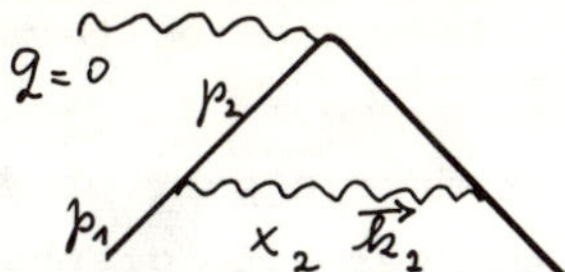

Fig. 7

As the external momentum in the calculation of $L^{(2)}$ we choose p_1, and not P . This is legitimate since p_1 is on the mass shell. Then ignoring the numerator structure which is the same as in the Feynman case the energy denominators are, for the unrenormalized graph

$$\frac{1}{E_0 - E_1} \quad \frac{1}{E_0 - E_2} \quad \frac{1}{E_0 - E_3} \quad \frac{1}{E_0 - E_4}$$

while for the counter term they are

$$\frac{1}{E_0 - E_1} \quad \frac{1}{E_0 - E_4} \quad \frac{1}{E_1 - E_2} \quad \frac{1}{E_1 - E_2}$$

It is easy to check that the difference between the two integrands now vanishes sufficiently rapidly as $k_2^2 \to \infty$.(I have not mentioned the infrared divergence problem which is the same as in the Feynman case.)

The essential idea then is to take the external momentum of the reducible part to be the momentum appearing in the unrenormalized graph. This leads to a modified denominator structure, so that the difference between the unrenormalized integrand and that of the counter term is a finite integral.

But life is not always that simple. Again in the magnetic moment calculation the Feynman graph

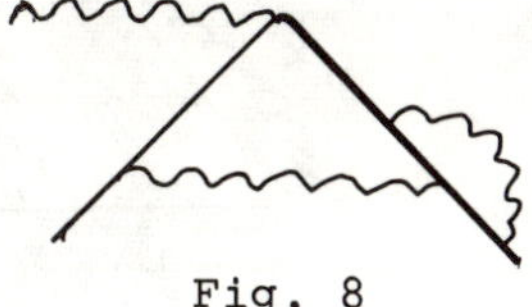

Fig. 8

has three time orderings

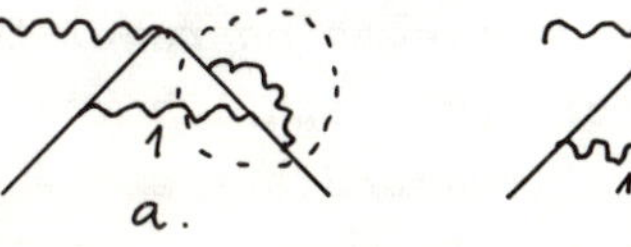

Fig. 9

The reducible piece is shown circled. In a counter term with photon 1 having $q = 0$, graphs   9. b and 9. c disappear, since a photon with $q = 0$ has   x = 0   and cannot produce a pair of fermions with only positive x. Thus the naive counter terms to graphs 9.b and 9. c vanish so one might expect the graphs to be finite. But 9. b is divergent. (9.c is finite because the divergent loop extends over three time intervals so that the energy denominators provide extra convergence.) Note that it will not do to subtract a counter term with a photon having x_1 , $\vec{k}_1$, since in the counter term energy must be conserved in the subgraph. Together with the mass shell condition for the fermions, this forces the photon to have $q = 0$ and not only $q^2 = 0$. The analysis in this case is then much more subtle. Full details can be found in reference[5].

In performing these renormalizations, we are subtracting infinite quantities, which is always a delicate procedure. The correct way to do so is to first regulate the integrals, rendering them finite, then subtract, and then let the regulators disappear. In the infinite momentum frame, since covariance is not manifest, one must be especially careful to regulate in an invariant manner. This can be achieved by using the Pauli-Villars regularization scheme.

The calculation of δm_e in second order is an excellent illustration of the subtleties of the limit $P \to \infty$. As we have already indicated our rule for incorporating fermions of negative x is not valid in this case so we revert to the older rules in which particles of negative x are treated explicitly. Then there are 2 graphs

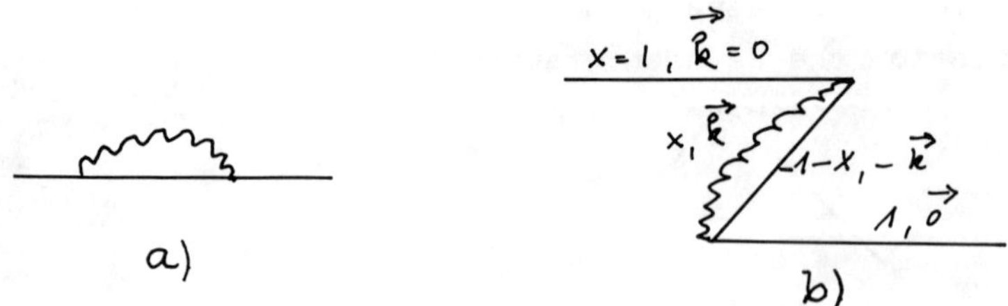

Fig. 10

As is well known these graphs are divergent and have to be regulated. A naive argument would say that upon regularization graph 10b vanishes, for since at least one of the particles in the intermediate state has x < 0   the energy denominator is just $1/\left[1 - \Sigma |x_i|\right]$

independent of the photon mass. Consequently subtracting a similar
integrand with a large photon mass will give identically zero.

Unfortunately the argument is wrong, because the limit
cannot be taken under the integral sign. One must integrate first
and only then let $P \to \infty$.

One can nevertheless obtain the correct result by defining the
$P = \infty$ rules more carefully. Ignoring the numerator structure for the
moment the denominators are, before $P \to \infty$,

$$\frac{1}{e_k} \quad \frac{1}{E_{-p-k}} \quad \frac{1}{E - e_k - E_{-p-k}} \tag{3.1}$$

where

$$e_k = \sqrt{(x P)^2 + k^2 + \lambda^2} \ ; \ E_{-p-k} = \sqrt{(1+x)^2 P^2 + k^2 + m^2} \ ; \ E = \sqrt{P^2 + m^2}$$

and λ is a small photon mass.

For large P we wrote

$$e_k = |x| P + \frac{k^2 + \lambda^2}{2 |x| P} \tag{3.2}$$

and disregarded the second term in (3.2) in the first term of (3.1).
This is legitimate provided that the function multiplying $\frac{1}{e_k}$ vanishes
at $x = 0$, so that the integral over x is well defined. In other words
as a distribution on functions vanishing at x = 0 we have

$$\frac{1}{e_k} \longrightarrow \frac{1}{|x| P} \qquad \text{for large P.}$$

But if the functions vanish at x = 0, we could also write

$$\frac{1}{e_k} \longrightarrow \frac{1}{|x| P} + C \, \delta(x) . \tag{3.3}$$

To fix the coefficient C, consider what happens if the function does
not vanish at x = 0.   Then the integral is not well defined and must
be regulated, by subtracting the contribution of a heavy photon. So
we must study

$$\lim_{P \to \infty} \left\{ \frac{P}{\sqrt{(x P)^2 + k^2 + \lambda^2}} - \frac{P}{\sqrt{(x P)^2 + k^2 + \Lambda^2}} \right\} \tag{3.4}$$

where Λ is the mass of the regulator photon. For x = 0 this
limit vanishes. But it is readily checked that as a distribution in
x,(3.4) tends to

$$- \ln \left(\frac{k^2 + \lambda^2}{k^2 + \Lambda^2} \right) \delta(x)$$

This is consistent with (3.3) if

$$C = - \ln \left(k^2 + \lambda^2 \right) .$$

One might argue that

$$\frac{P}{e_k} = \frac{1}{\sqrt{x^2 + (k^2 + \lambda^2)/P^2}}$$

is not a function of $(k^2 + \lambda^2)$, but of $(k^2 + \lambda^2)/P^2$, so that C should be
$-\ln\left[(k^2 + \lambda^2)/P^2\right]$. But on regularization the $\ln P^2$ terms cancel. Using then
as the energy term

$$P/e_k = 1/|x| - \ln \left(k^2 + \lambda^2 \right) \cdot \delta(x)$$

one shows that 10.b does give a contribution upon regularization and
when combined with 10.a it yields the Feynman result in terms of the
regulator mass.

It is clear that for amplitudes which are already finite, the
δ (x) cannot contribute and our previous rules are correct. So for
example in calculating the contribution of

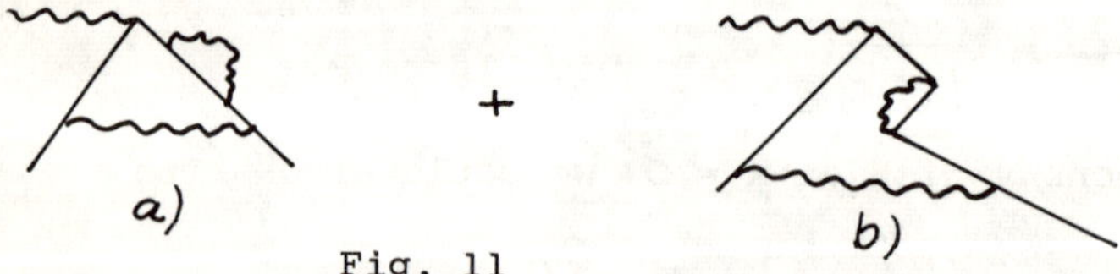

Fig. 11

to the magnetic moment of the electron, upon regulating we subtract

Fig. 12

and Figures 11b and 12a cancel identically. For renormalized
amplitudes only 10a remains in the self energy.

<u>Applications to Q E D</u>

Having understood the rules and the renormalization prescription we can calculate any quantity of interest in QED. To
compare these techniques with the usual ones, consider again the
problem of the magnetic moment of the electron with momenta as shown
in Figure 13.

Fig. 13

Assuming that the external photon has polarization $\mu = 0$, we find
that in second order there is only the time ordered graph. In fourth
order, there are 5 Feynman graphs, and 9 time ordered graphs
corresponding to them. In sixth order there are 40 distinct Feynman
graphs, 28 of which do not involve vacuum polarization. There are
between 1 and 15 timeordered graphs corresponding to each Feynman
graph.

Brodsky and I have done the fourth order calculation completely
and shown that it agrees with the usual answer[5]. This is a valuable
check on the validity of the renormalization prescription since all
three types of renormalization are required.

In sixth order we have calculated

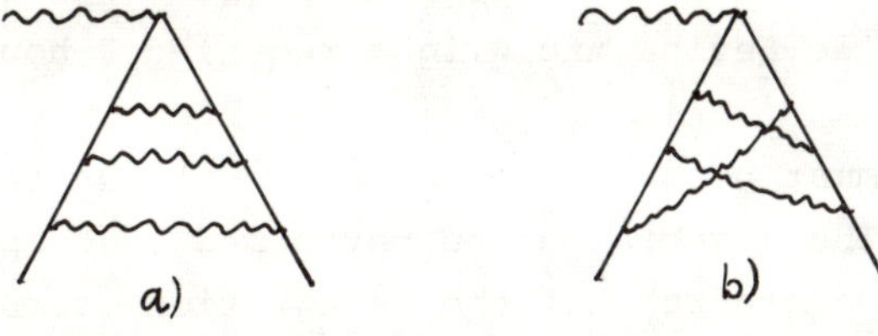

Fig. 14

Apart from the important feature of providing an independent
check on the usual calculation it might appear that there is no
advantage to our approach. The trace calculations are the same as in
the Feynman approach, but there are more time ordered graphs than
Feynman graphs. Thus although the trace calculation need be done only
once, there is still more algebra to perform in our approach.

Moreover, the dimensionality of our integral is always one greater
than that of the Feynman parameter integral.

The advantage of our approach comes in the numerical integrations to be performed. It turns out that the integrand is a much
smoother function of the infinite momentum variables (which we take to
be the X and k of the photon lines) than it is of the Feynman
parameters. Thus the numerical convergence is much faster. This is
quite significant because the difficulty of performing this calculation
is in the numerical integrations. The algebra is also done by computer
and the time this takes is small compared to the integration time.
Moreover the algebra need be done only once while the integrals must
be repeated if higher numerical precision is required.

To illustrate the practical difference between the two approaches
consider graph 14a suitably renormalized. (It has no overall infrared
divergence). For us it is a 6 dimensional integral, while in the usual
case it is a 5 dimensional one. (We are making use of some symmetry
properties of the graph.) There is only one time ordered graph corresponding to the given Feynman graph. It turns out that we were able
to obtain better than 1% accuracy by evaluating the integrand at
100,000 points, while Levine and Wright[7] required 2,000,000 points for
the same accuracy in the usual approach. Ours is the first independent
confirmation of the results of Levine and Wright, and we each find
that graph contributes

$$(1.77 \pm .01)\,\frac{\alpha^3}{\pi^3}$$

to $(g-2)/2$. The integration time for our problem was 4 minutes
on the SLAC IBM 360/91 while Levine and Wright required 2 hours on a
Univac 1108.

In all honesty it must be said that we have not had the same
success with graph 14b. The algebra can be performed by computer
(there are 8 different time orders) but the integration is much more
complicated than in the case of graph 14a. Our runs give results

$$.8\,\frac{\alpha^3}{\pi^3} \qquad \text{and} \qquad 1.3\,\frac{\alpha^3}{\pi^3}$$

but with large errors. The average of the 4 best runs was
$(1.11 \pm .23)\,\left(\frac{\alpha}{\pi}\right)^3$ based on evaluating the function at about
600,000 points, requiring about 2 hours on the SLAC machine. The
answer of Levine and Wright is $(.90 \pm .01)\left(\frac{\alpha}{\pi}\right)^3$ based on evaluating
their integrand at several million points.

<u>Covariant Approach At P = ∞</u>

Field theories at infinite momentum have been studied from a different point of view. Kogut and Soper[4] argued that the limit is a reformulation of the theory in which the equal time surface (in the regular frame) is replaced by a light-like surface i.e. v=c. Making the transformation

$$\tau = \frac{t+z}{2} \quad , \quad z' = \frac{t-z}{2}$$

they quantized the theory at equal τ . When passing from the Lagrangian to the Hamiltonian in this approach they found that the interaction Hamiltonian contained in addition to the usual piece a seagull term with the structure

$$e^2 \, \delta(\tau - \tau') \, \bar{\psi} \, \psi \, A_\mu A_\nu \, ,$$

that is, an instantaneous interaction involving 2 fermions and 2 photons. They then formulated OFPT for this theory and reproduced the rules we have been discussing. In this approach the limit $P \to \infty$ never appears, it has already been taken. The question of whether this theory is equivalent to the usual one is the question of whether the $P \to \infty$ limit is justified. Their approach was formulated in the Coulomb gauge which is difficult to renormalize.

A more covariant approach was developed by Chang, Root and Yan[8]. Starting with Schwinger's action principle, they "derived" the equal commutation relations which Kogut and Soper had guessed. They worked in the $\bar{\psi} \gamma_5 \psi \, \phi$ theory rather than in QED. They found that the Feynman propagator was identical to the usual one for spin zero, but differed for spin 1/2. But they were able to show that the extra term in the spin 1/2 propagator exactly cancelled the terms arising from the seagull in the interaction Hamiltonian, so that their theory formally agreed with the usual one for renormalized amplitudes. Their expressions for the renormalization constants differed from the usual Feynman ones.

These results are easily understandable in terms of ours. Because the Feynman propagator only involves free fields, and because all free particles have x > 0, the fermion propagator does not include fermions with negative x. These must be contained in the effective interaction Hamiltonian, and they are exactly the seagull term. We have seen that fermions of negative x extend only over one time

interval so that no other interaction can occur between its vertices.
One can then effectively assume that its vertices are simultaneous.
Formally one can also see this by noticing that the energy denominator
associated with a state containing a particle of negative x is
independent of the external energy, so that its Fourier transform
is a delta function in time. Moreover the seagull

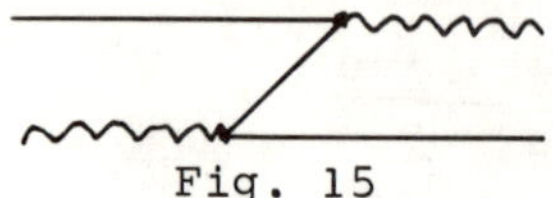

Fig. 15

clearly involves 2 external fermions and 2 photons.

In a theory of scalar particles with no derivative coupling,
our rules showed that at $P \rightarrow \infty$ there were no particles of negative x.
Thus the free propagator should agree with the Feynman result.

We can also understand why for example their expression
for δm_e does not agree with ours. As a field theory, it was
natural to interpret the seagull term as a normal ordered expression.
But that means that it will not contribute to δm_e since its
expectation value must be taken between states which have no photons.
But we have seen that the seagull does contribute to the usual Feynman
answer for δm_e, although its omission does not alter any renormalized
amplitude. Bouchiat et al[9] have shown that if one does not normal
order the seagull, the Feynman expression for δm_e is obtained.

APPENDIX

Convergence as $P \rightarrow \infty$

We restrict our attention to the ϕ^3 theories. First we show that in the terms we have kept there are no singularities at x=0 and the k^2 convergence is as in Feynman theories. We then indicate under what conditions the terms that are formally of order $1/p^2$ cannot be ignored.

We have kept only terms with $x > 0$. The factor $1/x$ associated with each line is compensated by the factor $\frac{k^2 + m^2}{x}$ appearing in the energy denominators of the intermediate states in which this line occurs. To count the divergence in k^2 , recall that the degree of divergence in the Feynman theory is, for a graph with V vertices and N internal lines

$$4N \quad - \quad 2N \quad - \quad 4(V-1)$$

$$\text{from } d^4k \qquad \text{from the} \qquad \text{from momentum}$$
$$\text{propagators} \qquad \text{conservation}$$

while in our rules it is

$$2N \quad - \quad 2(V-1) \quad - \quad 2(V-1)$$

$$\text{from } d^2k \qquad \text{from energy} \qquad \text{from momentum}$$
$$\text{denominators} \qquad \text{conservation}$$

which agrees with the Feynman result. (The problem is much more complicated for QED. It is well known for example that each of the graphs in Figure 10 is quadratically divergent, but their sum is only logarithmically divergent).[10]

We discarded contributions of particles with negative x because the energy denominators formally suppressed the graph by $1/p^2$. But the effective energy denominators

$$\frac{1}{p^2} \cdot \frac{1}{1 - \sum |x_i|}$$

does not contain the factor x/k^2 which we counted on in our previous analysis so that one could get divergence at x = o or $k^2 = \infty$. To study these recall that the energy denominator is really

$$\frac{1}{P^2} \; \frac{1}{1 - \sum |x_i| + \sum_{inc}\frac{k_i^2+m_i^2}{2x_iP} - \sum_{int}\frac{k_i^2+m_i^2}{2|x_i|P}}$$

If only x_1 is negative, then as it goes to zero the term is

$$\frac{1}{P^2} \; \frac{1}{2x_1 + 0\left(\frac{1}{x_1 P}\right)}$$

The $0\left(\frac{1}{x_1 P^2}\right)$ term cuts off the integral at $x_1 = 0\left(\frac{1}{P}\right)$. The contribution to the graph from small x_1 is then

$$\frac{1}{P^2} \int_{|x_1|>\frac{1}{P}} \frac{dx_1}{|x_1|} \; \frac{1}{2x_1} \approx 0\left(\frac{1}{P}\right)$$

and so is still negligible. By a similar argument the k^2 integral is cut off at P^2 (actually $k^2 = x\, P^2$). If the rest of the graph contributes a factor $\frac{1}{k^2}$ then the entire contribution is

$$\frac{1}{P^2} \int_{k^2<P^2} \frac{d^2k}{k^2} \approx \frac{\ln P^2}{P^2}$$

which is still negligible. If however there is no other factor the graph contributes

$$\frac{1}{P^2} \int_{k^2<P^2} d^2k^2 = 0(1)$$

and is <u>not</u> negligible. This only happens if the vector k does not occur in any other intermediate state, for otherwise the energy denominator of that state would have a factor $\frac{1}{k^2}$. But this can only happen in the graphs of Figure 10 or in any graph in which these are imbedded. These are self energy terms which must be regulated anyway, by subtracting the contribution of a heavy mass M. One verifies that after regularization the contribution of negative x can be discarded. (This is not so for vacuum graphs. See for example reference[11].)

In summary, our $P=\infty$ rules are valid for renormalized quantities but not for unrenormalized ones.

195

REFERENCES

1.   See e.g. S. Adler and R.Dashen, <u>Current Algebras</u>,
W.A. Benjamin, New York, 1968,Chapter IV.

2.   S. Weinberg, Phys.Rev. <u>150</u>, 1313 (1966).

3.   S. Drell, D. Levy and T.M. Yan, Phys.Rev. <u>187</u>, 2159 (1969);
<u>D1</u>, 1035, 1617 (1970); and T.M. Yan and S.Drell, Phys.Rev.<u>D1</u>,
2402 (1970).

4.   J.Kogut and D.Soper, Phys. Rev. <u>D1</u>, 2901 (1970)
J.D. Bjorken, J. Kogut and D. Soper, Phys. Rev. <u>D3</u>, 1382(1971).

5.   S.Brodsky and R. Roskies, to be published.

6.   See the second paper of reference 3.

7.   M. Levine, private communication. Their results for the
complete magnetic moment calculation in sixth order is published
in M.Levine and J. Wright, Phys. Rev. Lett. <u>26</u>, 1351(1971).

8.   S.J. Chang, R.G. Root and T.M. Yan, Univ. of Illinois preprints
ILL-(TH)-72-3, 72-4, Feb.(1972).

9.   C. Bouchiat, P. Fayet and N. Sourlas, Orsay preprint LPTHE 71/53,
October (1971).

10.   V. Weisskopf, Phys. Rev. <u>56</u>, 72 (1939).

11.   S.J. Chang and S.Ma, Phys. <u>180</u>, 1506 (1969).

SOME GENERAL ASPECTS OF MULTIPERIPHERAL DYNAMICS

M. TOLLER

CERN, Geneva

1. INTRODUCTION

The first multiperipheral models have been developed ten
years ago by ABFST and others[1-3]. Some general predictions of these
models, as Regge behaviour and Feynman scaling, are, at least qualitatively, in accord with the most recent results of accelerator and ISR
experiments. Other predictions, as the weak increase of correlations
with energy, will be tested soon at the ISR, permitting possibly a
choice between the multiperipheral ideas and other possibilities, as
the one or two fireball models and the diffractive excitation models,
which predict a strong increase of correlations with energy.

Other general features of the simplest multiperipheral models,
as the dominance of an isolated pomeron pole and its consequences
(factorization and short range of inclusive correlations), are expected
to hold only approximately and do not follow necessarily from multiperipheral models of a more general kind.

More detailed features, as the shape and the energy dependence
of the inclusive distribtuions, have been successfully interpreted
by means of particular multiperipheral models, making use of specific
phenomenological inputs.

In this situation, in order to facilitate the interpretation
of the forthcoming experimental data, it is useful to reanalyse in
detail the assumptions, the structure and the mathematical formalism
of multiperipheral dynamics. In these lectures, I shall summarize some
partial contributions in this direction, contained in some recent papers[4-6], adding when necessary a short exposition of previously known
results, in order to get, as far as possible, a self-contained treatment.

I shall deal only with a particular aspect of multiperipheral physics, which can be summarized as follows: make suitable assumptions on the production amplitudes and derive, by integration over the momenta and sum over the multiplicity, the total cross section and the inclusive distributions. We shall see that this problem is not so simple as it could seem.

The new contributions which will be treated with more details are essentially a proposed characterization of the multiperipheral amplitudes by means of a factorized upper bound[4,5] and some improvements in the $O(3,1)$ projection of the multiperipheral integral equation based on the properties of a semigroup[6].

2. <u>A CHARACTERIZATION OF MULTIPERIPHERAL PRODUCTION AMPLITUDES</u>

The mathematical formalism and the general results of multiperipherism are based on some general assumptions on the production amplitudes $M_n(P_A, P_B, P_{n+1}, \ldots, P_0)$ for the processes

$$A + B \longrightarrow (n+1) + (n) + \ldots + (1) + (0), \quad n = 0, 1, \ldots \quad (2.1)$$

We are assuming for simplicity that all the particles are identical and spinless, so that the functions M_n are symmetrical with respect to the permutations of the (n+2) final particles.

We introduce, as usual, the four-momentum transfers

$$Q_i = P_A - \sum_{j=0}^{i} P_j, \quad (2.2)$$

the corresponding invariants

$$t_i = (Q_i)^2, \quad i = 0, 1, \ldots, n \quad (2.3)$$

which, in the equal mass case we are considering are necessarily negative, and the subenergies

$$s_i = (P_{i+1} + P_i)^2 = (Q_{i-1} - Q_{i+1})^2, \quad i = 0, 1 \ldots n. \quad (2.4)$$

In order to get a complete set of (3n+2) invariants, we may consider also the subenergies

$$\sigma_i = (P_{i+1} + P_{i-1})^2, \quad i = 1, 2, \ldots, n. \quad (2.5)$$

We remark that in the definition of these invariants a special ordering of the final momenta has been assumed.

A first characteristic feature of the multiperipheral amplitudes is that they can be large only in the kinematical configurations in which, after a suitable permutation of the final particles, all the momentum transfers $t_0, \ldots, t_n$ are small (in absolute value). This requirement can be stated more precisely by means of the inequality

$$\left| M_m(P_A, P_B, P_{m+1}, \ldots P_0) \right| \leq \sup_{\pi} f_m\left(P_A, P_B, P_{\pi(m+1)}, \ldots, P_{\pi(0)}\right) \qquad (2.6)$$

where we have indicated by π a permutation of the $(n+2)$ final particles and f_n is a function which decreases in a suitable way when some momentum transfer t_i increases (in absolute value).

It is convenient to indicate by x a set of $(3n+2)$ kinematical invariants and by $\mathcal{O}_\pi x$ the set of invariants obtained from x by means of the permutation π of the final particles. Then eq. (2.6) takes the form

$$\left| M_n(x) \right| \leq \sup_{\pi} f_m\left(\mathcal{O}_\pi x\right). \qquad (2.7)$$

This condition is discussed in detail in ref.[4]. For our present purposes, it is sufficient to consider the weaker condition

$$\left| M_m(x) \right|^2 \leq \sum_{\pi} \left(f_m\left(\mathcal{O}_\pi x\right)\right)^2. \qquad (2.8)$$

We remark that, if this condition holds, it is always possible to decompose the modulus square of the amplitude as follows

$$\left| M_m(x) \right|^2 = \sum_{\pi} \mathcal{F}_m\left(\mathcal{O}_\pi x\right), \qquad (2.9)$$

where

$$0 \leq \mathcal{F}_m(x) \leq \left(f_m(x)\right)^2. \qquad (2.10)$$

A possible choice of the function $\mathcal{F}_n(x)$, which is not necessarily the most convenient one, is

$$\mathcal{F}_m(x) = \left| M_m(x) \right|^2 \left(f_m(x)\right)^2 \left[\sum_{\pi} \left(f_m\left(\mathcal{O}_\pi x\right)\right)^2 \right]^{-1}. \qquad (2.11)$$

In general, multiperipheral models are based on an explicit expression for the functions $\mathcal{F}_n(x)$, which have a peculiar factorization property, which was studied in the most general case by Chew, Goldberger and Low[7], who called it "short range correlation". In order to avoid confusion between this exclusive short range correlation and the inclusive short rang correlation, which is a completely different concept, we call this property "Q-factorization". A sequence of functions $\mathcal{F}_n(P_A, P_B, P_{n+1}, \ldots, P_0)$ is called Q-factorized of order k if all these functions can be written in the form

$$\mathcal{F}_n = \mathcal{B}(P_B, Q_n, \cdots Q_{n-k+2}) \, \mathcal{H}_k(Q_n, \cdots, Q_{n-k+1}) \cdots$$

$$\cdots \mathcal{H}_k(Q_{k-1}, \cdots, Q_0) \, \mathcal{A}(Q_{k-2}, \cdots, Q_0, P_A). \tag{2.12}$$

Remark that the functions $\mathcal{B}$, $\mathcal{H}_k$ and $\mathcal{A}$ do not depend on n, so that the whole sequence of functions $\mathcal{F}_n$ is described in terms of these three functions. In general , it is more convenient to express $\mathcal{B}$, $\mathcal{H}_k$ and $\mathcal{A}$ in terms of invariants rather than in terms of four-vectors.

Many physically important contributions to the amplitude, as multiple pion exchange or multiple reggeon exchange, are Q-factorized. However, one can show that any Q-factorized representation of the amplitudes is necessarily approximated.

For this reason, we prefer to found the multiperipheral formalism on a weaker assumption, which has some chance of being really true. Our assumption is that the functions f_n, which appear in the upper bounds (2.7), (2.8) and (2.10) are Q-factorized. For general purposes, it is sufficient to assume a simple Q-factorized upper bound, which takes into account only the decrease in the momentum transfers t_i and the polynomial boundedness in the sub-energies s_i . We shall use the explicit assumption

$$f_n(x) = c \prod_{i=0}^{n} \left[d(t_i) \left(\frac{s_i}{4m^2} \right)^\alpha \right] , \tag{2.13}$$

where $d(t)$ is a suitably decreasing function of $|t|$ and the exponent α is one or slightly larger in order to allow for logarithmic factors.

More complicated and restrictive upper bounds can be useful in order to take into account more detailed properties of the production amplitudes, but they are not necessary for the construction of a multiperipheral formalism.

In conclusion, the formalism that we shall develope is completely based on the assumptions (2.8) and (2.13) plus some condition on the function d(t). The more restrictive assumption (2.7) would be useful in the treatment of other problems, as the study of the elastic diffraction peak based on unitarity.

3. TOTAL CROSS SECTION AND MULTIPLICITY DISTRIBUTION

Substituting eq.(2.9) into the cross section formula, we see that the total cross section for a process with (n+2) final particles is given by

$$\sigma_n(s) = \frac{1}{2}(2\pi)^{-3n-2}\left[s(s-4m^2)\right]^{-1/2} \cdot$$

$$\cdot \int \mathcal{F}_n(P_A, P_B, P_{n+1}, \ldots, P_0)\, \delta^{(4)}\left(P_A + P_B - \sum_{i=0}^{n+1} P_i\right) \cdot$$

$$\left(2P_{n+1}^0\right)^{-1} d^3\underline{P}_{n+1} \ldots \left(2P_0^0\right)^{-1} d\underline{P}_0 \ .$$

Substituting into this formula the inequality (2.10) and eq.(2.13), after suitable majorizations we get (see ref.[5])

$$\sigma_n(s) \leq \pi H c^2 \left[s(s-4m^2)\right]^{-1}\left(\frac{s}{m^2}\right)^{2\alpha} \cdot$$

$$\cdot \frac{1}{n!}\left[H \log \frac{s}{m^2}\right]^n \ ,$$

where

$$H = (4\pi)^{-2} 4^{-\alpha} \int_{-\infty}^{0} (d(t))^2 \left(1 - \frac{\sqrt{-t}}{m}\right)^{8\alpha} dt$$

(we assume that this integral converges).

The inequality (3.2) contains a very important and characteristic feature of multiperipheral physics: due to the factor $(n!)^{-1}$, the cross section $\sigma_n(s)$ is a very fast decreasing function of n in the region $n \gg \log s$. Of course, $\sigma_n(s) = 0$ for $n > s^{1/2}m^{-1}$, but eq.(3.2) implies that $\sigma_n(s)$ is negligible for much smaller values of n (at high energy).

Starting from eq. (3.2) and from the very weak assumption that the total cross section $\sigma(s)$ does not decrease faster than any negative power of s, one can derive the inequalities

$$\langle n^\uparrow \rangle = \left(\sigma(s) \right)^{-1} \sum_n n^\uparrow \sigma_n(s) \leq c_p \left(\log \frac{s}{m^2} \right)^p \tag{3.4}$$

The detailed proof for p=1 can be found in ref.[5] and the generalization to arbitrary p is straightforward.

4. THE BALI-CHEW-PIGNOTTI VARIABLES

For the further developments of the formalism, it is convenient to describe the production amplitudes in terms of Bali-Chew-Pignotti (BCP) variables[8] . In order to introduce our notations, we give a short account of this argument.

If a is an element of the group SL(2C), namely a 2×2 complex matrix with determinant equal to one, we indicate by L(a) the corresponding 4 × 4 Lorentz transformation matrix. As well known, L(a) spans the proper orthochronous Lorentz group. We indicate respectively by u_z (μ) and a_z (χ) the elements of SL (2C) which correspond to a rotation of an angle μ around the z axis and to a boost of rapidity χ along the z axis. We use also similar notations with z replaced by x or y .

For each final particle, we introduce a frame of reference connected with a given arbitrary frame by means of the Lorentz transformation $L(a_i)$. The simplest procedure is to define the group elements a_i implicitly by means of the formulae

$$P_B = L(a_{m+1})(m,0,0,0)$$

$$Q_i = L(a_i)(0,0,0,\sqrt{-t_i})$$

$$Q_i = L(a_{i+1}\, a_z(\chi_{i+1}))(0,0,0,\sqrt{-t_i}), \quad i = 0,1,\ldots,n \tag{4.1}$$

$$P_A = L(a_0\, a_z(\chi_0))(m,0,0,0).$$

The condition that the outgoing particles have mass m permits to compute the quantities χ_i . In this way we get

$$\sinh x_{n+1} = (2m)^{-1}(-t_n)^{1/2}$$

$$\sinh x_0 = (2m)^{-1}(-t_0)^{1/2} \tag{4.2}$$

$$x_i = x(t_i, t_{i-1}), \quad i = 1, 2, \ldots n$$

where

$$\cosh x(t, t') = (m^2 - t - t')(4tt')^{-1/2}, \quad x(t, t') > 0. \tag{4.3}$$

If we put

$$g_i = a_z(-x_{i+1})\, a_{i+1}^{-1}\, a_i, \quad i = 0, 1, \ldots, n \tag{4.4}$$

from eq.(4.1) we see immediately that L(g_i) does not act on the z component of a four vector. It follows that g_i belongs to the subgroup SU(1,1) of SL(2C) and can be parametrized as follows

$$g_i = u_z(\mu_i)\, a_x(\xi_i)\, u_z(\nu_i),$$

$$0 \le \mu_i \le 4\pi, \quad \xi_i \ge 0, \quad 0 \le \nu_i \le 2\pi \tag{4.5}$$

The BCP variables t_i, g_i, (i=0,..., n) determine univocally all the kinematical invariants and therefore the amplitudes can be expressed in terms of them.

It is convenient in the following to express Q-factorized functions in terms of BCP variables. For instance, we shall replace the Q-factorized upper bound (2.13) by the following expression which is Q-factorized in a different way

$$f_n(x) = b(\xi_n, t_n)\, k(t_n, \xi_{n-1}, t_{n-1}) \cdots$$

$$\cdots k(t_1, \xi_0, t_0)\, a(t_0), \quad t_i < 0, \quad \xi_i \ge 0 \tag{4.6}$$

If we put for instance

$$b(\xi_n, t_n) = \left(\frac{t_n}{m^2}\right)^{1/4} c\,[d(t)]^{1/2}\left(\frac{4m^2 - t_n}{8m^2}\right)^{\frac{\alpha}{2}}(1 + \cosh \xi_n)^{\alpha},$$

$$k(t_{i+1}, \xi_i, t_i) = \left(\frac{t_i}{t_{i+1}}\right)^{1/4}\left[d(t_{i+1})\,d(t_i)\right]^{1/2} \cdot$$

$$\cdot \left(\frac{T(m^2, t_{i+1}, t_i)}{8m^2\sqrt{t_{i+1} t_i}}\right)^{\alpha}(1 + \cosh \xi_i)^{\alpha}, \tag{4.7}$$

$$a(t_0) = \left(\frac{m^2}{t_0}\right)^{1/4}\left[d(t_0)\right]^{1/2}\left(\frac{4m^2 - t_0}{8m^2}\right)^{\frac{\alpha}{2}},$$

$$T(a, b, c) = a^2 + b^2 + c^2 - 2ab - 2ac - 2bc, \qquad (4.8)$$

one can show that the expression (4.6) is always greater than the expression (2.13). In the following we shall use the less restrictive upper bound defined by eqs. (4.6) and (4.7).

5. OPERATOR Q-FACTORIZATION

The Q-factorization property (2.12) is essential in the mathematical treatment of the multiperipheral models, as it permits an iterative procedure which reduces the integration over the final momenta and the sum over the multiplicity to the solution of an integral equation. The group-theoretical analysis of this multi-peripheral equation provides a natural approach to complex angular momentum.

It could seem that, if one starts from the weaker assumption of a Q-factorized upper bound, all these formal developments have to be abandoned. Fortunately, it is not so, because one can show that, if a sequence of amplitudes satisfies a Q-factorized upper bound, these amplitudes possess an exact Q-factorized representation of the kind (2.12) with the new complication that the quantity $\mathcal{H}$ is an operator valued function, the quantities $\mathcal{B}$, $\mathcal{A}$ are vector valued functions and the right hand side of eq. (2.12) has to be interpreted as the matrix element of a product of operators.

The same result holds if we express our amplitudes in terms of invariants or of BCP variables. The special form of this result which we shall use and which is proven in ref.[4] is the following:

Proposition : If the sequence of continuous functions $\mathcal{F}_n(x)$ satisfies the inequality (2.10) with $f_n(x)$ given by eq. (4.6), for every n = 0, 1,...   we can write the operator Q-factorized representation

$$\mathcal{F}_n(g_n, t_n, \cdots, g_0, t_0) = \Big(\mathcal{B}(g_n, t_n), \; \mathcal{H}(t_n, g_{n-1}, t_{n-1}) \cdots$$
$$\cdots \; \mathcal{H}(t_1, g_0, t_0) \, \mathcal{A}(t_0) \Big) , \qquad (5.1)$$

where $\mathcal{A}$ is a vector belonging to a suitable Banach space, $\mathcal{H}$ is an operator in this space and $\mathcal{B}$ is a vector of the dual space. The norms of these quantities satisfy the conditions

$$\| \alpha(t_0) \| \leq \left(a(t_0) \right)^2$$

$$\| \mathcal{H}_k(t_{i+1}, g_i, t_i) \| \leq \left(k(t_{i+1}, \xi_i, t_i) \right)^2$$

$$\| \mathcal{B}(g_n, t_n) \| \leq \left(b(\xi_n, t_n) \right)^2 \tag{5.2}$$

The numerical Q-factorization can be considered as the special case in which the Banach space is one dimensional. For a more detailed discussion, see ref.[4] .

6.   THE CHEW-de TAR MULTIPERIPHERAL INTEGRAL EQUATION

Now we use the Q-factorized representation (5.1) in order to derive the multiperipheral integral equation, following essentially a very elegant procedure introduced by Chew and de Tar[9] . For simplicity, we consider the case of numerical Q-factorization, but it is easy to see that exactly the same procedure holds for operator Q-factorized amplitudes.

We start from the total cross section formula (3.1). The analogous treatment for r particle inclusive distributions can be found in ref.[4] .   If we introduce the new variables $a_0, \ldots, a_{n+1}$ by means of eq. (4.1), the integral (3.1) can be written in the form[4]

$$\sigma_n(s) = \frac{1}{2} (2\pi)^{-3n-2} \left[s(s - 4m^2) \right]^{1/2} \int \mathcal{F}_n(x) \cdot$$

$$\cdot \frac{2\pi^2}{m^2} \delta_+^3 (b_B^{-1} a_{n+1})(4\pi)^{-1} \left(T(m^2, m^2, t_n) \right)^{1/2} d^6 a_{n+1} \cdot$$

$$\cdot \frac{2\pi^2}{|t_n|} \delta_-^3 (g_n)(4\pi)^{-1} \left(T(m^2, t_n, t_{n-1}) \right)^{1/2} dt_n \, d^6 a_n \ldots \tag{6.1}$$

$$\ldots \frac{2\pi^2}{|t_0|} \delta_-^3 (g_0)(4\pi)^{-1} \left(T(m^2, t_0, m^2) \right)^{1/2} dt_0 \, d^6 a_0 \cdot$$

$$\cdot \frac{2\pi^2}{m^2} \delta_+^3 (a_2(-x_0) a_0^{-1} b_A),$$

where the group elements b_A and b_B have the property

$$p_A = L(b_A)(m, 0, 0, 0),$$

$$p_B = L(b_B)(m, 0, 0, 0), \tag{6.2}$$

and the elements g_i are given by eq. (4.4). We have indicated by $\delta_\pm^3(a)$ a measure on SL(2C) defined by

$$\int_{SL(2C)} f(a)\,\delta_\pm^3(a)\,d^6a = \int_{H_\pm} f(h)\,d^3h \qquad (6.3)$$

where H_+ stands for the subgroup SU(2), H_- is the subgroup SU(1,1) and d^3h is the invariant measure on these subgroups.

If we introduce the Q-factorized expression for $\mathcal{F}_n$ and we put

$$\underline{B}(a, t_n) = \frac{1}{8\pi|t_n|}\left[T(m^2, m^2, t_n)\right]^{1/2}\delta_-^3(a_z(-\chi_{m+1})a)\cdot$$
$$\cdot B(a_z(-\chi_{m+1})a, t_n), \qquad (6.4)$$

$$\underline{\mathcal{H}}(t', a, t) = \frac{1}{16\pi^2|t|}\left[T(m^2, t, t')\right]^{1/2}\delta_-^3(a_z(-\chi(t,t'))a)\cdot \qquad (6.5)$$
$$\cdot \mathcal{H}(t', a_z(-\chi(t,t'))a, t),$$

$$\underline{\mathcal{Q}}(t_o) = \frac{1}{4\pi}\left[T(m^2, t_o, m^2)\right]^{1/2}\mathcal{Q}(t_o), \qquad (6.6)$$

the eq. (6.1) takes the form

$$\sigma_m(s) = \frac{1}{2}\left[s(s-4m^2)\right]^{-1/2}\int \frac{2\pi^2}{m^2}\,\delta_+^3(b_B^{-1}a_{m+1})\cdot \qquad (6.7)$$
$$\cdot \underline{B}(a_{m+1}^{-1}a_m, t_m)\,\underline{\mathcal{H}}(t_m, a_m^{-1}a_{m-1}, t_{m-1})\cdots$$
$$\cdots \underline{\mathcal{H}}(t_1, a_1^{-1}a_o, t_o)\,\underline{\mathcal{Q}}(t_o)\,\frac{2\pi^2}{m^2}\,\delta_+^3(a_z(-\chi_o)a_o^{-1}b_A)\cdot$$
$$\cdot d^6a_{m+1}\cdots d^6a_o\,dt_m\cdots dt_o.$$

We remark that the integrals over the group elements $a_{n+1}, \ldots, a_0$ are of the convolution type. Summing over n the expression (6.7), we get the total cross section $\sigma(s)$. It is clear that it can be expressed in terms of the quantity

$$\underline{Q}(t', a, t) = \underline{\mathcal{H}}(t', a, t) + \int \underline{\mathcal{H}}(t', a', t'')\,\underline{\mathcal{H}}(t'', a'^{-1}a, t)\,d^6a'\,dt'' + \ldots \qquad (6.8)$$

As shown in ref.[4] , also the r particle inclusive distributions
can be written as finite sums of integrals containing the expression
(6.8). In conclusion, we see that the most difficult step in the
calculation of the total cross section and of the inclusive distributions is the treatment of the series (6.8). It already contains, as
we shall see, the complex angular momentum singularities which determine the asymptotic behaviour of the measurable quantities.

　　　　We see easily that the series (6.8) is, if it converges, the
perturbative solution of the multiperipheral integral equation

$$\underline{R}(t',a,t) = \underline{\mathcal{K}}(t',a,t) +$$

$$+ \int \underline{\mathcal{K}}(t',a',t'')\, \underline{R}(t'',a'^{-1}a,t)\, d^6a'\, dt''. \tag{6.9}$$

One has to keep in mind that $\underline{R}$ is uniquely defined by eq. (6.8),
while the equation (6.9) could also have different solutions.

7.　　　　DIAGONALIZATION OF CONVOLUTION PRODUCTS ON SL(2C)

　　　　In the eqs. (6.8) and (6.9) we find integrations over the
"radial" variables t, t'...   and over the "angular" variables which
are represented by the group elements a, a'...   . The integrations
over the angular variables, can be interpreted as convolution products.
We remember that the convolution of two functions defined on a group
(e.g. SL(2C))   is defined by

$$(F_1 * F_2)(a) = \int_{SL(2C)} F_1(a')\, F_2(a'^{-1}a)\, d^6a' \tag{7.1}$$

If, as in our case, the quantities F_i are measures ($\ast$) or distributions, eq. (7.1) has to be interpreted as

$$\int (F_1 * F_2)(a)\varphi(a)\, d^6a = \int F_1(a')\, F_2(a'')\, \varphi(a'a'')\, d^6a'\, d^6a'' \tag{7.2}$$

where $\varphi(a)$ is a test function.

($\ast$) We indicate a measure on SL(2C) by the notation F(a) d^6a, where F
is a generalized function, which for the sake of brevity will also be
called a measure.

As well known, the Fourier or Laplace transformations transform the convolution product into an ordinary product. If $\mathcal{D}$ (a) is a continuous representation of SL(2C) by means of bounded operators in a Hilbert or Banach space, if we define the projection integrals

$$\mathcal{D}(F) = \int_{SL(2c)} F(a)\, \mathcal{D}(a)\, d^6 a \, , \tag{7.3}$$

from eq. (7.2) and from the representation property

$$\mathcal{D}(a)\, \mathcal{D}(a') = \mathcal{D}(aa') \tag{7.4}$$

we have

$$\mathcal{D}(F_1 \times F_2) = \mathcal{D}(F_1)\, \mathcal{D}(F_2) \, . \tag{7.5}$$

Applying this procedure to the equations (6.8) or (6.9), we get the corresponding projected partial wave equations, which are in general simpler and have interesting properties. This procedure has been discussed in detail in refs.[9-11] .

The critical point of this treatment is the convergence of the integral (7.3). A sufficient condition for the existence of an integral of an operator valued function of this kind, is

$$\int |F(a)| \, \| \mathcal{D}(a) \| \, d^6 a \, < \, \infty \, . \tag{7.6}$$

In order to discuss this condition, we have to remember some properties of the irreducible representations of SL(2C) and of the corresponding norms. A complete account can be found in refs.[12-14] .

We choose as representation space the space of the measurable functions of the complex variable z such that the expression

$$\|f\|_p = \left[\int |f(z)|^p \left(1 + |z|^2 \right)^{-2-p(\operatorname{Re}\lambda - 1)} d^2 z \right]^{\frac{1}{p}} \, , \quad p \geqslant 1 \tag{7.7}$$

$$d^2 z = d\operatorname{Re}z \, d\operatorname{Im}z \tag{7.8}$$

is finite. With respect to this norm, the representation space is a

Banach space, and even an Hilbert space if p = 2. The different representations are labelled by the complex parameter λ and by the integral or half integral parameter M . The representation operators are defined by

$$[\mathcal{D}^{M\lambda}(a)f](z) = (a_{12}z + a_{22})^{\lambda - M - 1} \cdot$$

$$\cdot (\bar{a}_{12}\bar{z} + \bar{a}_{22})^{\lambda + M - 1} f(z_a), \tag{7.9}$$

where

$$z_a = \frac{a_{11}z + a_{21}}{a_{12}z + a_{22}} . \tag{7.10}$$

From eq. (7.10) we have

$$d^2z = |a_{11} - a_{12}z_a|^{-4} d^2z_a \tag{7.11}$$

and from eq. (7.7) and (7.9) we get

$$\| \mathcal{D}^{M\lambda}(a)f \|_p = \left[\int |f(z_a)|^p \cdot \right.$$

$$\left. \cdot \left(|a_{11} - a_{12}z_a|^2 + |a_{22}z_a - a_{21}|^2 \right)^{-2 - p(\mathcal{R}e\,\lambda - 1)} d^2z_a \right]^{\frac{1}{p}} . \tag{7.12}$$

From the definition of the norm of an operator, we have finally

$$\| \mathcal{D}^{M\lambda}(a) \|_p = \sup_{z} \left(\frac{1 + |z|^2}{|a_{11} - a_{12}z|^2 + |a_{22}z - a_{21}|^2} \right)^{\nu} \tag{7.13}$$

where

$$\nu = \mathcal{R}e\,\lambda - 1 + 2p^{-1} . \tag{7.14}$$

If we use the parametrization

$$a = u_1 \, a_z(\xi)\, u_2 \,, \quad u_1, u_2 \in SU(2) \tag{7.15}$$

we obtain from eq. (7.13) after some calculation

$$\| \mathcal{D}^{M\lambda} \|_p = \exp |\nu \xi| \geqslant 1 . \tag{7.16}$$

We see that the condition (7.6) can be satisfied only if F
represents a bounded measure. If we apply this condition to the
kernel (6.5) at fixed values of t and t' , we see, using eq.(6.3),
that the function $\mathcal{H}_k$ (t', g, t) must be integrable on SU(1,1). This
is ensured by the bound (5.2), with k given by eq.(4.7), only if
$\alpha < -1/2$, which is a condition too restrictive for physical purposes.

Of course, one has to remember that the condition (7.6) is
not necessary for the existence of the integral (7.3). Nevertheless,
one can see that there is no hope to project a physically interesting
kernel on the representations $\mathcal{D}^{M\lambda}$.

Fortunately, the analysis given in refs.[10,11] of the Chew-
de Tar equation (6.9) and the treatment given in refs.[15-17] of the
simpler multiperipheral equations of the ABFST type suggest that there
is some way to avoid these difficulties.

In the following we summarize a systematic approach developed
in ref.[6] . It is based on the remark that the existence of the
projection integral (7.3) is strictly related with the existence of
the convolution integral (7.1). In fact, if the main property of a
Laplace transformation is to transform a convolution product into an
ordinary product, defined only for functions which satisfy some
conditions which ensure the existence of the convolution products.
More exactly, the domain of definition of a Laplace transform should
be a convolution algebra, i.e. a linear space of functions or measures
which is closed under the convolution product. Different "natural"
Laplace transforms should correspond to different convolution algebras.
The bounded measures form a convolution algebra and the projection on
the representations $\mathcal{D}^{M\lambda}$, with p chosen in such a way that
v = 0,   is the corresponding Laplace transform.

Now we note that the convolutions which appear in eq. (6.8)
are well defined for physical reasons even if the kernel $\underline{\mathcal{H}_k}$ is not
a bounded measure. As we shall see in the next Section, the existence
of these convolutions is due to the fact that the kernel $\mathcal{H}_k$, due to
the positivity of the mass and the energy of physical states, vanishes
outside a given subset of SL(2C). Analysing this property we shall
find a convolution algebra to which the kernel belongs and the corresponding natural Laplace transform.

8.   A SEMIGROUP CONTAINED IN SL(2C)

From eq. (6.5) we see that the kernel $\mathcal{H}$ does not vanish only
if its argument a is an element of SL(2C) of the form

$$ a = a_z(x)\, g \quad , \quad x > 0 \quad , \quad g \in SU(1,1) \qquad (8.1) $$

It is easy to show that the elements of the form (8.1) have
the properties

$$ L_{33}(a) \geq 1 \quad , \quad L_{03}(a) \geq 0 \quad , \quad L_{30}(a) \geq 0 \qquad (8.2) $$

We indicate by S the set of the elements of SL(2C) which satisfy
these inequalities. If a and b belong to S, using well known
properties of the Lorentz matrices and the Schwarz inequality, one can
show that

$$ L_{33}(ab) \geq L_{33}(a) L_{33}(b) \geq 1 \quad , $$

$$ L_{30}(ab) \geq L_{33}(a) L_{30}(b) \geq 0 \quad , \qquad (8.3) $$

$$ L_{03}(ab) \geq L_{03}(a) L_{33}(b) \geq 0 . $$

This means that also the product ab belongs to S, and therefore S
is a semigroup.

We see also that the set S° defined by the inequalities

$$ L_{33}(a) > 0 \quad , \quad L_{30}(a) > 0 \quad , \quad L_{03}(a) > 0 \qquad (8.4) $$

is a semigroup. One can show [6] that S° is the interior of S and
that S is the closure of S° .   The elements of the kind (8.1)
belong to S° . Also the elements of the more general form

$$ a = g' a_z(x)\, g \quad , \quad x > 0 \quad , \quad g, g' \in SU(1,1) \qquad (8.5) $$

belong to S° . One can show [6] that all the elements of S° can
be written in the form (8.5).

In conclusion, we have seen that, for fixed values of t' and
t, the kernel $\mathcal{H}$ is a measure with support in S. It is clear from
eq. (7.2) that the convolution of two measures with support in S, if
it exists, has support in S. To transform these convolution products

into ordinary products, one can use a projection on a representation
of the smigroup S.   We remark that a representation of SL(2C) gives
by restriction a representation of S, but not all the representations
of S can be extended to the whole group SL(2C).

9.   LINEAR REPRESENTATIONS OF THE SEMIGROUP S

In this Section we define and study an important class of
Banach space representations of S .   First of all, we consider the
transformation (7.10) and we show that if a $\in$ S   and $|z| \geq 1$,
we have $|z_a| \geq 1$. This is clear in the special case

$$a = a_z(\chi) = \begin{pmatrix} \exp \frac{\chi}{2} & 0 \\ 0 & \exp(-\frac{\chi}{2}) \end{pmatrix}, \quad \chi > 0 \qquad (9.1)$$

In the case

$$a = \begin{pmatrix} \alpha & \beta \\ \bar{\beta} & \bar{\alpha} \end{pmatrix} \in SU(1,1) \qquad (9.2)$$

we see immediately that the transformation (7.10) maps the circle
$|z| = 1$ onto itself and has therefore the required property.
It is easy to extend the proof to any element of S^o , as it can be
decomposed according to eq. (8.5). The general result follows from
a continuity argument.

From the property proven above, it follows that if a $\in$ S the
operator $\mathcal{O}^{M\lambda}$ (a) defined by eq. (7.9) transforms a function f(z)
which vanishes outside the circle $|z| < 1$ into a function which
has the same property. Therefore we can define the operator $\mathcal{B}^{M\lambda}$ (a)
which is the restriction of $\mathcal{O}^{M\lambda}$ (a) to a space of functions
which vanish for $|z| \geq 1$. It can also be defined directly by means
of the formula

$$[\mathcal{B}^{M\lambda}(a) f](z) = (a_{12} z + a_{22})^{\lambda - M - 1} \qquad (9.3)$$

$$\cdot (\bar{a}_{12} \bar{z} + \bar{a}_{22})^{\lambda + M - 1} f(z_a), \quad a \in S$$

In this case it is useful to introduce, instead of the norms
(7.7), the norms

$$\|f\|_{pr} = \left[\int_{|z|<1} |f(z)|^p \left(1 - |z|^2\right)^{-rp} d^2z \right]^{\frac{1}{p}} \tag{9.4}$$

$$p \geq 1 , \quad r \geq 0 .$$

Proceeding as in the derivation of eq. (7.13), we get the operator norms

$$\|B^{M\lambda}(a)\|_{pr} = \sup_{|z| \leq 1} \left[\left(\frac{1 - |z|^2}{|a_{11} - a_{12}z|^2 - |a_{22}z - a_{21}|^2} \right)^2 \cdot |a_{11} - a_{12}z|^{2(r-\nu)} \right] = S_{r\nu}(a) , \quad a \in S \tag{9.5}$$

where ν is given by eq. (7.14). It is easy to show that this expression is finite.

We have obtained in this way a class of representations of S by means of bounded operators in Banach spaces. For p=2 we get Hilbert space representations. Representations which are labelled by the same values of M and λ but correspond to different values of p or r are not essentially different, as they coincide on some dense subspaces of the representation spaces.

From the representation property, it follows that the norms (9.5) satisfy the inequality

$$S_{r\nu}(ab) \leq S_{r\nu}(a) S_{r\nu}(b) . \tag{9.6}$$

The explicit calculation of these norms is very difficult, but it is easy to get the following partial results

$$S_{r\nu}(\mu_z(\mu)) = 1 \tag{9.7}$$

$$S_{r\nu}(a_z(x)) = \exp(-\nu x) , \quad x \geq 0 \tag{9.8}$$

$$\tag{9.9}$$

$$S_{r\nu}(a_x(\xi)) = \exp|(\nu-r)\xi|$$

From eqs. (9.6.)-(9.9) we get

$$S_{\nu\nu}(a) = \exp(-\nu x) , \quad a \in S^{\circ} , \tag{9.10}$$

where a is parametrized as in eq. (8.5). In particular we have

$$\sup_{|z| \leq 1} \left(\frac{1 - |z|^2}{|a_{11} - a_{12} z|^2 - |a_{22} z - a_{21}|^2} \right) = \exp(-\chi) \tag{9.11}$$

and introducing this result into eq. (9.5) we obtain the inequality

$$\mathscr{S}_{\lambda+c,\,\nu+c}(a) \leq e^{-\chi c} \mathscr{S}_{\lambda\nu}(a) \quad , \quad c \geq 0 \; . \tag{9.12}$$

We are mainly interested in group elements of the kind (8.1) and we shall only need the following result, which can be obtained from eq. (9.5) after complicated calculations.

$$\mathscr{S}_{\lambda\nu}\left(a_z(\chi) g\right) \leq e^{(\nu-\lambda)\xi} \, \overline{\Phi}_{\lambda\nu}(\chi), \quad \nu \leq \lambda < 2\nu \tag{9.13}$$

where   g is parametrized as in eq. (4.5) and

$$\overline{\Phi}_{\lambda\nu}(\chi) = O\left(\chi^{2(\nu-\lambda)}\right) \quad , \quad \chi \to 0 \tag{9.14}$$

$$\overline{\Phi}_{\lambda\nu}(\chi) = O\left(e^{-\nu\chi}\right) \quad , \quad \chi \to \infty \tag{9.15}$$

10. <u>CONVOLUTION ALGEBRAS AND LAPLACE TRANSFORM ON   S</u>

Now we consider a measure F(a) with support in S. A sufficient condition for the existence of the projection integral

$$\mathcal{B}^{\mu\lambda}(F) = \int_S \mathcal{B}^{\mu\lambda}(a) F(a) d^6a \tag{10.1}$$

is that the quantity

$$\| F \|_{\lambda\nu} = \int_S |F(a)| \mathscr{S}_{\lambda\nu}(a) d^6a \tag{10.2}$$

is finite.

The measures with support in S such that the norm (10.2) is finite form a Banach space M_{rv} . They form also a convolution algebra and we have the following inequality

$$\| F_1 * F_2 \|_{rv} = \int_S \left| \int_S F_1(a') F_2(a'^{-1} a) \, d^6 a' \right| g_{rv}(a) \, d^6 a \leq \tag{10.3}$$

$$\leq \int_S \int_S | F_1(a') | | F_2(a'') | \, g_{rv}(a') \, g_{rv}(a'') \, d^6 a' d^6 a'' =$$

$$= \| F_1 \|_{rv} \| F_2 \|_{rv} \; .$$

We have used the inequality (9.6). The inequality (10.3) shows that M_{rv} is a Banach algebra with respect to convolution.

The measures belonging to M_{rv} have the fundamental property

$$B^{r\lambda}(F_1 * F_2) = B^{r\lambda}(F_1) \, B^{r\lambda}(F_2) \; . \tag{10.4}$$

In order to show that this formalism is suitable for the treatment of multiperipheral equations, we have to show that the kernel (6.5) for fixed values of t, t' belongs to the algebra M_{rv} with some choice of the parameters r and v. From the inequalities (5.2) and (9.13) we get

$$\| \mathcal{H}(t',t) \|_{rv} \leq \frac{1}{16 \pi^2 |t|} \left[T(m^2, t, t') \right]^{1/2} \cdot \tag{10.5}$$

$$\cdot \int_0^\infty \left(k(t', \xi, t) \right)^2 e^{(v-r)\xi} \Phi_{rv}(\chi(t', t)) \tfrac{1}{2} \sinh \xi \, d\xi \leq$$

$$\leq 2^{-6} \pi^{-2} (2m)^{-4\alpha} (t \, t')^{-\alpha - 1/2} \left[T(m^2, t, t') \right]^{1/2 + 2\alpha} \cdot$$

$$\cdot \, d(t) \, d(t') \, \Phi_{rv}(\chi(t', t)) (r - v - 2\alpha - 1)^{-1} ,$$

if the parameters r and v satisfy the inequalities

$$0 < 2\alpha + 1 < r - v < v \; . \tag{10.6}$$

Comparing with eq. (7.14), we see that, with a proper choice of the parameters p and r, the projected kernel exists in the half plane

$$Re \; \lambda > 2\alpha \; . \tag{10.7}$$

This is just the result expected from previous treatments[9-11, 18]. Moreover, the results proven above show that all the convolutions which appear in eq. (6.8) exist and belong to M_{rv} if r and v satisfy the condition (10.6).

11.   DISCUSSION OF THE MULTIPERIPHERAL EQUATION

In order to complete our analysis, we have to discuss the integration over the radial variables and the convergence of the series which appears in eq.(6.8) . For this purpose it is convenient to consider the kernel $\mathcal{H}$ and the resolvent $\mathcal{R}$ as functions of the variables t, t' with values in the convolution algebra M_{rv}. We always assume that these functions satisfy some measurability condition. Then the existence of the integrals over t" ... is ensured if the following sufficient condition is satisfied

$$N_{\rho v}(\underline{\mathcal{H}}) = \left[\int \| \mathcal{H}(t',t) \|^2_{\rho v} \, dt \, dt' \right]^{1/2} < \infty . \tag{11.1}$$

This expression has all the properties of a norm and satisfies also the inequality

$$N_{\rho v}(\underline{\mathcal{H}}_1 \underline{\mathcal{H}}_2) \leq N_{\rho v}(\underline{\mathcal{H}}_1) N_{\rho v}(\underline{\mathcal{H}}_2), \tag{11.2}$$

where we have used the notation

$$(\underline{\mathcal{H}}_1 \underline{\mathcal{H}}_2)(t',t) = \int \underline{\mathcal{H}}_1(t',t'') * \underline{\mathcal{H}}(t'',t) \, dt''. \tag{11.3}$$

It follows that if

$$N_{\rho v}(\underline{\mathcal{H}}) < 1, \tag{11.4}$$

the series (6.8) converges and we have

$$N_{\rho v}(\underline{\mathcal{R}}) \leq N_{\rho v}(\underline{\mathcal{H}}) \left[1 - N_{\rho v}(\underline{\mathcal{H}}) \right]^{-1}. \tag{11.5}$$

Moreover, $\mathcal{R}$ satisfies the integral equation (6.9).

In order to see if the condition (11.1) is satisfied, we have just to use eqs. (10.5), (9.14) and (9.15). If the function d(t) decreases for large $|t|$ faster than any negative power of $|t|$, one can easily show that the integral (11.1) converges if the condition (10.6) is satisfied. The case in which d(t) decreases as a power of

$|t|$ is treated in ref.[6].

In order to prove eq. (11.4), we remark that from eqs. (6.5), (9.12) and (10.2) it follows

$$\left\| \underline{\mathcal{H}}(t,t') \right\|_{r+c,\,v+c} \leq exp\left(-c\,\chi(t,t')\right) \left\| \underline{\mathcal{H}}(t,t') \right\|_{r\,v}, \quad c>0 \qquad (11.6)$$

and from eq. (11.1) we get

$$\lim_{c\to\infty} N_{r+c,\,v+c}\left(\underline{\mathcal{H}}\right), = 0 \qquad (11.7)$$

if r and v satisfy the condition (10.6).

It follows that, if v is sufficiently large and r is properly chosen, the resolvent $\underline{\mathcal{R}}$ defined by eq. (6.8) has the property

$$N_{r\,v}\left(\underline{\mathcal{R}}\right) < \infty \qquad (11.8)$$

and satisfies the eq. (6.9). Moreover, if Re λ is sufficiently large and p, r are suitably chosen, the Laplace transform $\mathcal{B}^{M\lambda}\left(\underline{\mathcal{R}}(t',t)\right)$ exists and satisfies the "partial wave" equation

$$\mathcal{B}^{M\lambda}\left(\underline{\mathcal{R}}(t,t')\right) = \mathcal{B}^{M\lambda}\left(\underline{\mathcal{H}}(t,t')\right) + \qquad (11.9)$$

$$+ \int \mathcal{B}^{M\lambda}\left(\underline{\mathcal{H}}(t,t'')\right) \mathcal{B}^{M\lambda}\left(\underline{\mathcal{R}}(t'',t)\right) dt'' .$$

One can show that $\mathcal{B}^{M\lambda}(a)$ is an entire operator valued function of λ and that the operators $\mathcal{B}^{M}\left(\underline{\mathcal{H}}(t',t)\right)$ are analytic in λ in the half plane defined by eq. (10.7). The operator $\mathcal{B}^{M\lambda}\left(\underline{\mathcal{R}}(t',t)\right)$ is analytic in the smaller half plane mentioned above, but, if one has a specific model, one can often use eq. (11.9) to get its analytic continuation in a larger region and to study the nature of its singularities.

Summarizing, we have shown that very general assumptions are sufficient for a completely satisfactory definition of the multi-peripheral integral equation and of its partial wave projection. In this procedure, the four-dimensional complex angular momentum λ and the other Lorentz quantum number M appear in a natural way.

In order to complete the treatment summarized above, one has to invert the Laplace transform $\mathcal{B}^{M\lambda}\left(\underline{\mathcal{R}}(t',t)\right)$, to get the resolvent $\underline{\mathcal{R}}$, and to derive the explicit expressions of the asymptotic behaviours of the observable quantities in terms of the singularities in the complex λ plane.

A problem which is equivalent to the inversion of the transformation (10.1) has been treated in ref.[11], while the analogous problem for the three-dimensional case is treated in ref.[19]. We think, however, that some work is still necessary in order to get a completely clear treatment.

REFERENCES

1.   D. Amati, S. Fubini and A.Stanghellini;
        Nuovo Cimento **26**, 896 (1962).

2.   L. Bertocchi, S. Fubini and M. Tonin;
        Nuovo Cimento **25**, 626 (1962).

3.   C. Ceolin, F. Duimio, S. Fubini and R.Stroffolini;
        Nuovo Cimento **26**, 247 (1962).

4.   A. Bassetto, L. Sertorio and M. Toller; CERN Preprint
        TH.1470 (1972), to be published in Nuovo Cimento.

5.   A. Bassetto, L. Sertorio and M. Toller; Lettere al
        Nuovo Cimento **4**, 73 (1972).

6.   S. Ferrara, G. Mattioli, G. Rossi and M. Toller;
        CERN TH Preprint in preparation.

7.   G.F. Chew, M.L. Goldberger and F.E. Low;
        Phys. Rev. Letters **22**, 208 (1969).

8.   N.F. Bali, G.F. Chew and A. Pignotti;
        Phys.Rev. Letters **19**, 614 (1967) and Phys.Rev. **163**, 1572(1967).

9.   G.F. Chew and C. de Tar; Phys. Rev. **180**, 1577 (1969).

10.   A.H. Mueller and I.J. Muzinich; Ann.Phys. (N.Y.) **57**, 20 (1970)

11.   M. Ciafaloni and C. de Tar; Phys.Rev. **D1**, 2917 (1970)-

12.   W. Rühl; "The Lorentz Group and Harmonic Analysis",
        New York, 1970.

13.   I.M. Gel'fand, M.I. Graev and N.Ya. Vilenkin; "Generalized
        Functions", Vol. 5, New York and London, 1966.

14.   M.A. Naimark; "Linear Representations of the Lorentz Group",
        London, 1964.

15.   S. Nussinov and J. Rosner; J.Math.Phys. **7**, 1670 (1966).

16.   H.D.I. Abarbanel and L.M. Saunders;
        Phys. Rev. $\underline{D2}$, 711 (1970), and Ann. Phys. (N.Y.) $\underline{69}$,583(1972)
17.   A. Bassetto and M. Toller; CERN Preprint TH. 1499, (1972)·
18.   S. Pinski and W.I. Weisberger; Phys. Rev. $\underline{D2}$, 1640 and
        2365 (1970).
19.   M. Ciafaloni, C. de Tar and M.N. Misheloff;
        Phys. Rev. $\underline{188}$, 2522 (1969).

LECTURE NOTES ON DUAL AMPLITUDES

I.T. DRUMMOND

CAMBRIDGE, England

1.   **DUAL AMPLITUDES**

The multiparticle dual model is a generalization of Veneziano's Betafunction model for two-particle scattering[1]. In the case of the elastic scattering of spinless particles A of mass m

$$A(p_1) + A(p_2) \rightarrow A(p_3) + A(p_4)$$

the amplitude is

$$T = B(-\alpha(s), -\alpha(t)) + (s \leftrightarrow u) + (t \leftrightarrow u) \tag{1.1}$$

where $\quad \alpha(s) = \alpha_0 + \alpha' s$ with

$$s = (p_1 + p_2)^2, \quad t = (p_3 - p_2)^2, \quad u = (p_1 - p_3)^2 \tag{1.2}$$

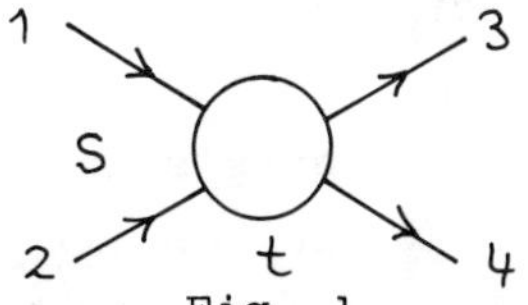

1   3

S   t

2   4

Fig. 1

The important feature of the amplitude is that it represents a solution of Finite Energy Sum Rules with complete saturation of the absorptive part with zero width resonances. This follows because $B(-\alpha(s), -\alpha(t))$ has Regge behaviour in the cut s-plane and can be expressed as a sum over poles (dispersion relation)

$$B(-\alpha(s), -\alpha(t)) = \begin{cases} \displaystyle\sum_0^\infty \frac{1}{n - \alpha(s)} \frac{1}{n!} \frac{\Gamma(n + \alpha(t) + 1)}{\Gamma(\alpha(t) + 1)} \\[2ex] \displaystyle\sum_0^\infty \frac{1}{n - \alpha(t)} \frac{1}{n!} \frac{\Gamma(n + \alpha(s) + 1)}{\Gamma(\alpha(s) + 1)} \end{cases} \tag{1.3}$$

$$B(-\alpha(s),-\alpha(t)) \approx \begin{cases} [-\alpha(s)]^{\alpha(t)}\,\Gamma(-\alpha(t)), & s\to\infty \\ [-\alpha(t)]^{\alpha(s)}\,\Gamma(-\alpha(s)), & t\to\infty \end{cases}$$

$$(1.4)$$

The equivalence of the two expressions in eq. (1.3) is the principle of Duality for scattering amplitudes in the narrow width resonance approximation. We say the two channels (or variables) s and t are dual.

The residue of the nth pole in s is a polynomial of degree n in t and therefore represents a set of intermediate states with angular momenta $j \le n$.

2. <u>THE N-POINT FUNCTION</u>

If we use the notation $\alpha_{12} = \alpha(s)$ the Beta function has the representation

$$B(-\alpha_{12},-\alpha_{23}) = \int_0^1 dI\; u_{12}^{-\alpha_{12}-1}\, u_{23}^{-\alpha_{23}-1} \qquad (2.1)$$

where

$$u_{12} = 1 - u_{23} \tag{2.2}$$

$$dI = du_{12} \text{ or } du_{23} \tag{2.3}$$

The important property of $B(-\alpha_{12}, -\alpha_{23})$ namely that it does not have simultaneous poles in the <u>dual</u> channels is guaranteed by the "<u>duality constraint</u>" on the integration variables u_{12}, u_{23} expressed in eq.(2.3). This observation is the basis of subsequent generalizations.

For example the five point function[2] is obtained by supposing it is a sum of $\frac{1}{2}(5-1)! = 12$ planar pieces in each of which the external particles have a definite order. A typical one is shown in Fig.2. The contribution B_5, is allowed to have

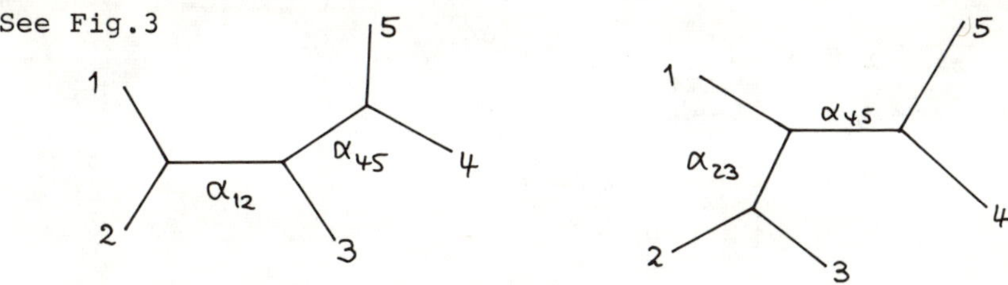

Fig.2

poles in the channels (α_{12}, α_{23}, ..., α_{15}). The channels which can have simultaneous poles can be determined from the Feynman diagrams which can be drawn consistent with the order of the external particles. See Fig.3

Fig. 3

Thus α_{12} and α_{45} may simultaneously have poles. Such channels or variables are called non-dual. A list of duality relations among channels is

$$
\begin{array}{lllll}
\alpha_{12} & \text{dual to} & \alpha_{23} & \text{and} & \alpha_{15} \\
\alpha_{23} & " \quad " & \alpha_{12} & " & \alpha_{34} \\
\alpha_{34} & " \quad " & \alpha_{23} & " & \alpha_{45}
\end{array}
$$

etc.

These variables may not have simultaneous poles. The amplitude corresponding to Fig.2 is

$$
B_5 = \int dI \prod_{i,j} u_{ij}^{-\alpha_{ij}-1} \tag{2.4}
$$

where the integrand has one factor for each channel with poles. The non-trivial part of the problem is to choose the duality constraints on the u_{ij} so that simultaneous poles do not appear in dual channels.

We shall require that when $u_{12} = 0$ <u>both</u> the dual variables u_{23} and u_{15} are unity and when <u>either</u> u_{23} or u_{15} is zero, then u_{12} is unity. This is achieved by imposing the constraint

$$
u_{12} = 1 - u_{23}\,u_{15} \tag{2.5}
$$

The other constraints are

$$u_{23} = 1 - u_{12} u_{34}$$
$$u_{34} = 1 - u_{23} u_{45}$$
$$u_{45} = 1 - u_{34} u_{15} \qquad (2.6)$$
$$u_{15} = 1 - u_{12} u_{45}$$

It is not obvious but turns out to be the case nevertheless, that the constraints eqs. (2.5) and (2.6) are mutually compatible. They leave two degrees of freedom which may be chosen to be the variables associated with the internal lines of any allowed Feynman diagram. A solution to the constraint equations is

$$u_{12} = u \,, \qquad u_{45} = v$$
$$u_{23} = (1-u)/(1-uv)$$
$$u_{34} = (1-v)/(1-uv) \qquad (2.7)$$
$$u_{15} = 1 - uv$$

The integration measure is chosen to make the integrand cyclically symmetric. It is

$$dI = \frac{du_{12} du_{45}}{1 - u_{12} u_{45}} = \frac{du_{23} du_{45}}{1 - u_{23} u_{45}} \qquad (2.8)$$

We can now easily check that the residue of the pole at α_{12} is

$$B_4(-\alpha_{45}, -\alpha_{34}) = \int_0^1 du_{45} \, u_{45}^{-\alpha_{45}-1} \, u_{34}^{-\alpha_{34}-1} \qquad (2.9)$$

Since, when $u_{12} = 0$ the duality constraints imply $u_{23} = u_{15} = 1$ and

$$u_{34} = 1 - u_{45} \qquad (2.10)$$

See Fig.4

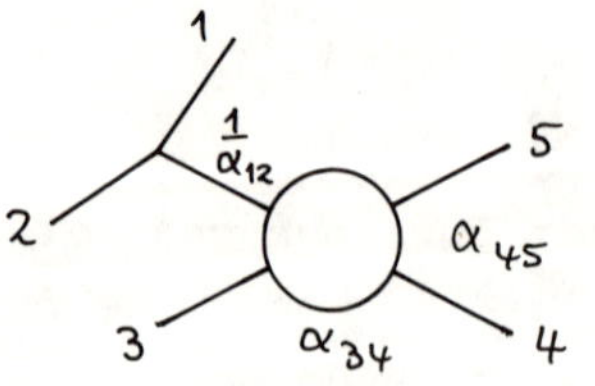

Fig. 4

The theory therefore has a welcome internal consistency.

The generalization to the N-body case[3] is "straightforward". Once again we assume there are $\frac{1}{2}(N-1)!$ different contributions one for each order of the external lines. Fig. 5 shows a typical one

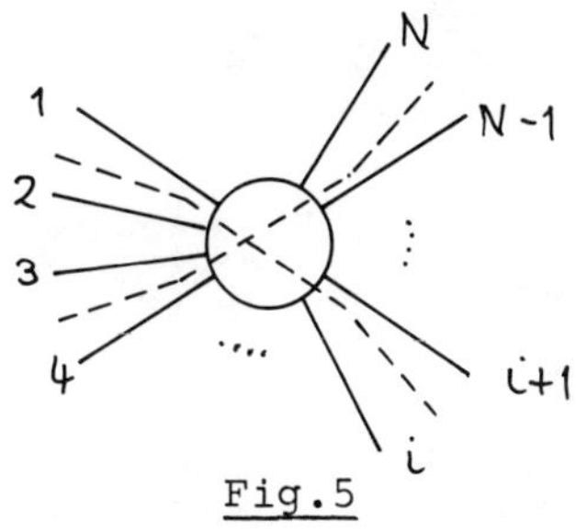

<u>Fig.5</u>

The channels which may have poles correspond to <u>partitions</u> of the lines
into two groups. Two partitions are dual if the lines determining them
(see Fig. 5) cross. It is easy to verify that two channels dual in
this sense cannot have simultaneous poles.

$$T = \int dI \; \widetilde{\prod_{P}} \; u_P^{-\alpha_P - 1} \tag{2.11}$$

where the product is overall partitions P. The duality constraints
are

$$u_P = 1 - \widetilde{\prod_{\bar{P}}} \; u_{\bar{P}} \tag{2.12}$$

where $\bar{P}$ is any partition dual to P. Although it is not obvious, these constraints leave $N-3$ independent variables which may
be taken to be those associated with the internal lines of the multiperipheral diagram of Fig. 6.

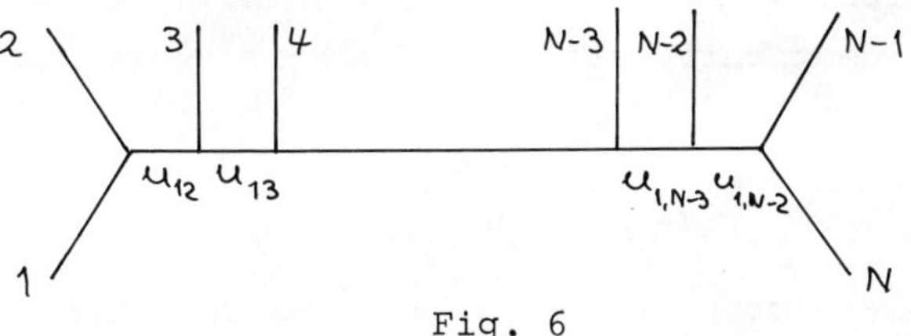

<u>Fig. 6</u>

If we denote a partition by (i,j) where i and j are the end lines in
the group not containing N then the independent variables are u_{ij}
($i = 2, \dots N-2$). Cyclic symmetry requires

$$dI = \widetilde{\prod_{i=2}^{N-2}} du_{1i} \Big/ \widetilde{\prod_{2}^{N-3}} (1 - u_{1i} u_{1\,i+1}) \tag{2.13}$$

The constraints (2.12) are solved by the relations

$$u_{ij} = \frac{(1 - x_{ij})(1 - x_{i-1,j+1})}{(1 - x_{i-1,j})(1 - x_{i,j+1})} \tag{2.14}$$

where $x_{ij} = u_{1i} u_{1,i+1} \cdots u_{1,j-1}$.

Finally if we express the amplitude entirely in terms of the multi-peripheral set of variables we find

$$A_N = \int \prod_{i=2}^{N-2} du_{1i} \, u_{1i}^{-\alpha_{1i} - 1} \, (1 - u_{1i})^{\alpha_0 - 1}$$
$$\times \prod_{2 \leq i < j \leq N-1} (1 - x_{ij})^{-2 p_i p_j} \tag{2.15}$$

To achieve precisely this result we need to have all internal trajectories identical and the external particle to lie on the leading trajectory.

3. <u>KOBA-NIELSEN VARIABLES</u>

The difficulty of the approach used so far is the large number of variables (Chan variables) involved. Koba and Nielsen[4] found a way round this difficulty. They parametrized the Chan variables u_{ij} in terms of N variables ($z_1 \ldots z_N$) in such a way that the duality constraints are automatically satisfied. Their expression is

$$u_{ij} = \frac{(z_j - z_i)(z_{j+1} - z_{i-1})}{(z_j - z_{i-1})(z_{j+1} - z_i)} \tag{3.1}$$

The Chan variables are cross-ratios in the K-N variables. When these latter variables all lie on a circle in the complex plane the u_{ij} are real. If they are positioned round the circle in the order of the external particles ($z_1 \rightarrow z_2 \rightarrow \ldots \rightarrow z_N \rightarrow z_1$) then

$$0 \leq u_{ij} \leq 1$$

for each u_{ij} .

For the 4-point function

$$\mu_{12} = \frac{(z_2 - z_1)(z_3 - z_4)}{(z_2 - z_4)(z_3 - z_1)}$$

$$\mu_{23} = \frac{(z_3 - z_2)(z_4 - z_1)}{(z_3 - z_1)(z_4 - z_2)} \tag{3.2}$$

It follows immediately that the duality constraint $1 - \mu_{23} = \mu_{12}$, is satisfied. Again, it is quite easy to check the constraints for the 5-point function.

The particular circle we choose in the complex plane is arbitrary. The reason is that our two circles can be mapped into one another by a Möbius transformation and the cross-ratios are <u>invariant</u> under this proceedure. A Möbius transformation is of the form

$$z \rightarrow \xi = Mz = \frac{\alpha z + \beta}{\gamma z + \delta}$$

We can easily check that

$$\xi_1 - \xi_2 = \frac{(z_1 - z_2)\Delta}{(\gamma z_1 + \delta)(\gamma z_2 + \delta)} \tag{3.3}$$

where $\Delta = \alpha\delta - \beta\gamma$. The invariance of the cross-ratios follows immediately. In passing we note that eq. (3.3) implies that

$$d\xi = \frac{dz\,\Delta}{(\gamma z + \delta)^2} \tag{3.4}$$

However, once we have fixed on a particular circle there is still a three parameter subgroup of Möbius transformations which maps the circle into itself. This is most immediately evident if we choose the real axis as the $K - N$ circle in which case the relevant subgroup is that of real Möbius transformations.

These transformations combine in the following way. If

$$Mz = \frac{\alpha z + \beta}{\gamma z + \delta}$$

$$Nz = \frac{az + b}{cz + d} \tag{3.5}$$

then

$$MNz = \frac{Az + B}{Cz + D} \tag{3.6}$$

where

$$\begin{pmatrix} A & B \\ C & D \end{pmatrix} = \begin{pmatrix} \alpha & \beta \\ \gamma & \delta \end{pmatrix}\begin{pmatrix} a & b \\ c & d \end{pmatrix} . \tag{3.7}$$

Thus the group of real Möbius transformations is SL(2R). This invariance group plays an important role in the theory. In the same way, if we had chosen the unit circle as the K - N circle, the relevant subgroup would have been SU(1,1).

Let $(x_1 \ldots \ x_N)$ be the N K-N variables on the real axis. The invariance of the Chan variables under the three parameter group of real Möbius transformations allows us to fix <u>three</u> of the K - N variables at arbitrary points leaving correctly just N - 3 degrees of freedom for the Chan variables. We will choose $(x_1, x_{N-1}, x_N) = (0, 1, \infty)$. In that case

$$u_{1i} = x_i / x_{i+1} \tag{3.8}$$

which can be inverted to yield

$$x_i = u_{1i}\, u_{1,i+1} \cdots u_{1,N-2} \tag{3.9}$$

For the other u_{ij} it is easily seen that

$$u_{ij} = \frac{(1 - x_i/x_j)(1 - x_{i-1}/x_{j+1})}{(1 - x_i/x_{j+1})(1 - x_{i-1}/x_j)} \tag{3.10}$$

Since $x_i/x_j = x_{ij}$ as defined in section 2 we see that eq.(3.10) reduces to eq.(2.12). Hence we have shown the duality constraints are indeed satisfied by the formalism.

We want now to write the dual amplitude in the form

$$A_N = \int dI_N(x)\, F_N(x, p) \tag{3.11}$$

where both the measure dI_N and the integrand F_N are invariant under Möbius transformations.

$$\begin{aligned} dI_N(Mx) &= dI_N(x) \\ F_N(Mx, p) &= F_N(x, p) \end{aligned} \tag{3.12}$$

The reason is that both were originally expressed in terms of Chan variables. We would like to treat the particles in a cyclically symmetric way. A possible measure which satisfied (3.12) is (see eqs.(3.3) and (3.4)

$$\prod_{i=1}^{N} \frac{dx_i}{|x_i - x_{i+1}|} \, , \qquad (x_{N+1} = x_1) \tag{3.13}$$

However, it has the disadvantage of that it involves too many integration variables, N instead of N-3.   It turns out that it would also lead to an infinite result. The way to overcome both difficulties is to devide out of the measure the invariant volume element dR of the real Möbius group under which the measure is invariant. The measure we want is

$$dI_N(x) \; = \; \left[\prod_{i=1}^{N} \frac{dx_i}{|x_i - x_{i+1}|} \right] \Big/ dR \tag{3.14}$$

We can understand the meaning of eq.(3.14) and the definition of $dI_{N'}$ in the following way. Express the general set of K-N variables $\{x\}$ in terms of a basic set $\{\bar{x}\}$ in which three (x_a, x_b, x_c) say, are <u>fixed</u> by the formula

$$x \; = \; R \, \bar{x} \tag{3.15}$$

The three parameters in R bridge the gap between the numbers of degrees of freedom in the two groups. A general theorem tells us that when we calculate the Jacobian of the transformation $\{x\} \leftrightarrow \{R, \bar{x}\}$ we find

$$\prod_{i=1}^{N} \frac{dx_i}{|x_i - x_{i+1}|} \; = \; dR \, dI_N(\bar{x}) \tag{3.16}$$

This calculation defines $dI_N(x)$.

A practical method of calculating is first to note that if $(x_a, x_b, x_c) \; = \; R \, (\bar{x}_a, \bar{x}_b, \bar{x}_c)$ then

$$dR \; = \; \frac{dx_a \, dx_b \, dx_c}{|(x_a - x_b)(x_b - x_c)(x_c - x_a)|} \tag{3.17}$$

and then use eq.(3.17) in eq.(3.14) treating all factors including differentials as algebraic quantities. Afterwards x_a, x_b and x_c are fixed at the chosen values.

It turns out that the integrand we want is

$$F_N(x,p) = \widetilde{\prod_i} |x_i - x_{i+1}|^{\alpha_o} \widetilde{\prod_{1 \leq i < j \leq N}} |x_i - x_j|^{-2p_i \cdot p_j} \tag{3.18}$$

It is easily checked (remember eqs. (3.3)) that

$$F_N(Rx,p) = F_N(x,p) \widetilde{\prod_{i=1}^{N}} \left| \frac{\Delta}{(\gamma x_i + \delta)^2} \right|^{p_i \sum_{j=1}^{N} p_j} \tag{3.19}$$

Since $\sum_j p_j = 0$ (momentum conservation) F_N is Möbius invariant.

If we calculate by letting $(x_1, x_{N-1}, x_N) \to (0, 1, \infty)$ then

$$dI_N(x) \to \widetilde{\prod_2^{N-2}} dx_i \bigg/ \widetilde{\prod_1^{N-2}} |x_i - x_{i+1}|$$

$$F_N(x,p) \to \widetilde{\prod_{i=1}^{N-1}} |x_i - x_{i+1}|^{\alpha_o} \widetilde{\prod_{1 \leq i < j \leq N-1}} |x_i - x_{i+1}|^{-2p_i \cdot p_j} \tag{3.20}$$

Hence our general formula is

$$B_N = \int \widetilde{\prod_{i=2}^{N-2}} dx_i \, \widetilde{\prod_1^{N-1}} |x_i - x_{i+1}|^{\alpha_o - 1} \widetilde{\prod_{1 \leq i < j \leq N-1}} |x_i - x_j|^{-2p_i \cdot p_j} \tag{3.21}$$

The next step is to use eqs. (3.8) and (3.9) in which case it easily follows that eq. (3.21) is the same as eq. (2.13). From now on we shall consider the special case $\alpha_o = 1$ which is especially simple and interesting.

4. <u>FACTORIZATION</u>

We will approach factorization indirectly by first of all expressing the Veneziano amplitude in terms of operators[5,6,7] and then exhibiting the scalar product structure of the pole residues (that is factorization) directly.

Introduce $\{a_{n\mu}\} \longrightarrow [a_{n\mu}, a_{m\nu}^{\dagger}] = -\delta_{mn}\, g_{\mu\nu}$

$$m, n = 1, 2, 3, \ldots$$

$$g_{\mu\nu} = \begin{pmatrix} 1 & & & \\ & -1 & & 0 \\ & & -1 & \\ 0 & & & -1 \end{pmatrix} \tag{4.1}$$

Next we introduce a generalized "position" operator [7]

$$Q_{\mu}(x) = Q_{\mu}^{\circ}(x) + Q_{\mu}^{(+)}(x) + Q_{\mu}^{(-)}(x) \tag{4.2}$$

where

$$Q_{\mu}^{(+)}(x) = \sum_{1}^{\infty} \sqrt{\frac{2}{n}}\; a_{n\mu}^{\dagger}\, x^{n}$$

$$Q_{\mu}^{(-)}(x) = \sum_{1}^{\infty} \sqrt{\frac{2}{n}}\; a_{n\mu}\, x^{-n}$$

$$Q_{\mu}^{\circ}(x) = q_{\mu} + 2i\, p_{\mu}\, \ln x \tag{4.3}$$

where

$$[q_{\mu}, p_{\nu}] = i g_{\mu\nu} \qquad \left(p_{\mu} = -i \frac{\partial}{\partial q^{\mu}} \right) \tag{4.4}$$

The zero mode operators q, p are like the overall position and total momentum of the system. We also introduce a ground state $|0\rangle$

$$a_{n\mu}|0\rangle = 0$$

$$p_{\mu}|0\rangle = 0 \tag{4.5}$$

The Hilbert space we are interested in is then spanned by the occupation number states

then

$$|\lambda, k\rangle = \widetilde{\prod_{n}} \frac{(a_{n}^{\dagger})^{\ell_n}}{\sqrt{\ell_n!}}\; e^{ikq}\, |0\rangle \tag{4.6}$$

$$p\,|\lambda, k\rangle = k\,|\lambda, k\rangle$$

and if

$$H = -\sum_{1}^{\infty} n\, a_{n}^{\dagger} a_{n}$$

(generalized number operator)

then

$$H|\lambda, k\rangle = \left(\sum_{1}^{\infty} n\, \ell_n \right) |\lambda, k\rangle \;.$$

Now consider the commutator

$$\left[Q_\mu^{(-)}(x), Q_\nu^{(+)}(y) \right]$$

$$= \; 2 \sum_{mn} \frac{1}{\sqrt{mn}} \, x^{-m} y^n \, \left[a_{m\mu}, a_{n\nu}^\dagger \right] \tag{4.7}$$

$$= \; -2 g_{\mu\nu} \sum_{1}^{\infty} \frac{1}{n} \left(\frac{y}{x} \right)^n$$

$$= \; 2 g_{\mu\nu} \, \ln \left(1 - \frac{y}{x} \right)$$

Furthermore

$$e^{i k \cdot Q^{(-)}(x)} \; e^{i k' \cdot Q^{(+)}(y)}$$

$$= \; e^{i k' \cdot Q^{(+)}(y)} \; e^{i k \cdot Q^{(-)}(x)} \; e^{\left[i k Q^{(-)}(x), \, i k' Q^{(+)}(y) \right]}$$

$$= \; e^{i k' \cdot Q^{(+)}(y)} \; e^{i k \cdot Q^{(-)}(x)} \; e^{-2 k k' \ln \left(1 - \frac{y}{x} \right)}$$

$$= \; e^{i k' \cdot Q^{(+)}(y)} \; e^{i k \cdot Q^{(-)}(x)} \; \left(1 - \frac{y}{x} \right)^{-2 k \cdot k'} \tag{4.8}$$

The formula for the Veneziano integrand is then

$$F_N(x,p) = \langle 0 | \, \frac{V(k_N, x_N)}{x_N} \, \cdots \, \frac{V(k_1, x_1)}{x_1} \, | 0 \rangle \tag{4.9}$$

where

$$V(k_i, x_i) = \; : e^{i k_i \, Q(x_i)} :$$

$$= \; e^{i k_i q - 2 k_i \cdot p \, \ln x_i} \; e^{i k_i \cdot Q^{(+)}(x_i)} \; e^{i k_i \cdot Q^{(-)}(x_i)} \tag{4.10}$$

We evaluate the expression (4.9) by pulling all the creation operators
to the left and all the annihilation operators to the right. Whenever
we interchange two factors

$$e^{i k_j \cdot Q^{(-)}(x_j)} \; e^{i k_i \cdot Q^{(+)}(x_i)}$$

we throw out a C-number factor $(1 - x_i / x_j)^{-2 k_i \cdot k_j}$. Finally the
vacuum annihilates the a-operators. We obtain

$$F_N = \langle 0| \prod_i e^{ik_i \cdot Q^{(0)}(x_i)} |0\rangle \prod_i \frac{1}{x_i} \prod_{1 \le i < j \le N} (1 - x_i/x_j)^{-2k_i k_j} \tag{4.11}$$

When we evaluate the zero modes and remember to use momentum conservation and $k_i^2 = -1$, which arises because $\alpha_0 = 1$, we obtain a factor

$$\prod_{i=1}^{N} x_i^{-k_i^2} \prod_{i<j} x_j^{-2k_i k_j} \tag{4.12}$$

The result is

$$F_N = \prod_{i<j} |x_i - x_j|^{-2k_i k_j}$$

Our expression for the Veneziano amplitude in operator form is

$$A_N = \int \frac{1}{dR} \prod_i dx_i \; \langle 0| \frac{V(k_N, x_N)}{x_N} \cdots \frac{V(k_1, x_1)}{x_1} |0\rangle \tag{4.13}$$

We exploit the Möbius invariance of the integrand to set $x_1 \to 0$, $x_{N-1} \to 1, x_N \to \infty$. It follows that

$$dR \approx \frac{dx_1 \, dx_{N-1} \, dx_N}{x_N^2} \tag{4.14}$$

Now

$$\frac{V(k_1, x_1)}{x_1} |0\rangle = e^{ik_1 \cdot Q^{(+)}(x_1)} e^{ik_1 q} |0\rangle$$
$$\longrightarrow |k_1\rangle \tag{4.15}$$

$$\langle 0| \frac{V(k_N, x_N)}{x_N} \approx \frac{1}{x_N^2} \langle 0| e^{ik_N q} e^{ik_N Q^{(-)}(x_N)}$$
$$\longrightarrow \frac{1}{x_N^2} \langle -k_N|$$

$$A_N = \int \prod_2^{N-2} dx_i \; \langle -k_N| \frac{V(k_{N-1}, x_{N-1})}{x_{N-1}} \cdots \frac{V(k_2, x_2)}{x_2} |k_1\rangle$$

$$\tag{4.16}$$

If we remember the properties of $H = -\sum n a_n^{\dagger} a_n$ we know that

$$x^{H} a_n^{\dagger} x^{-H} = x^{n} a_n^{\dagger} \tag{4.17}$$

$$x^{H} a_n x^{-H} = x^{-n} a_n$$

so, taking care of the zero modes as well we have

$$V(k,x) = x^{H-p^2} V(k,1) \, x^{-H+p^2} \tag{4.18}$$

If we use the notation $V(k) = V(k,1)$, $L_o = H - p^2$ and put $t_i = x_i / x_{i+1}$, we can express the amplitude A_N as

$$A_N = \int \prod_i^{N-2} \frac{dt_i}{t_i} \langle -k_N | V(k_{N-1}) \, t_{N-2}^{L_o-1} \, V(k_{N-2}) \, t_{N-3}^{L_o-1} \cdots$$
$$\cdots t_2^{L_o-1} \, V(k_2) | k_1 \rangle \tag{4.19}$$

Thus using the formula

$$\frac{1}{L_o - 1} = \int_0^1 dt \, t^{L_o - 2} \tag{4.20}$$

we find finally

$$A_N = \langle -k_N | V(k_{N-1}) \frac{1}{L_o-1} V(k_{N-2}) \frac{1}{L_o-1} \cdots \frac{1}{L_o-1} V(k_2) | k_1 \rangle \tag{4.21}$$

The eigenstates of the propagator $\frac{1}{L_o-1}$ are the states $|\lambda, k\rangle$. It can be spectrally decomposed as

$$\frac{1}{L_o-1} \approx \sum_{\lambda} \frac{|\lambda, k\rangle \langle -k, \lambda|}{\ell_{\lambda} - k^2 - 1} \tag{4.22}$$

where k is the momentum flowing through the given propagator. The amplitude can be written therefore as

$$A_N = \sum_{\lambda} \frac{\langle \Psi' | \lambda k \rangle \langle -k \lambda | \Psi \rangle}{\ell_{\lambda} - k^2 - 1} \tag{4.23}$$

where

$$|\Psi\rangle = V(k_i) \frac{1}{L_o-1} \cdots V(k_2) | k_1 \rangle \tag{4.24}$$

and similarly with $\langle \psi' |$. The poles in the amplitude are now evident as is the factorization of the residues. For any given ℓ_λ there corresponds only a finite number of states. The amplitude for coupling an excited state $| \lambda, k \rangle$ to i ground state particles is one factor in the residue that is

$$\langle -k, \lambda | \, V(k_i) \, \frac{1}{L_0 - 1} \cdots \, V(k_2) | k_1 \rangle \qquad (4.25)$$

This amplitude is a superposition of tensor functions of the $\{ k_i \}$ multiplied by invariant amplitudes which are B_{i+1} functions with displaced trajectories.

5. GAUGE CONDITIONS AND PHYSICAL STATES

As a preparation for what follows let us return to eq.(4.13)

$$A_N = \int \frac{1}{dR} \prod_{i=1}^{\infty} dx_i \, \langle 0 | \prod_i \frac{V(k_i, x_i)}{x_i} | 0 \rangle \qquad (5.1)$$

Although we know by calculating, that the product of integrand and measure in eq. (5.1) is Möbius invariant nevertheless it would be gratifying to show that it is, directly. We will then see some of the general machinery of dual models in action.

We want a method of performing Möbius transformations[7] on $\frac{V(k,x)}{x}$. In fact we know already how to send $x \longrightarrow \lambda x$

$$\lambda^{L_0} \, \frac{V(k,x)}{x} \, \lambda^{-L_0} = \lambda \left[\frac{V(k, \lambda x)}{\lambda x} \right] \qquad (5.2)$$

So L_o is the infinitesimal generator of expansions. The other generators of independent Möbius transformations are

$$L_1 = -i\sqrt{2}\, p \cdot a_1 - \sum \sqrt{n(n+1)} \, a_n^\dagger \cdot a_{n+1}$$

$$L_{-1} = +i\sqrt{2}\, p \cdot a_1^\dagger - \sum \sqrt{n(n+1)} \, a_{n+1}^\dagger a_n \qquad (5.3)$$

They are called the Gliozzi generators[8]. We easily check the following commutation relations

$$[L_1, L_{-1}] = 2L_0$$

$$[L_{\pm 1}, L_o] = \pm L_{\pm 1} \qquad (5.4)$$

These are just the commutation relations of the generators of the real Möbius group which is rather like angular momentum. We also have

$$[L_{-1}, a_n] = -\sqrt{n(n+1)}\, a_{n+1}$$
$$[L_{-1}, a_n^\dagger] = \sqrt{n(n-1)}\, a_{n-1}^\dagger$$
$$[L_{+1}, a_n] = -\sqrt{n(n-1)}\, a_{n-1}$$
$$[L_{+1}, a_n^\dagger] = \sqrt{n(n+1)}\, a_{n+1}^\dagger$$

(5.5)

Using these commutation relations we can easily evaluate the effect of these generators on $Q_\mu(x)$.

$$e^{\alpha L_{-1}} Q(x)\, e^{-\alpha L_{-1}} = \sum_0^\infty \frac{\alpha^P}{P!}\, [L_{-1}, Q(x)]_{(P)} = Q(x+\alpha)$$

(5.6)

$$e^{\alpha L_{+1}} Q(x)\, e^{-\alpha L_{+1}} = Q\left(\frac{x}{1-\alpha x}\right)$$

The infinitesimal form of these equations is

$$[L_n, Q(x)] = x^{n+1}\frac{d}{dx} Q(x)$$

(5.7)

with $n = 0, \pm 1$. The vertex $V(k, x)$ is apart from normal ordering simply a function of $Q(x)$. The corresponding commutation relations are

$$[L_n, V(k,x)] = x^n \left(x\frac{d}{dx} - n k^2\right) V(k,x)$$

$$= x^n \left(x\frac{d}{dx} + n\right) V(k,x)$$

(5.8)

The extra term comes because of the normal ordering in V. When eqs. (5.8) are integrated we find[7]

$$U(M)\, \frac{V(k,x)}{x}\, U^{-1}(M) = \left[\frac{(\gamma x + \delta)^2}{\Delta}\right]^{-1} \frac{V(k, Mx)}{Mx}$$

(5.9)

where $U(M) = e^{\vec{n}\cdot\vec{L}}$ is operator corresponding to the Möbius transformation M. Now

$$\widetilde{\prod_i} dx_i = \widetilde{\prod_i} \left[d(Mx_i)\frac{(\gamma x_i + \delta)^2}{\Delta}\right]$$

(5.10)

from which we can conclude that (integrand) x (measure) is Möbius invariant simply in terms of the operator formalism.

We shall need a generalization of the Gliozzi generators, they are for $n \geq 1$,

$$L_n = -i\sqrt{2n}\, p\, a_n - \sum_{\tau=1}^{\infty} \sqrt{\tau(\tau+n)}\, a_\tau^\dagger a_{\tau+n}$$

$$-\frac{1}{2}\sum_1^{n-1} \sqrt{\tau(n-\tau)}\, a_\tau a_{n-\tau}$$

$$L_{-n} = L_n^\dagger \tag{5.11}$$

They are called Virasoro generators and obey the algebra[9]

$$[L_m, L_n] = (m-n)L_{m+n} + \frac{d}{12}\delta_{m,-n}\, m(m^2-1) \tag{5.12}$$

where d is the dimensions of space time. They commute with the vertex as in eq. (5.8).

We shall use these operators to show that the physical states appearing in the factorization of pole residues span only a subspace of occupation number space. At the same time the ghosts in the model arising from the zero components of the creation operators are eliminated [10,11,12]. Consider

$$[L_0 - L_n, V(k,x)] = \left[(1-x)\times\frac{d}{dx} - n\right] V(k,x)$$

that is $x = 1$,

$$[L_0 - L_n, V(k)] = -n V(k) \tag{5.13}$$

If we put $W_n = [L_0 - L_n + n - 1]$ then

$$W_n V(k) = V(k)[L_0 - 1 - L_n]$$

$$W_n V(k)\frac{1}{L_0-1} = V(k)\left[1 - L_n\frac{1}{L_0-1}\right] = V(k)\left[1 - \frac{1}{L_0+n-1}L_n\right]$$

that is

$$W_n V(k)\frac{1}{L_0-1} = V(k)\frac{1}{L_0+n-1}W_n \tag{5.14}$$

Eq.(5.14) shows that

$$W_n|\Psi\rangle = V(k_i)\frac{1}{L_0+n-1}\cdots\frac{1}{L_0+n-1}V(k_2)(L_0-1-L_n)|k_1\rangle,$$

and for $n \geq 1$, $L_n|k_1\rangle = (L_0-1)|k_1\rangle = 0$. Hence

$$W_n|\Psi\rangle = 0 \tag{5.15}$$

It follows that states of the form $\langle s| = \langle \chi |W_n$ do not couple to ground state scalars[10],

$$\langle s|\psi\rangle = \langle \chi|W_n|\psi\rangle = 0$$

We call them __spurious states__. At a given pole residue we have the mass-shell condition

$$\langle s|(L_0-1) = 0$$

that is

$$\langle \chi|W_n(L_0-1) = 0$$

Using the commutation relations of the L_n's we see that this requires

$$\langle \chi|(L_0+n-1) = 0$$

Consequently __at a pole residue__ the spurious states can be written

$$\langle s| = \langle \chi|L_n \tag{5.16}$$

The physical states $|\phi\rangle$ lie in the subspace orthogonal to the spurious states, that is

$$\langle s|\phi\rangle = 0 \tag{5.17}$$

or

$$\langle \chi|L_n|\phi\rangle = 0$$

which careful consideration shows to have the consequence

$$L_n|\phi\rangle = 0 \tag{5.18}$$

These together with

$$(L_0-1)|\phi\rangle = 0$$

are the physical state constraints or gauge conditions. In fact we only need eq.(5.18) for $n = 1$ and 2, the remainder are implied as a consequence of eq.(5.12). Consider now the 1st excited level.

The general state has the form

$$|\phi\rangle = \alpha\, a_1^{\dagger}|k\rangle \tag{5.19}$$

with $k^2 = 0$.
We see easily $L_2|\phi\rangle$ vanishes automatically and

$$k \cdot \alpha = 0$$

follows from

$$L_1 |\phi\rangle = -i\sqrt{2}\, k \cdot a_1\, \alpha \cdot a_1^\dagger |k\rangle = 0 \qquad (5.20)$$

The solutions are $\alpha = \alpha_{\text{transverse}}$, $\alpha = k$.

The transverse excitations are physical <u>positive norm states.</u> The longitudinal excitation appears to be physical but has zero norm! Fortunately it is spurious because

$$|\phi\rangle = k \cdot a_1^\dagger |k\rangle \propto L_{-1}|k\rangle \qquad (5.21)$$

We see that our equations were not strong enough to exclude <u>zero norm spurious states</u>. The important point however is that all the other physical states have <u>positive norm.</u>

At the second excited level the general state has the form

$$|\phi\rangle = (\alpha^{\mu\nu} a_{1\mu}^\dagger a_{1\nu}^\dagger + \beta^\mu a_{2\mu}^\dagger)|k\rangle$$

with $k^2 = 1$. The effective parts of the gauge operators acting on this state are

$$L_1 \approx -i\sqrt{2}\, p \cdot a_1 - \sqrt{2}\, a_1^\dagger \cdot a_2$$
$$L_2 \approx -i\, 2 p \cdot a_2 - \tfrac{1}{2}\, a_1 \cdot a_1 \qquad (5.22)$$

Evaluating the gauge conditions we find

$$\beta_\mu = -2i\, \alpha_{\mu\nu} k^\nu$$
$$\mathrm{tr}\,\alpha = -2i\, k \cdot \beta \qquad (5.23)$$

If we take $k = (1,0,0,0)$ then

$$\beta_\mu = -2i\, \alpha_{0\mu}$$
$$\alpha_{00} = \tfrac{1}{5}\sum_{i=1}^{3} \alpha_{ii} \qquad (5.24)$$

We find then the general state satisfying the gauge condition is

$$|\phi\rangle = \sum_{ij}(\alpha_{ij} - \tfrac{1}{3}\delta_{ij}\sum_k \alpha_{kk})\, a_i^\dagger a_j^\dagger |k\rangle$$
$$+ \sum_k \alpha_{kk}\left(\tfrac{1}{3}\sum_i a_i^\dagger a_i^\dagger + \tfrac{1}{5}(a_{10}^{\dagger\,2} - 2i\, a_{20}^\dagger)\right)|k\rangle$$
$$+ \sum_i \beta_i (a_{2i}^\dagger + i\, a_{1i}^\dagger a_{10}^\dagger)|k\rangle$$
$$(5.25)$$

with β_i and α_{ij} arbitrary.

The last contribution is of zero norm.

Again it is spurious being proportional to $L_{-1} \sum_i \beta_i a_{1i}^{\dagger} |k\rangle$.

The first two contributions of spin 2 and 0 respectively have both positive norm[10].

It turns out though it is quite a long story that the gauge conditions do exclude all negative norm states from the physical space provided the space time dimension $d \leqq 26$ [12].

6.   THE NEVEU SCHWARZ MODEL

We modify the theory by introducing new <u>anticommuting</u> operators $\{b_{m\mu}\}$ in addition to the $\{a_{n\mu}\}$. We have

$$\{b_{m\mu}, b_{n\nu}^{\dagger}\} = -\delta_{mn} g_{\mu\nu} \tag{6.1}$$

For notational convenience $m, n = \frac{1}{2}, \frac{3}{2}, \ldots$ ... and we put $b_{-m} = b_m^{\dagger}$.
Now we introduce $H_\mu(z)$ analoguous to $Q_\mu(z)$

$$H_\mu(z) = \sum_{-\infty}^{+\infty} b_{m\mu} z^{-m} \tag{6.2}$$

The Virasoro generators become

$$L_n = L_n^{(a)} + L_n^{(b)} \tag{6.3}$$

where $L_n^{(a)}$ are the ones previously discussed and

$$L_n^{(b)} = :\frac{1}{2} \sum_{-\infty}^{+\infty} (r - \frac{n}{2}) b_r b_{n-r}: \tag{6.4}$$

For example

$$L_o = -\sum_1^{\infty} r b_r^{\dagger} b_r - \sum_1^{\infty} n a_n^{\dagger} a_n - p^2 \tag{6.5}$$

Once again our propagator turns out to be $\dfrac{1}{L_o - 1}$.

We see that the factorizing states lie in a space spanned by occupation number states of $a^{\dagger}$ and $b^{\dagger}$ excitations. Since the b excitations yield $\frac{1}{2}$-integer contributions to the (mass)^2 the trajectories are separated by $\frac{1}{2}$-integers instead of integers.

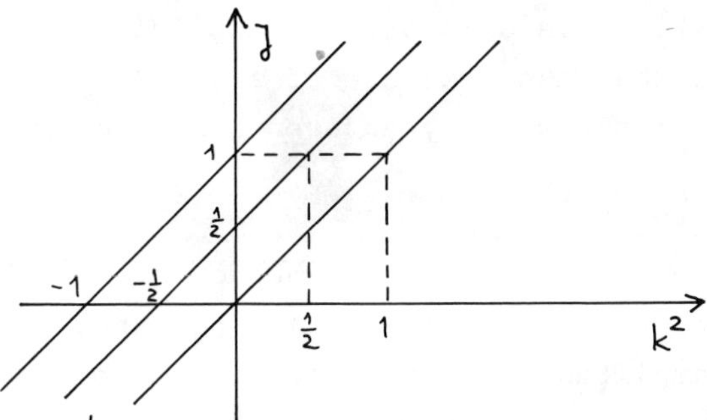

The state $k b_{\frac{1}{2}}^{+} |k\rangle$ with $k^2 = -\frac{1}{2}$ corresponds to the "pion". It is unfortunately still a tachyon. However the other apparent tachyon at $k^2 = -1$ which is formally the ground state of the theory turns out not to couple to systems of pions and hence may be excluded from the theory. We have thus moved our tachyon nearer to zero! Indeed a more complicated version of the theory exists which contains no tachyon.

The amplitude for coupling n pions is constructed by means of a formula similar to eq.(5.1). We have

$$A_n = \int \frac{1}{dR} \prod dz_i \ \langle 0| \frac{W(k_n, z_n)}{z_n} \cdots \frac{W(k_1, z_1)}{z_1} |0\rangle \tag{6.6}$$

where

$$W(k, z) = k H(z) V(k, z) , \quad k^2 = -\frac{1}{2} \tag{6.7}$$

The acceptability of the above integrand and the duality of the resulting amplitude depends on its Möbius invariance. This is guaranteed (for $n = 0, \pm 1$) by the commutation relations

$$[L_n, W(k, z)] = z^n (z \frac{d}{dz} + n) W(k, z) \tag{6.8}$$

This last equations holds because of the result (c.f. eq.5.7)

$$[L_n, H_\mu(z)] = z^n (z \frac{d}{dz} + \frac{1}{2} n) H_\mu(z) \tag{6.9}$$

and the requirement $k^2 = -\frac{1}{2}$. The $\{L_n\}$ still obey the commutation relations eq.(5.12) with the c-number terms slightly modified. Following exactly the same path as before we deduce that

$$A_n = \langle -k_n | k_n \cdot b_{\frac{1}{2}} W(k_{n-1}) \frac{1}{L_0 - 1} \cdots \frac{1}{L_0 - 1} W(k_2) k_1 \cdot b_{\frac{1}{2}}^{+} |k_1\rangle \tag{6.10}$$

240

where $W(k) = W(k, 1)$. This demonstrates explicitly the factorization discussed above. Note that the presence of a factor linear in the b's in each vertex implies that only an <u>even</u> number of pions can be coupled together.

The decoupling of the apparent tachyon at $k^2 = -1$ from systems of pions or what is the same thing, its non-appearance as an intermediate in many-pion amplitudes can be demonstrated by means of extra gauge operators

$$G_m = \sum_{-\infty}^{+\infty} \alpha_n \, b_{m-n} \tag{6.11}$$

where $\alpha_n = i\sqrt{n}\, a_n$, $\alpha_{-n} = -i\sqrt{n}\, a_n^\dagger$, $\alpha_0 = \sqrt{2}\, p$

The new gauge operators obey the anticommutation relations

$$\{G_m, G_n\} = 2L_{m+n} + c\text{-number} \tag{6.12}$$

The formula which permits us to derive the desired result is

$$-\sqrt{2}\,\{G_m, W(k)\} = (L_0 + m - \tfrac{1}{2})\, V(k) - V(k)(L_0 - \tfrac{1}{2}) \tag{6.13}$$

For the derivation of this result see Ref.[13]. Now the pion state with momentum k can be written

$$\sqrt{2}\, k\, b_{\frac{1}{2}}^\dagger\, |k\rangle = G_{-\frac{1}{2}}\, |k\rangle$$

Therefore the n pion amplitude is

$$A_n = \langle -k_n|\, G_{\frac{1}{2}}\, W(k_{n-1})\frac{1}{L_0 - 1} \cdots \frac{1}{L_0 - 1}\, W(k_2)\, G_{-\frac{1}{2}}\, |k_1\rangle \tag{6.14}$$

Now we use eq.(6.13) with $m = -\tfrac{1}{2}$ and the commutator

$$[L_0, G_{-\frac{1}{2}}] = \tfrac{1}{2} G_{-\frac{1}{2}} \tag{6.15}$$

to deduce that

$$\frac{1}{L_0 - 1}\, W(k)\, G_{-\frac{1}{2}} = -G_{-\frac{1}{2}}\, \frac{1}{L_0 - \frac{1}{2}}\, W(k) - \frac{1}{\sqrt{2}}\, V(k) \tag{6.16}$$
$$+ \frac{1}{L_0 - 1}\, V(k)(L_0 - \tfrac{1}{2})$$

The first term is the important one since it is easily shown that the

others give a rise to zero contributions to the amplitude. Consequently
we find (remembering that n is even)

$$A_n = \langle -k_n | G_{\frac{1}{2}} G_{-\frac{1}{2}} W(k_{n-1}) \frac{1}{L_0 - \frac{1}{2}} \cdots \frac{1}{L_0 - \frac{1}{2}} W(k_2) | k_1 \rangle \qquad (6.17)$$

that is

$$A_n = \langle -k_n | W(k_{n-1}) \frac{1}{L_0 - \frac{1}{2}} \cdots \frac{1}{L_0 - \frac{1}{2}} W(k_2) | k_1 \rangle \qquad (6.18)$$

This last equation shows that the effective propagator is $(L_0 - \frac{1}{2})^{-1}$ rather
than $(L_0 - 1)^{-1}$ hence the original tachyon at $k^2 = -1$ has disappeared
along with all its excitations.

The precise meaning of the various occupation number states
which are used to factorize has now changed however. As an intermediate
state in expression (6.18) the pion is represented by the ground state
. There is a zero mass vector meson (the ρ) represented
by $b_{-\frac{1}{2}}^{\mu} | k \rangle$. The orbital excitations contain the f^0 -meson
with $(\text{mass})^2 = 1$ namely $b_{-\frac{1}{2}}^{\mu} a_{-1}^{+\nu} | k \rangle$. There is also an ω -type
meson represented by $\varepsilon_{\mu\nu\sigma\tau} k^{\nu} b_{-\frac{1}{2}}^{\sigma} b_{-\frac{1}{2}}^{\tau} | k \rangle$.

This has $(\text{mass})^2 = \frac{1}{2}$ and is not degenerate with the ρ -meson, an unphysical feature which cannot be removed from this type of model.

The gauge operators $\{ G_m \}$ can also be used to verify the
absence of ghosts in the factorization.

7. OFF SHELL AMPLITUDES

An obvious problem in dual models is to construct Green functions or off-shell amplitudes to describe the coupling of[14-17] photons
to strongly interacting particles. Some progress in this direction has
been achieved recently. The theory for many photon amplitudes as it
is presently formulated is quite complicated. However, it is rather
easy and quite enlightening to discuss the case of one photon amplitudes.

We will return to our original model with $\alpha_0 = 1$ and consider
the coupling of a single off-shell scalar photon (momentum k)
to a number of ground state scalars. The trick is to associate a

<u>complex</u> K - N variable with the off-shell leg keeping real variables for the on shell-legs. The amplitude is ($\mathcal{I}m\, z \geq 0$)

$$A_{n+1}(k,p) = \int \frac{1}{dR} \frac{d^2z}{|\mathcal{I}m z|^2} \prod_i dx_i \, |\mathcal{I}m z|^{-k^2} \prod_i |z - x_i|^{-2k\cdot p_i}$$

$$\times \prod_{i<j} |x_i - x_j|^{-2p_i\cdot p_j}$$

$$(7.1)$$

It is easy to verify that this integrand is invariant, for arbitrary k^2 , under real Möbius transformations. The poles in k^2 arise from divergences at $\mathcal{I}m\, z = 0$. It is easy to see that the residue at $k^2 = -1$ is just

$$A_{n+1}(k,p) = \int \frac{1}{dR}\, dx \prod_{i=1}^{n} dx_i \prod_{i=1}^{n} |x - x_i|^{-2kp_i}$$

$$\times \prod_{i<j} |x_i - x_j|^{-2p_i\cdot p_j}$$

$$(7.2)$$

which is the K - N expression for the amplitude coupling $n+1$ scalars. It is not hard to verify that the off-shell amplitude has the desired pole structure of intermediate resonances. If we consider the case of $n = 3$ and exploit the Möbius invariance to put $(x_1, x_2, x_3) = (0, 1, \infty)$ we find

$$A_4(k, p_1, p_2, p_3) = \int d^2z \, |\mathcal{I}m z|^{-k^2-2} \, |z|^{-2p_1\cdot k} \, |1-z|^{-2p_2 k}$$

$$(7.3)$$

This expression can actually be evaluated[15] to yield

$$A_4 = \frac{\sqrt{\pi}}{2} \frac{\Gamma(-\tfrac{1}{2}\alpha(k^2))\, \Gamma(-\tfrac{1}{2}\alpha(s))\, \Gamma(-\tfrac{1}{2}\alpha(t))\, \Gamma(-\tfrac{1}{2}\alpha(u))}{\Gamma(-\tfrac{1}{2}(\alpha(s)+\alpha(t)))\, \Gamma(-\tfrac{1}{2}(\alpha(t)+\alpha(u)))\, \Gamma(-\tfrac{1}{2}(\alpha(u)+\alpha(s)))}$$

$$(7.4)$$

This is just the Virasoro amplitude multiplied by $\Gamma(-\tfrac{1}{2}\alpha(k^2))$ which yields the external mass poles. When we take the residue at $\alpha(k^2) = 0$ the amplitude coincides with the <u>sum</u> of the three Veneziano Beta-functions.

  This reveals the intrinsically non-planar character of the amplitude we have described. Neveu and Scherk[17] have constructed a <u>planar</u> version of this amplitude which in the case of A_4 simply involves restricting the Z-integration in eq. (7.3) further so that

$\mathcal{I}m\,z \geqq 0$, $\mathcal{R}e\,z \leqq 0$. This produces a term with only poles in $\alpha(s)$, $\alpha(t)$ and $\alpha(k^2)$ but not in $\alpha(u)$. It does not have a simple explicit form like eq.(7.4). It has the rather unconventional feature that the asymptotic behaviour is controlled by two kinds of Regge exchanges. The conventional kind of unit slope together with a companion trajectory of half the slope which intersects the original at the first nonsense point $\alpha(t) = -1$. These two terms in the asymptotic behaviour are ($s \to \infty$, t - fixed)

$$A_4\,(\text{planar}) \approx \tfrac{1}{2}(-\alpha(s))^{\alpha(t)}\,\Gamma(-\alpha(t))\,B(-\tfrac{1}{2}\alpha(k^2),\tfrac{1}{2}(1+\alpha(t)))$$
$$-(-\alpha(s))^{\tfrac{1}{2}(\alpha(t)-1)}\,2^{-\tfrac{1}{2}(\alpha(t)+1)}\,\Gamma(\tfrac{1}{2}(1-\alpha(t)))(1+\alpha(t))^{-1}$$

$$(7.5)$$

The second term is the companion which has an effective trajectory $\bar{\alpha}(t) = \tfrac{1}{2}(\alpha(t) - 1)$. Notice that the two terms have compensating poles at the nonsense point $\alpha(t) = -1$ and the companion term does not have poles in $\alpha(k^2)$, so it vanishes when the external leg is placed back on the mass shell.

It is possible to extend the idea to vector currents[18] and many photon amplitudes[17]. Whether the theory is finally successful in producing a representation of current algebra remains at present an interesting open question !

REFERENCES

1.   G. Veneziano, Nuov. Cim. __57A__, 190 (1968).

2.   K. Bardakci and H. Ruegg, Phys. Letts. __28B__, 342 (1968);
        C.G. Goebel and B. Sakita, Phys. Rev. Letts. __22__,257 (1969).

3.   Chan Hong-Mo and Tsuo Sheung Tsun, Phys.Letts. __28B__, 425(1969).

4.   Z. Koba and H.B. Nielsen, Nucl. Phys. __B12__, 517 (1969).

5.   S. Fubini and G. Veneziano, Nuov. Cim. __64A__, 811 (1969);
        K. Bardakci and S. Mandelstam, Phys. Rev. __184__, 1640 (1969).

6.   S. Fubini, D. Gordon and G. Veneziano, Phys.Letts.__29B__,679(1969);

Y. Nambu, Enrico Fermi Institute Preprint COO 264-507;
L. Susskind, Phys.Rev. $\underline{D1}$, 1182 (1970).

7. S. Fubini and G. Veneziano, Nuov. Cim. $\underline{67A}$,29 (1970);
K.Bardakci and M.B. Halpern, Phys. Rev. $\underline{183}$, 1456(1970);
L. Clavelli and P. Ramond, Phys. Rev. $\underline{D3}$, (1971) 988.

8. F. Gliozzi, Nuov. Cim. Letts. $\underline{2}$, 846 (1969);
C.B. Chiu, S. Matsuda and C. Rebbi, Phys. Rev. Letts.$\underline{23}$, 1526 (1969).

9. M.A. Virasoro, Phys. Rev. $\underline{D1}$, 2933 (1970).

10. E. Del Giudice and P. Di Vecchia, Nuovo. Cim. $\underline{70A}$,579 (1970).

11. R.C. Brower and C.B. Thorn, Nucl. Phys. $\underline{B31}$, 163 (1971).

12. R.C. Brower, MIT Preprint, Spectrum Generating Algebra and No Ghost Theorem for the Dual Model;
P. Goddard and C.B. Thorn, CERN Preprint TH 1493.

13. A. Neveu and J.H. Schwarz, Nucl. Phys. $\underline{B31}$, 86 (1971);
Phys.Lett. $\underline{35B}$, 529 (1971).

14. H. B. Nielsen, L. Susskind and A.B. Kraemmer, Nucl.Phys. $\underline{B28}$, 34 (1971).

15. I.T. Drummond, Nucl.Phys. $\underline{B35}$, 269 (1971).

16. C. Rebbi, Nuov.Cim. Letts.
Nuovo.Cim.Letts. $\underline{1}$, 967 (1971).

17. A. Neveu and J. Scherk, TH 1481 - CERN;
E. Cremmer and J. Scherk, TH 1449- CERN.

18. K. Kikkawa and B. Sakita, Dual Resonance Model of Vector Currents, Preprint - City College of New York;
R.C. Brower and L. Susskind, Current Algebra and the Dual Parton Model.

CURRENTS ON THE LIGHT CONE

H. Leutwyler

Institut für theoretische Physik
BERN, Switzerland

These lectures are meant as an elementary introduction to some
of the topics in light cone physics. In an attempt to discuss the
foundation of the subject as thoroughly as possible we devote the major
part of the paper to electron-proton scattering. Related processes
such as electron-neutron scattering or neutrino-nucleon interactions
are not treated because the extension of the methods discussed here to
these processes is straightforward and may be found in the literature[1].
We adapt a rather conservative point of view and discuss only those
applications which do not involve dynamical assumptions beyond those
inherent in light cone physics. More speculative developments concerning e.g. the algebraic structure of the current commutators on the
light cone are only touched upon. The interested reader will find more
material on these topics in the references.

If possible we quote review papers rather than original literature in order not to confuse the student with an embarrassingly large
number of references. A very readable review which treats problems
similar to those discussed here is contained in the lectures by Stichel[2].
As further supplementary reading we mention the paper by Leutwyler and
Otterson which contains more material on some of the topics reviewed
here.(This paper is quoted as LO).

1.   KINEMATICS OF ELECTRON SCATTERING

Why is electron-proton scattering theoretically more rewarding
than proton-proton scattering? The small number of experimental papers
on the process $e + p \rightarrow e + $ anything has given rise to such

an inflation of theoretical papers because part of this process is described by Quantum Electrodynamics. According to QED the process takes place mainly via one-photon-exchange (lowest order contribution in perturbation theory); Fig.1.

Both the interaction of the photon with the electron and the propagation properties of the virtual photon may be taken from QED and only the interaction of the photon with the hadronic states $|p,s\rangle$ and $|f,\text{out}\rangle$ is unknown:

$$T_f = e^2\,\bar{u}(k')\,\gamma_\mu\,u(k)\,\frac{g^{\mu\nu}}{q^2}\,\langle f\,\text{out}\,|\,j_\nu(0)\,|\,p,s\rangle \tag{1.1}$$

In this expression for the T-matrix-element $s = \pm\frac{1}{2}$ denotes the spin direction of the incoming proton and $q = k - k'$ stands for the momentum of the virtual photon. In the cross section the square of the unknown matrix element $\langle f\,\text{out}\,|\,j_\nu\,|\,p,s\rangle$ comes in, together with a sum over all final states f. This sum may be expressed in terms of the quantity

$$W_{\mu\nu}^+(p,q,s) = \frac{1}{4\pi}\int d^4x\, e^{iqx}\langle p,s\,|\,j_\mu^+(x)\,j_\nu(0)\,|\,p,s\rangle \tag{1.2}$$

In fact, inserting a complete set of intermediate states in this expression we obtain

$$W_{\mu\nu}^+(p,q,s) = 4\pi^3\sum_f \delta^4(p+q-P_f)\langle p,s\,|\,j_\mu(0)\,|\,f\,\text{out}\rangle\cdot$$

$$\cdot\langle p,s\,|\,j_\nu(0)\,|\,f\,\text{out}\rangle^* \tag{1.3}$$

We restrict ourselves to electron scattering on an unpolarized proton target for which the average over the two proton spin directions is relevant

$$W_{\mu\nu}^+(p,q) = \frac{1}{2}\sum_{s=\pm\frac{1}{2}} W_{\mu\nu}^+(p,q,s) \tag{1.4}$$

This quantity is lorentzinvariant and, furthermore, satisfies $q^\mu W_{\mu\nu}^+ = q^\nu W_{\mu\nu}^+ = 0$ on account of current conservation. The most general expression with these properties reads

$$W_{\mu\nu}^{+}(p,q) = \{ q_\mu q_\nu - g_{\mu\nu} q^2 \} V_1^{+} +$$

$$+ \{ (P_\mu q_\nu + P_\nu q_\mu) \nu^2 - g_{\mu\nu} \nu^2 - P_\mu P_\nu q^2 \} V_2^{+} \qquad (1.5)$$

where V_1^{+} and V_2^{+} are invariant functions of p and q, i.e. functions of p^2, $pq \equiv \nu$ and q^2. Since p^2 is the square of the proton mass which we choose as our mass unit, $p^2 = 1$, we have only two relevant variables

$$V_i^{+} = V_i^{+}(\nu, q^2) \qquad ; \qquad i = 1, 2$$

The cross section for the process $e + p \longrightarrow e +$ anything may therefore be expressed in terms of only two unknown functions of two variables. The explicit expression reads

$$\frac{d^3\sigma}{d\Omega \, dE'} = \frac{\alpha^2}{2 E^2 \sin^2\frac{\Theta}{2}} \left\{ q^2 V_1^{+} + (\nu^2 + 2 E E' \cos^2\tfrac{\Theta}{2}) V_2^{+} \right\} \qquad (1.6)$$

Here E and E' denote the initial and final energies of the electron and Θ is the scattering angle [everything taken in the Laboratory frame , p = (1,0,0,0)] , $d\Omega = \sin\Theta \, d\Theta \, d\varphi$ is the solid angle into which the electron is scattered. In terms of electron energies and scattering angle we finally have $\nu = E - E'$, $q^2 = -4 E E' \sin^2\tfrac{\Theta}{2}$. [The normalization of the proton state occuring in the definition of $W_{\mu\nu}^{+}$ is taken as $\langle p', s' | p, s \rangle = (2\pi)^3 2 P_0 \, \delta^3(\vec{p}' - \vec{p}) \, \delta_{s's}$].

Instead of the functions V_1^{+} and V_2^{+} one often uses the quantities W_1^{+} and W_2^{+} defined by

$$W_{\mu\nu}^{+}(p,q) = \left(\frac{q_\mu q_\nu}{q^2} - g_{\mu\nu} \right) W_1^{+} + \left(P_\mu - \frac{q_\mu \nu}{q^2} \right) \left(P_\nu - \frac{q_\nu \nu}{q^2} \right) W_2^{+} \qquad (1.7)$$

These quantities are related to V_1^{+} and V_2^{+} by

$$W_1^{+} = q^2 V_1^{+} + \nu^2 V_2^{+}$$

$$W_2^{+} = -q^2 V_2^{+} \qquad (1.8)$$

A further set, the Hand cross sections σ_T, σ_L is also used frequently

$$\sigma_T = \frac{4\pi^2\alpha}{K} (q^2 V_1^{+} + \nu^2 V_2^{+}) \qquad ; \qquad K = \nu + \frac{q^2}{2}$$

$$\sigma_L = \frac{4\pi^2\alpha}{K} (-q^2)(V_1^{+} + V_2^{+}) \qquad (1.9)$$

The quantity σ_T (σ_L) is the cross section for scattering of transversely (longitudinally) polarized virtual photons.

The crucial point here is that the cross section for the process
$e + p \longrightarrow e +$ anything involves only two unknown functions.
Measuring the cross section amounts to measuring these functions, i.e.
measuring the Fourier transform of the quantity $\langle P | j_\mu(x)\, j_\nu(0) | P \rangle$.
This is an extremely interesting object as it involves the product of
local fields. Local fields are normally not accessible to direct experimental investigation. They come in here because the process is
dominated by exchange of virtual photons whose mass may be varied by
looking at different energies and angles of the outcoming electron.
Ordinary cross sections, say the cross section for $\pi + p \rightarrow \pi + p$,
are related to the square of a matrix element whose momenta are all on
the mass shell of the particles involved. The matrix element describing
$\pi + p \longrightarrow \pi + p$ e.g. is given by the function

$$T(p';k'|p,k) = (m_\pi^2 - k^2)(m_\pi^2 - k'^2)\int d^4x\, e^{ik'x}\langle p'|T\varphi(x)\,\varphi(0)|p\rangle$$

evaluated at $k^2 = k'^2 = m_\pi^2$. Since the mass shell is a set of measure zero in momentum space, the T-matrix contains very little informa-
tion about the behaviour of the matrix element $\langle p'|T\varphi(x)\,\varphi(0)|p\rangle$
in coordinate space. The same remark applies to inclusive hadronic
cross sections say $p + p \longrightarrow \pi +$ anything. Measuring this cross
section one obtains information about the matrix element

$$T(p_1,p_2,q) = \int d^4x\, e^{-iqx}\langle p_1,p_2 \text{ in }|j_\pi(x)\, j_\pi(0)|p_1,p_2 \text{ in}\rangle$$

only at $q^2 = m_\pi^2$ which is again not enough to determine the behaviour
of the local fields $j_\pi(x)$ in coordinate space.

To close this kinematical section we look at two limiting
cases in electron scattering. The process $e + p \longrightarrow e +$ anything
contains in particular the _elastic_ process $e + p \longrightarrow e + p$. This
contribution corresponds to the proton intermediate state in the matrix
element

$$\langle p,s|j_\mu(x)\,j_\nu(0)|p,s\rangle = (2\pi)^{-3}\sum_{s'}\int \frac{d^3p'}{2p^{0'}} e^{i(p-p')x}\langle p,s|j_\mu|p',s'\rangle\langle p's'|j_\nu|p,s\rangle$$

$$+ \text{ inelastic contributions}$$

The elastic contribution is determined by the form factors
$G_E(q^2)$, $G_M(q^2)$ of the proton. Explicitly, the elastic contribution
to V_1^+ and V_2^+ reads

$$V_1^+(\nu, q^2) = \frac{\nu}{\nu^2 - q^2} \left\{ G_E^2 - G_M^2 \right\} \delta(q^2 + 2\nu) \qquad + \text{inel.contr.}$$

$$V_2^+(\nu, q^2) = \frac{1}{\nu(\nu^2 - q^2)} \left\{ \nu^2 G_M^2 - q^2 G_E^2 \right\} \delta(q^2 + 2\nu) + \text{ " \quad " } \tag{1.10}$$

A different limiting case results if one considers virtual photons that are almost <u>real</u>, $q^2 \to 0$. In this limit of electron scattering the unknown hadronic part of the scattering amplitude is the same as for the process real photon + p $\to$ anything. Indeed, the total cross section for absorption of real photons by unpolarized protons is given by

$$\sigma_\gamma(\nu) = 4\pi^2 \alpha \, \nu \, V_2^+(\nu, 0) \tag{1.11}$$

Only the function V_2^+ is relevant here since there are no longitudinally polarized real photons ($\sigma_T(\nu, 0) = \sigma_\gamma(\nu)$; $\sigma_L(\nu, 0) = 0$).

The <u>support</u> of the functions V_1^+ and V_2^+ as defined by the Fourier transform (1.2) is shown in Fig.2.

V_1^+ and V_2^+ vanish outside the shaded region (except for the elastic contribution at $q^2 + 2\nu = 0$). Since virtual photons emitted by an electron have always spacelike momenta, the cross section for electron scattering determines the functions V_1^+ and V_2^+ only in the physical region $q^2 \leq 0$. There is no direct experimental information about the behaviour of these functions in the unphysical region $q^2 > 0$. We will see, however, that causality fills in this gap and determines the functions V_1^+ and V_2^+ essentially uniquely from their values in the physical region.

2.   <u>CURRENT COMMUTATOR</u>

In the following we will work with the current commutator rather than with the product of two currents

$$W_{\mu\nu}(p, q) = \frac{1}{4\pi} \int d^4x \, e^{iqx} \langle p | [j_\mu(x), j_\nu(0)] | p \rangle \tag{2.1}$$

(In this abbreviated notation an average over spin directions is understood). Of course this quantity is uniquely determined by the matrix element of the product:

$$W_{\mu\nu}(p, q) = W_{\mu\nu}^+(p, q) - W_{\nu\mu}^+(p, -q) \tag{2.2}$$

In terms of the invariant decomposition

$$W_{\mu\nu}(p,q) = \{q_\mu q_\nu - g_{\mu\nu} q^2\} V_1 + \{(p_\mu q_\nu + p_\nu q_\mu)\nu + \cdots\} V_2 \qquad (2.3)$$

we have

$$V_i(\nu,q^2) = V_i^+(\nu,q^2) - V_i^+(-\nu,q^2) \quad ; \quad i = 1,2 \qquad (2.4)$$

Clearly, the invariant functions $V_1(\nu,q^2)$ and $V_2(\nu,q^2)$ associated with the current commutator are odd functions of ν . Their support is indicated in Fig.3.

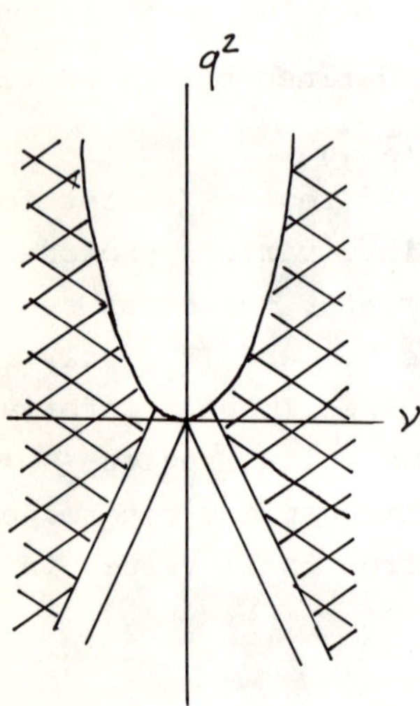

Fig. 3

What is remarkable in the present case is that the current commutator also uniquely determines the operator product. This is because the structure functions V_i^+ vanish for $\nu < 0$. We therefore have, as an inversion of (2.4):

$$V_i^+(\nu,q^2) = \Theta(\nu) V_i(\nu,q^2) \qquad (2.5)$$

The current commutator contains precisely the same information as the operator product. The commutator however has an additional very important property: it is <u>causal</u>, i.e. $[j_\mu(x), j_\nu(o)] = o$ for $x^2 < o$.

Actually, this basic property of the current operator has not been tested in the world of strongly interacting particles. It represents one of the few basic theoretical principles abstracted from QED with some chance of being correct in the hadronic world. Causality is indeed the mildest of the assumptions we will go through in these lectures. The light cone philosophy is based on the faith that the currents, i.e. the sources of the electromagnetic and weak interactions have simple properties which other fields, say the interpolating spin $\frac{1}{2}$ field which is supposed to connect the in- and out-states of the proton may not exhibit. Since the proton is very probably some sort of composite system there is no reason to expect that its dynamics may be described in terms of a single local field describing the motion of the entire system. The experience we have gained with composite systems such as atoms or nuclei in any case suggest that this is an unreasonable hope. Although it may be possible to construct a local field with the quantum numbers of say the ground state of the hydrogen atom this field is of no relevance for the dynamics of the system. The essential point here is, however, that the hydrogen atom still admits of a local electromagnetic current which is of fundamental significance as it describes its interaction with external electromagnetic fields or with photons. We will in the following assume that the electromagnetic current <u>is</u> a local field; some of the consequences we will derive may then be used to test this hypothesis.

We thus assume that the Fourier transform of $W_{\mu\nu}(P,q)$ vanishes for $x^2 < 0$. As was discussed in the lectures by Tavkhelidze[3] this assumption implies that the functions

$$\widetilde{V}_i(x) = (2\pi)^{-3} \int d^4q\; e^{-iqx}\, V_i(\nu, q^2) \qquad (2.6)$$

are causal, i.e. vanish for $x^2 < 0$. Clearly, causality of the functions $\widetilde{V}_i(x)$ is sufficient to insure causality of the current commutator, since this quantity takes the form of a derivative of $\widetilde{V}_1$ and $\widetilde{V}_2$:

$$\tfrac{1}{2}\langle P | [j_\mu(x), j_\nu(0)] | P \rangle = \{g_{\mu\nu}\Box - \partial_{\mu\nu}\}\widetilde{V}_1(x) + \{P_\mu P_\nu \Box + \cdots\}\widetilde{V}_2(x) \qquad (2.7)$$

What is not trivial is to show that causality of $\widetilde{V}_i(x)$ is also necessary. In fact, as was pointed out by Meyer and Suura[4], this is not the case if one considers the commutator of charged currents, say the commutator $\langle P | [j_\mu^+(x), j_\nu^-(0)] | P \rangle$ of the isovector currents – the corresponding invariant functions $\widetilde{V}_i(x)$ are not causal.

Let us at this point come back to a technical detail which we skipped in the foregoing. The matrix element $\langle P | [j_\mu(x), j_\nu(0)] | P \rangle$ is not well-defined because the current commutator (as well as the operator product) contains a c-number contribution:

$$[j_\mu(x), j_\nu(0)] = \langle 0 | [j_\mu(x), j_\nu(0)] | 0 \rangle + [j_\mu(x), j_\nu(0)]_{conn} \qquad (2.8)$$

In the forward matrix element the c-number contribution (disconnected piece) gives rise to an infinity $\sim \delta^3(\vec{p} - \vec{p})$. In order to deal with well-defined quantities we have in all our previous equations to replace the product $j_\mu(x) \, j_\nu(0)$ or the commutator $[\, j_\mu(x), j_\nu(0) \,]$ by the corresponding connected pieces. This change does not affect the relation between the matrix element of the current product and the cross section for electron scattering. This is because the Fourier transform of the disconnected piece contributes only for $q^2 > 0$:

$$\int d^4x \, e^{iqx} \langle 0| \, j_\mu(x) \, j_\nu(0) \, |0\rangle = 0 \qquad q^2 < 4\,m_\pi^2$$

(The lowest value of q^2 at which the disconnected piece contributes is $q^2 = 4\,m_\pi^2$, corresponding to the two-pion intermediate state.)

The change does affect, however, <u>the positivity conditions</u> satisfied by the functions V_1 and V_2 . These conditions, which guarantee that the cross sections σ_T, σ_L are positive, arise from the fact that V_1^+ and V_2^+ are defined in terms of sums of squares of current matrix elements as explicitly shown in (1.3). This expression guarantees that the quantity $y^\mu y^{\nu *} W_{\mu\nu}^+(p, q)$ is positive for any vector y^μ and one easily shows that this is the case if and only if the conditions

$$V_1^+ + V_2^+ \geq 0$$
$$q^2 \, V_1^+ + \nu^2 V_2^+ \geq 0 \tag{2.9}$$

are satisfied. Actually, as mentioned above, the expression (1.3) is valid only for $q^2 < 4\,m_\pi^2$ because we are considering only the connected piece. As a consequence, the positivity conditions (2.9) hold only for $q^2 < 4\,m_\pi^2$.

3.   <u>DO THE DATA DETERMINE THE CURRENT COMMUTATOR UNIQUELY?</u>

As mentioned in the first section, the data on electron scattering immediately determine the functions V_1 and V_2 only in the spacelike region $q^2 \leq 0$. We now wish to show that causality essentially determines the rest. This claim is based on a theorem by Bogoliubov and Vladimirov[5] which states the following.

<u>Theorem</u>: Let $\tilde{V}(x)$ be a tempered distribution and denote by $V(q)$ its Fourier transform. If
- $\tilde{V}(x)$ vanishes for $x^2 < 0$

253

- V(q) vanishes for $q^2 < 0$

then V(q) is of the form

$$V(q) = \varepsilon(q^\circ) \sum_{n=0}^{N} P_n(q)\, \rho^{(n)}(q^2) \tag{3.1}$$

where $P_n(q)$ is a polynomial in q and $\rho^{(n)}(s) = 0$ for $s < 0$.
It is easy to see that every such function has the two properties required. The non-trivial part of the theorem is the statement that the form (3.1) is already the most general expression with these properties.

The application of this theorem to V_1 and V_2 is immediate. Since the data in principle determine these quantities up to a causal function that vanishes for $q^2 < 0$ we conclude that the data determine V_1 and V_2 up to functions of the type (3.1). Supposed that we find one function V_1 which agrees with the data in $q^2 < 0$ and whose Fourier transform is causal. Then the most general function with these two properties is

$$V_1'(\nu, q^2) = V_1(\nu, q^2) + \varepsilon(\nu) \sum_{n=0}^{N} \nu^n \rho_1^{(n)}(q^2) \tag{3.2}$$

In other words, V_1 is determined by the data for all values of q^2 up to a polynomial in ν whose coefficients may be functions of q^2. Clearly, if we have the additional information that V_1 tends to zero as $\nu \to \infty$ for both spacelike and timelike q then the data determine it uniquely; in general, however, causality is not quite sufficient.

The polynomials which are not fixed by electron scattering manifest themselves in a variety of applications, in particular they appear in fixed q^2 dispersion relations as subtraction "constants". The reader is referred to LO for a detailed discussion of these problems. There is good reason to believe that the function V_2 actually tends to zero as $\nu \to \infty$ for both spacelike and timelike q and, therefore there is no ambiquity in its causal extension - V_2 is uniquely determined by the data on electron scattering. The situation is different for V_1, however, in which case there is no experimental ground for pinning down an ambiguity of the type (N = 0 in (3.2))

$$V_1'(\nu, q^2) = V_1(\nu, q^2) + \varepsilon(\nu)\, \rho_1(q^2) \tag{3.3}$$

There are theoretical prejudices which suggest that the amplitude should not contain a "fixed pole at $\alpha = 0$ " in the absorptive part, but

there are really no convincing arguments against their presence. A speculative model with rather strong experimental predictions for the high
energy behaviour of longitudinally polarized virtual photons which
would nail down the ambiguity in V_1 may be found in L O.

In the following we will simply accept that V_1 is fixed by the
electron scattering data only up to the ambiguity exhibited in (3.3).
This ambiguity manifests itself in x-space as follows

$$\tilde{V}_1(x)' = \tilde{V}_1(x) + \frac{1}{i}\int_0^\infty ds\ \varrho_1(s)\,\Delta(x; s) \tag{3.4}$$

Note that the additional term is a function of x^2 only whereas $\tilde{V}_1(x)$
depends on both x^2 and px.

We now turn to a different problem: Is it possible to pres-
cribe the values of V_1 and V_2 in the spacelike region at will?
<u>Does there always exist a continuation to timelike values of q such
that the Fourier transform</u> of these functions <u>is causal?</u> The answer
to this question is no. This is the content of a theorem due to
P.Otterson[18] which states that the conditions in the Bogoliubov -
Vladimirov theorem may be relaxed. It suffices that the Fourier transform V (q) of the causal tempered distribution $\tilde{V}(x)$ vanishes in some
strip $a < q^2 < b \leqq o$; this already implies that V(q) vanishes in the
entire spacelike domain and hence it must be of the form (3.1) For
a proof see L O . This means that it suffices to know the cross
sections σ_T , σ_L in some arbitrarily narrow interval of q^2 , say
between $q^2 = -1$ and $q^2 = -1.1$. This information already determines
the cross sections for all spacelike momenta. If the measurements outside the interval do not agree with the predicted values we have a vio-
lation of causality.

Unfortunately, this theorem is of interest more in principle
than in practice, since the cross sections would have to be known to
very high precision to make the causal continuation outside the interval
we started with sufficiently stable. The situation is analogous to the
case of the elastic proton form factors where it is also sufficient
to know, say, $G_E(q^2)$ in an arbitrarily small interval to determine the
entire function. The reader is referred to the lectures by Ciulli[6]
for a comprehensive exposition of stability problems of this type.
Note that in contrast to the form factors the structure functions
$V_i(\nu, q^2)$ are not analytic; they are instead absorptive parts of analytic functions.

4. LIGHT CONE SINGULARITIES

The asymptotic behaviour of an amplitude in momentum space reflects its singularities in coordinate space. There are many different ways of going to infinity in a fourdimensional space. We briefly discuss the three particular high energy limits indicated in Fig. 4.

B: Bjorken limit; $\nu \to \infty$, $q^2 \to \infty$, $\frac{q^2}{\nu}$ finite

R: Regge limit; $\nu \to \infty$, q^2 fixed

BJL: Bjorken-Johnson-Low limit; $q^0 \to \infty$, $\vec{q}$ fixed.

Qualitatively, these limits reflect the behaviour of the amplitude in coordinate space in the regions indicated in Fig. 5.

The Borken limit shows the behaviour of the amplitude on the light cone. The structure of the amplitude in the Regge limit is governed by its behaviour at $x^0 \to \infty$ for fixed x^2. Finally, the equal time behaviour is dual to the asymptotic behaviour in the BJL limit.

In the following we only specify the relation between the Bjorken limit and the singularities of the amplitude on the light cone in some detail. For a mathematical treatment of other limits in momentum space see ref.[7].

We limit our discussion of the relation between light cone singularities and Bjorken limit to a very special class of light cone singularities, the so-called canonical singularities. The prototype of a canonical behaviour on the light cone is a singularity of the type

$$\tilde{V}(x) = \frac{1}{2\pi i} \, \epsilon(x^0) \, \delta(x^2) \, g(x) + R(x) \tag{4.1}$$

where the remainder $R(x)$ is less singular than the leading term. The next smoother canonical singularity is a discontinuity on the light cone:

$$\tilde{V}(x) = \frac{i}{4\pi} \, \epsilon(x^0) \, \theta(x^2) \, h(x) + R(x) \tag{4.2}$$

A more general class of non-canonical light cone singularities is described by amplitudes of the type

$$\tilde{V}_\lambda(x) = \frac{1}{2\pi i} \, \epsilon(x^0) \, \Gamma(\lambda+1)^{-1} \, (x^2)_+^\lambda \, g(x) + R(x)$$

$$(4.3)$$

where λ is an arbitrary real parameter. (For a definition of the distribution z_+^λ see Gel'fand and Shilov[8]. The singularity is called canonical if the parameter λ is an integer. (The distribution z_+^λ has a pole at $\lambda = -1$ with the residue $\delta(z)$) Non-canonical light cone singularities have recently attracted much attention in connection with conformally invariant field theory. We will stick to canonical singularities because we are dealing with <u>currents.</u> Current algebra demands that the leading singularities of current commutators are canonical at least on the tip of the cone. Of course, it might happen, that away from the tip a non-canonical term takes over. We do not know of any evidence indicating that this actually happens, but we also have to admit that (except for the Thirring model) there is no evidence for the existence of an interacting field theory whose current commutators possess canonical leading light cone singularities, either. In my opinion the question of whether or not there is such a relativistic theory of interacting fields is one of the most interesting unsolved problems in field theory. It may be that to assume canonical leading light cone singularities for currents means to exclude any interaction.

5.   <u>S C A L I N G</u>

We now consider an amplitude $\tilde{V}(x)$ whose leading light cone singularity is of the type (4.1). We claim that the corresponding Fourier transform

$$V(q) = \frac{1}{2\pi} \int d^4x \, e^{iqx} \, \tilde{V}(x)$$

$$(5.1)$$

obeys the <u>scaling law</u>

$$V(k + E n) \to \frac{1}{2E} \, F(k, n) ; \quad E \to \pm \infty$$

$$(5.2)$$

provided we take n to be a lightlike vector, $n^2 = 0$. Note that we do not attempt here to prove the inverse, i.e. to show that scaling implies a specific light cone singularity. Some statements concerning this problem may be found in L O. To derive the scaling law (5.2) from the postulated behaviour on the light cone we make use of the following trick[11].

We reduce the problem to the one-dimensional case by considering the
quantity

$$\tilde{v}(\tau) = \int dx\, \tilde{f}(x)\, \tilde{V}(x)\, \delta(nx - \tau) \tag{5.3}$$

where n is a fixed, lightlike vector and $\tilde{f}(x)$ is some smooth test
function. The contribution of the leading light cone singularity is a
smooth function of τ except at $\tau = 0$ where the lightlike
plane $nx = \tau$ just touches the light cone. One easily verifies that
this contribution is discontinuous at $\tau = 0$. The discontinuity is
determined by the values of the functions $\tilde{f}(x)$ and $g(x)$ along the
ray $x = \lambda n$, at which the plane touches the cone:

$$\tilde{v}(+o) - \tilde{v}(-o) = \frac{1}{2i} \int_{-\infty}^{+\infty} d\lambda\, \tilde{f}(\lambda n)\, g(\lambda n) \tag{5.4}$$

(Here we have assumed that the remainder $R(x)$ is sufficiently smooth
such that its contribution to $\tilde{v}(\tau)$ is continuous.)

Let us now look at the Fourier transform of $\tilde{v}(\tau)$:

$$v(E) = \int_{-\infty}^{+\infty} d\tau\, e^{iE\tau}\, \tilde{v}(\tau) \tag{5.5}$$

It is well-known (<u>Riemann-Lebesgue-Lemma</u>) that a discontinuity of
$\tilde{v}(\tau)$ at the origin manifests itself in the asymptotic behaviour
of its Fourier transform:

$$v(E) \rightarrow \frac{i}{E} \left\{ \tilde{v}(+o) - \tilde{v}(-o) \right\} \; ; \qquad E \rightarrow \pm\infty \tag{5.6}$$

What remains to be done is to express this scaling law in terms of $V(q)$.
Inserting Fourier transforms

$$\tilde{f}(x) = \frac{1}{2\pi} \int d^4k\, e^{ikx}\, f(k)$$

in the definition of $\tilde{v}(\tau)$ one finds

$$v(E) = \int d^4k\, f(k)\, V(k + En) \,. \tag{5.7}$$

The scaling law (5.6) therefore reads, more explicitly

$$\lim_{E \rightarrow \pm\infty} 2E \int d^4k\, f(k)\, V(k+En) = \int d^4k\, F(k,n) f(k) \tag{5.8}$$

with

$$F(k,n) = \frac{1}{2\pi} \int_{-\infty}^{+\infty} d\lambda\, e^{i\lambda kn}\, g(\lambda n) \tag{5.9}$$

As a by-product we have thus found that the scaling function F(k,n) is the Fourier transform of the coefficient g(x) of the leading light cone singularity. Note that F depends on k only through its projection onto n, F = F(kn,n).

There is a price to pay for the simplification achieved by reducing the problem to the one-dimensional case: the scaling law (5.2) is valid only in the sense of a limit relation between distributions as is explicitly indicated in (5.8): The amplitude V (k + En) has to be smeared with a test function f(k) <u>before</u> one goes to the limit E $\rightarrow \infty$. This remark may seem to be of relevance for the purists only. Actually this is by no means the case as we will see below.

Let us now apply our results to an amplitude which depends on q only through the two invariants $\nu =$ pq and q^2. If q = k + En we have

$$\nu = pk + Epn$$
$$q^2 = k^2 + 2Ekn \tag{5.10}$$

Clearly, as E $\rightarrow \infty$ both these invariants tend to infinity while their ratio approaches the constant limit

$$\frac{q^2}{2\nu} \rightarrow \frac{kn}{pn}$$

The limit E $\rightarrow \infty$ is thus precisely the Bjorken limit. As E varies, the invariants q^2 and ν run along the straight line

$$q^2 = -2\xi\nu - \eta \tag{5.11}$$

in the (ν, q^2) -plane. The parameters ξ and η are determined by k and n. In particular, we have $\xi = -kn/pn$ and $\eta \geq \xi^2$.

In this invariant notation the scaling law (5.2) reads

$$V(\nu, -2\xi\nu - \eta) \rightarrow \frac{1}{2\nu} F(\xi) \; ; \; \nu \rightarrow \pm\infty \tag{5.12}$$

From the support properties of V it immediately follows that the scaling function F (ξ) vanishes outside the interval $-1 \leq \xi \leq 1$.

In the present case the function g(x) depends only on the scalar px and the relation (5.9) takes the form

$$g(px) = \int_{-1}^{+1} d\xi \, F(\xi) \, e^{i\xi px} \tag{5.13}$$

We emphasize one important aspect of canonical leading light cone singularities: the limits $\nu \to +\infty$ and $\nu \to -\infty$ of the quantity νV must be the same. If we consider, e.g. a positive value of ξ then the limit $\nu \to +\infty$ is in the spacelike, physical region of q, whereas the limit $\nu \to -\infty$ is in the unphysical region $q^2 > 0$. A canonical light cone singularity not only demands scaling in the physical as well as the unphysical region, but it requires that the behaviour of the function νV in these two physically different domains be described by the same scaling function! In particular, the limit $\nu \to -\infty$ for a value of $\xi > 1$ must vanish despite the fact that this limit is within the support of V .

Finally, let us illustrate the significance of the statement that the scaling law is valid only in the sense of distribution theory. In terms of the invariant notation introduced above this statement means that the scaling law (5.12) holds only after smearing with a smooth test function in ξ and η .

This warning concerns the following pitfall. If one puts $\xi = 0$ in (5.12) then the limit $\nu \to \infty$ amounts to the Regge limit ($q^2 = -\eta = $ fixed). It thus appears that the leading light cone singularity specifies the behaviour of the amplitude in the Regge limit. This is of course not the case. We have pointed out at the beginning of this section that the Regge limit is controlled by a different region in x-space. Indeed, the prescribed canonical light cone singularity does not exclude that the function $V(\nu, q^2)$ for example behaves like $V \to \nu^\alpha \beta(q^2)$ for fixed q^2, $\nu \to \infty$, (α may very well be positive). Although, formally, the limit $\nu \to \infty$ at fixed q^2 is a special case of the scaling law (5.12) it is illegal to simply put $\xi = 0$. The scaling law is not guaranteed to hold pointwise in ξ , η . It is very easy to construct explicit amplitudes with a fixed q^2 behaviour of the type indicated above which at the same time do have a canonical leading light cone singularity[9]. These amplitudes indeed satisfy the scaling law (5.12) (with a scaling function F (ξ) that is singular at $\xi = 0$) but only after smearing in ξ, η . Fortunately, the fixed q^2 limit seems to be the only case for which the distribution character of the scaling law is essential; for $\xi \neq 0$ the scaling law seems to hold pointwise in ξ, η .

It is straightforward to extend the analysis to <u>different types of leading light cone singularities</u>. Let us briefly discuss an amplitude whose leading light cone singularity is a discontinuity rather than a δ-function,

$$\tilde{V}(x) = \frac{i}{4\pi} \, \epsilon(x^0) \, \Theta(x^2) \, h(px) \; + \; R(x) \qquad (5.14)$$

This case may easily be reduced to the δ-function singularity analyzed before by considering

$$\tilde{V}^*(x) = p^\mu \partial_\mu \tilde{V}(x) = \frac{i}{2\pi} \, \epsilon(x^0) \, \delta(x^2) \, px \, h(px) \; + \; R^*(x) \qquad (5.15)$$

Applying our previous results we find that the corresponding behaviour in the Bjorken limit is given by the scaling law

$$V(\nu, -2\xi\nu - \eta) \;\longrightarrow\; \frac{1}{2\nu^2} \, G(\xi) \qquad (5.16)$$

Again, the scaling function $G(\xi)$ is related to the coefficient of the leading light cone singularity

$$px \, h(px) = i \int_{-1}^{+1} d\xi \, G(\xi) \, e^{i\xi px} \qquad (5.17)$$

6.   <u>LEADING LIGHT CONE SINGULARITIES OF THE CURRENT COMMUTATOR</u>

Experimentally, the functions W_1 and W_2 seem to satisfy the scaling laws[10]

$$\begin{aligned}
W_1 &\simeq F_1(\xi) \\
\nu W_2 &\simeq F_2(\xi)
\end{aligned} \qquad ; \quad \xi = \frac{-q^2}{2\nu} \qquad (6.1)$$

in the "deep" inelastic region $q^2 < -1$, $q^2 + 2\nu > 3$. In terms of the functions V_1 and V_2 which we are using here, these scaling laws read

$$V_1 \simeq \frac{1}{2\nu\xi} \, F_L(\xi) \qquad (6.2)$$

$$V_2 \simeq \frac{1}{2\nu^2\xi} \, F_2(\xi) \qquad (6.3)$$

where

$$F_L(\xi) = - F_1(\xi) + \frac{1}{2\xi} \, F_2(\xi) \qquad (6.4)$$

The scaling function F_2 is rather well measured and shows a behaviour as indicated in Fig.6. The quantity F_1 is also quite well known and, qualitatively, behaves like $F_1 \simeq (2\xi)^{-1} F_2$. In fact the data are consistent with $F_L(\xi) \equiv 0$ (F_1 is a measure for the ratio σ_L/σ_T ; in the Bjorken limit we have $\sigma_L/(\sigma_L + \sigma_T) \to 2\xi F_L/F_2$). However, within the accuracy available till now it is still possible that F_1 deviates from $(2\xi)^{-1} F_2$ by an amount of the order of 20% – 30%. In the following we first discuss the situation that occurs if F_L turns out to be different from zero: suppose that V_1 satisfies the scaling law (6.2) with $F_L \neq 0$. According to our discussion in section 5 such a scaling law is expected to hold if the leading light cone singularity of $\tilde{V}_1$ is of the type

$$\tilde{V}_1(x) = \frac{1}{2\pi i}\, \epsilon(x^0)\, \delta(x^2)\, g_1(px) + R_1(x) \tag{6.5}$$

and we have the relation

$$g_1(px) = \int_{-1}^{+1} d\xi\, \frac{F_L(\xi)}{\xi}\, e^{i\xi px} \tag{6.6}$$

Similarly, the scaling law (6.3) results if the leading light cone singularity of $\tilde{V}_2$ is a step function:

$$\tilde{V}_2(x) = \frac{i}{4\pi}\, \epsilon(x^0)\, \Theta(x^2)\, h_2(px) + R_2(x) \tag{6.7}$$

Suppose now that (6.5) and (6.7) are indeed the correct leading light cone singularities of the amplitudes V_1 and V_2. What is the corresponding singularity structure of the current commutator? To work this out we return to (2.7). Clearly the leading singularity of the commutator arises from V_1 and we have

$$\langle p | [j_\mu(x), j_\nu(0)]_c | p \rangle = \frac{i}{\pi}\, g_1(px)\, \partial_{\mu\nu}\{\epsilon(x^0)\delta(x^2)\} + \text{less singular} \tag{6.8}$$
$$\text{terms}$$

This expression for the leading light cone singularity describes, in particular, the leading singularity of the current commutator near the <u>tip</u> of the cone. It is a simple matter to work out the equal-time limit with the result

$$\langle p | [j_0(x), j_0(0)]_c | p \rangle = \langle p | [j_i(x), j_k(0)]_c | p \rangle = 0$$
$$\langle p | [j_0(x), j_i(0)]_c | p \rangle = 2i\, g_1(0)\, \partial_i \delta^3(\vec{x}) \tag{6.9}$$

The commutator between space- and time-components of the current contains a <u>Schwinger term</u> given by

$$g_1(0) = \int_{-1}^{1} d\xi\, \frac{F_L(\xi)}{\xi} \tag{6.10}$$

Note that this expression concerns only the operator part of the

Schwinger term as we are discussing only the connected part of the current commutator.

Some remarks concerning the validity of the Schwinger term sum rule (6.10) are in order here. We first note that the scaling function F_L/ξ which appears on the right hand side may be a rather singular object at $\xi = 0$. In fact, if the Pomeron couples to the longitudinal scaling function one expects F_L/ξ to behave like ξ^{-2} at $\xi = 0$. Nevertheless, the integral (6.10) makes sense because the scaling law defines the quantity F_L/ξ not as a function but as a distribution for which integrals like (6.10) are unambiguously specified[9]. Next, we mention that the ambiguity $S_1(q^2)$ in the causal extension of the function V_1 discussed in section 3 affects both the scaling function F_L/ξ and the Schwinger term $g_1(0)$ in such a manner as to keep (6.10) intact. The change (3.3) in V_1 amounts to the change

$$F_L'(\xi)/\xi = F_L(\xi)/\xi + \delta(\xi)\int_0^\infty ds\, S_1(s) \tag{6.11}$$

in the scaling function. At the same time the Fourier transform of V_1 is transformed according to (3.4). The contribution of the additional term to the coefficient of the leading light cone singularity is a constant :

$$g_1'(px) = g_1(px) + \int_0^\infty ds\, S_1(s) \tag{6.12}$$

The sum rule (6.10) clearly remains true.

The significance of this result, unfortunately, is that the Schwinger term is not really a measurable quantity, unless one finds a way to nail down the ambiguity in the causal extension of V_1.

Another somewhat disappointing feature of the Schwinger term sum rule is connected with positivity. The positivity conditions mentioned in section 2 imply that the quantity F_L/ξ is positive. One is therefore tempted to conclude - and this erroneous conclusion may indeed be found e.g. in ref.[11] - that the Schwinger term $g_1(0)$ can be zero only if the entire integrand F_L/ξ vanishes. An obvious counter example to this conclusion is provided by adding a term of the type $S_1(q^2)$ to V_1, which is sufficiently large and negative. Such a term escapes the positivity conditions because its support may be taken to lie entirely in the region $q^2 > 4 m_\pi^2$ where the positivity conditions do not hold. Why is the conclusion wrong? The reason is easy to see if one recalls that the scaling "function" F_L/ξ is defined by a limit operation on a distribution. Positivity holds provided we smear the amplitudes V_1 , V_2 with positive test functions whose support is contained in $q^2 < 4m_\pi^2$, $\nu > 0$. One verifies that test functions with this property vanish for $\xi \leq 0$ and one is therefore entitled to conclude that F_L / ξ is positive only on test functions whose support excludes $\xi = 0$. An example of a distribution which is positive on all such functions but which nevertheless gives a negative value to the integral (6.10) is $F_L / \xi = \theta(1 - \xi^2) |\xi|^{-2}$ for which $\int d\xi \, F_L / \xi = -2$.

7.   <u>ABSENCE OF THE WORST SINGULARITY</u>

We have mentioned that the experimental uncertainties in the determination of V_1 leave the very attractive possibility open that the scaling function $F_L (\xi)$ vanishes. Let us now discuss this possibility in more detail. We assume that V_1 not only has no singularity of the type $\delta(x^2)$ on the light cone, but that it is leading singularity is instead of the stepfunction-type:

$$\widetilde{V}_1(x) = \frac{i}{4\pi} \, \epsilon(x^0) \, \theta(x^2) \, h_1(px) + R_1(x) \tag{7.1}$$

(There is, of course, the possibility that the strength of the leading singularity lies somewhere between $\delta(x^2)$ and $\theta(x^2)$ say $\widetilde{V}_1 (x) \sim (x^2)_+^{-1/2}$. We restrict ourselves to a discussion of the most optimistic and most interesting hypothesis). If (7.1) indeed describes the leading singularity of $\widetilde{V}_1(x)$ correctly we have the scaling law

$$V_1 (\nu, -2\xi\nu - \eta) \;\longrightarrow\; \frac{1}{2\gamma^2} \, \frac{G_1(\xi)}{\xi} \tag{7.2}$$

This scaling law states that the ratio σ_L / σ_T of longitudinal to transverse cross sections not only tends to zero in the Bjorken limit, but scales according to

$$\frac{\sigma_L}{\sigma_T} \;\longrightarrow\; \frac{(-q^2)}{\gamma^2} \left\{ 1 + \frac{G_1(\xi)}{F_2(\xi)} \right\} \tag{7.3}$$

If this scaling law fails to be verified by the data then all we are saying from here on is for entertainment only. Since under the above assumptions the functions V_1 and V_2 have the same singularities on the light cone they contribute equally to the leading singularity of the current commutator which may then be written in the form

$$\langle p | [j_\mu(x), j_\nu(0)]_C | p \rangle = \frac{1}{i} \{ b_\mu(x) \partial_\nu D(x) + b_\nu(x) \partial_\mu D(x)$$
$$- g_{\mu\nu} b^s(x) \partial_s D(x) \} \qquad \text{+ less singular} \atop \text{terms}$$

$$(7.4)$$

where

$$D(x) = \frac{1}{2\pi} \epsilon(x^0) \delta(x^2) \tag{7.5}$$

$$b_\mu(x) = x_\mu h_1(px) + 2 p_\mu p x h_2(px)$$

This singularity structure is to be compared with the result (6.8) found previously for the more singular case $F_L \neq 0$. It is interesting to note that the structure (7.4) is characteristic of a current associated with spin $\frac{1}{2}$ particles, whereas the more singular structure (6.8) holds for spinless particles. In fact, let φ and ψ denote free fields of spin 0 and spin $\frac{1}{2}$ respectively. The corresponding currents

$$J_\mu^a = i : \varphi^*(x) \overleftrightarrow{\partial}_\mu t^a \varphi(x) :$$
$$K_\mu^a = : \bar\psi(x) \gamma_\mu t^a \psi(x) : \tag{7.6}$$

have the following leading singularities on the light cone (we consider only the connected parts):

$$[J_\mu^a(x), J_\nu^b(y)]_C = \frac{1}{i} B^{ab}(x,y) \partial_{\mu\nu} D(x-y) + \text{ less singular terms}$$
$$[K_\mu^a(x), K_\nu^b(y)]_C = \frac{1}{i} \{ B_\mu^{ab}(x,y) \partial_\nu D(x-y) + B_\nu^{ab}(x,y) \partial_\mu D(x-y)$$
$$\tag{7.7}$$
$$- g_{\mu\nu} B_s^{ab}(x,y) \partial^s D(x-y) + i \epsilon_{\mu\nu\rho\sigma} \tilde B^{abs}(x,y) \partial^\sigma D(x-y) \}$$
$$\text{+ less singular terms} \qquad (7.8)$$

The coefficients of the leading singularities are bilocal fields:

$$B^{ab}(x,y) = : \varphi^*(x) t^a t^b \varphi(y) : + : \varphi^*(y) t^b t^a \varphi(x) :$$

$$B_\mu^{ab}(x,y) = i : \bar\psi(x) \gamma_\mu t^a t^b \psi(y) : - i : \bar\psi(y) \gamma_\mu t^b t^a \psi(x) :$$

$$\tilde B_\mu^{ab}(x,y) = : \bar\psi(x) \gamma_\mu \gamma_5 t^a t^b \psi(y) : + : \bar\psi(y) \gamma_\mu \gamma_5 t^b t^a \psi(x) :$$

The functions $g_1(px)$ and $b_\mu(x)$ in (6.8) and (7.4) correspond to matrix elements of these bilocal fields (The spin averaged forward matrix element of $\tilde{B}_\mu$ vanishes.) Qualitatively the question of whether or not the worst singularity $\sim \partial_{\mu\nu} D(x)$ in the current commutator is absent therefore amounts to the question of whether or not the photon sees any pointlike <u>scalar</u> constituents in the proton. If it interacts only with pointlike spin $\frac{1}{2}$ "partons" then one expects the current to exhibit the smoother singularity (7.4) on the light cone.

Needless to say that these arguments are very vague. If the fields φ and ψ are not free but interacting fields then it is still possible to formally work out the leading light cone singularities of the current by means of the canonical formalism adapted to lightlike rather than spacelike planes[12]. One finds that for renormalizable interactions the leading singularities remain canonical; the interaction only affects the coefficients of the leading singularities. It is however well-known that these formal canonical manipulations are not reliable. Perturbation theory reveals that renormalizable interactions produce logarithmic corrections in the behaviour of the fields on the light cone[13]. As mentioned earlier, the question of whether or not there is a theory of interacting fields with canonical singularities in the current commutator is entirely open.

If one is willing to accept that the leading singularities of the weak and electromagnetic currents are indeed of the type (7.8) then interesting algebraic structures may be set up. Some of these algebras were discussed in the lectures; I do not include this material here as there is abundant literature on the subject[14]. The "light cone algebras" may be regarded as an extension of the familiar equal-time current algebra to the light cone. As in the case of old-fashioned current algebra one may attempt to saturate the light cone algebras and to find explicit representations of the commutation relations. The representation theory of light cone commutation relations has very interesting features with the fact that neither Haag's nor Coleman's theorems are valid for lightlike planes[15]. Work on explicit representations of light cone algebras is still in a very primitive stage. It is known that some of the infinite component field models which saturate local current algebra at infinite momentum also saturate the algebra of currents on the light cone[16], but it is not known whether this algebra admits of any non-trivial solutions which do not show the unphysical features familiar from infinite component field equations.

8.   CAUSALITY

Finally we briefly mention a different topic intimately related to the material discussed here. We have seen that a specific singularity structure on the light cone implies scaling laws in momentum space. In this connection the causality property of the functions $\widetilde{V}_1 (x)$ and $\widetilde{V}_2 (x)$ is irrelevant. The scaling laws merely state that these quantities possess a specific jump on the light cone. If one adds an arbitrary smooth function to $\widetilde{V}_1(x)$ or to $\widetilde{V}_2(x)$ the scaling laws still hold, even if this function does not vanish outside the light cone. The additional information that the functions $\widetilde{V}_i (x)$ are causal manifests itself in the form of sum rules[11]. As an example we quote the sum rules satisfied by $V_2 (\nu , q^2)$:

$$\int_{-\infty}^{\infty} d\nu \, V_2 (\nu_1 , -2 \xi \nu - \eta) = 0$$

$$\int_{-\infty}^{\infty} d\nu \, \nu \, V_2 (\nu_1 , -2 \xi \nu - \eta) = \frac{1}{2} P \int_{-1}^{1} \frac{d\xi'}{\xi' - \xi} \frac{F_2(\xi')}{\xi'} \tag{8.1}$$

These sum rules are necessary and sufficient conditions for V_2 to be the Fourier transform of a causal object. For a detailed discussion of these sum rules, in particular for an analysis of their fixed q^2 limit (Cornwall-Corrigan-Norton sum rule) the reader is referred to L O. We emphasize that the leading light cone singularity guarantees that these sum rules are valid only in the sense of distributions in ξ , η . Similarly, the light cone guarantees that the dispersion relation for the Compton amplitudes

$$V_i^{ret} (\nu_1 , -2 \xi \nu - \eta) = \frac{1}{2\pi i} \int_{-\infty}^{\infty} \frac{d\nu}{\nu' - \nu - i\varepsilon} V_i (\nu_1' , -2 \xi \nu - \eta) \tag{8.2}$$

is valid without subtractions. The light cone does not control their fixed q^2 limits, however, which in general do require subtractions. These matters are discussed at length in L O.

The sum rules (8.1) cannot be confronted with experiment without further ado since they require knowledge of the function $V_2(\nu , q^2)$ for all q^2 , including the unphysical region. It is however possible to rewrite them in a form which allows to test saturation in the physical region only. Such a test has been performed in ref.[17]. For more details and a comparison of these sum rules with the Finite Energy Sum Rules proposed by Bloom and Gilman and by Rittenberg and Rubinstein we again refer to L O.

REFERENCES

1. Y. Georgelin, H. Sazdijan and J. Stern, Nuovo Cim., in press.

2. P. Stichel, Contribution to "Ecole Internationale de la Physique des Particules Elementaires", Basko Polje-Makarska (Yougoslavia), 1971.

    LO H. Leutwyler and P. Otterson, _Theoretical Problems in Deep Inelastic Scattering_, Proceedings of the Frascati meeting on Outlook for Broken Conformal Symmetry, 1972, to be published.

3. A.N. Tavkhelidze, Lectures given at the Kaiserslautern Summer School, 1972.

4. J.W. Meyer and H. Suura, Phys.Rev. $\underline{160}$, 1366 (1967).

5. N.N. Bogoliubov and V.S. Vladimirov,Nauchn.Dokl.Vysshei Shkoly $\underline{3}$, 26 (1958). For a generalization to many variables see J. Bros, H. Epstein and V. Glaser, Comm. Math. Phys. $\underline{6}$, 77 (1967).

6. S. Ciulli, Lectures given at the Kaiserslautern Summer School, 1972.

7. V.F. Müller and W. Rühl, Nucl.Phys. B $\underline{30}$, 513 (1971) Fritz Schwarz, preprint Kaiserslautern, April 1972; Jour.Math. Phys., October 1972.

8. I.M. Gel'fand and G.E. Shilov, Generalized Functions, Academic Press, 1968.

9. Y. Georgelin, H. Leutwyler and J. Stern, _Some Mathematical Aspects of Light Cone Physics_, to be published; Y. Frishman and S. Yankielowicz, _Scaling Laws and Equal Time Commutators_, preprint Weizmann Institute, Rehovoth, 1972.

10. For a review of experimental data see, e.g. H.W. Kendall, Report at the International Symposium on Electron and Photon Interactions, Cornell University, 1971.

11. H. Leutwyler and J. Stern, Nuclear Physics $\underline{B\ 20}$, 77 (1970).

12. F. Jackiw, Springer Tracts in Modern Physics, $\underline{62}$, 1 (1972) For an elegant treatment of canonical light cone commutators on the basis of Dirac's method for dealing with constraints see L. Bányai and L.A. Mezincescu, preprint Institute of Physics, Bucharest, 1972.

13. R.A. Brandt and G. Preparata, Nuclear Physics, in press.

14. See ref.[12]; J.M. Cornwall, and R. Jackiw, Phys.Rev.$\underline{D4}$, 367, (1971); H. Fritzsch and M.Gell-Mann, Tracts in Mathematics

and the Natural Sciences, Vol. $\underline{2}$, 1, Gordon and Breach,
(1971); H. Leutwyler, <u>Superlocal Sources</u> in "Magic without
Magic: John Archibald Wheeler", ed. by J.R. Klauder, Freeman
Co. San Francisco, on press since 1970.

15.   F. Rohrlich, Acta Physica Austriaca, Suppl VIII, 277 (1971);
J.R. Klauder, H. Leutwyler, and L. Streit, Nuovo Cim. $\underline{66A}$,
536,(1970).

16.   H. Bebie, V. Gorgé, and H. Leutwyler, Ann. Physics, in press;
H. Kleinert, Proceedings of the Frascati meeting on Outlook
for Broken Conformal Symmetry, 1972, to be published.

17.   H. Leutwyler and J. Stern, Phys. Letters, $\underline{31B}$, 458 (1970).

18.   This theorem is a special case of the Convex Hull Theorem
(V.C. Vladimirov, Dokl.Akad.Nauk, SSSR $\underline{134}$,251 (1960);
H.J. Borchers, Nuovo Cim. $\underline{19}$, 787, (1961)).
I am indebted to Prof. Vladimirov for pointing this out.

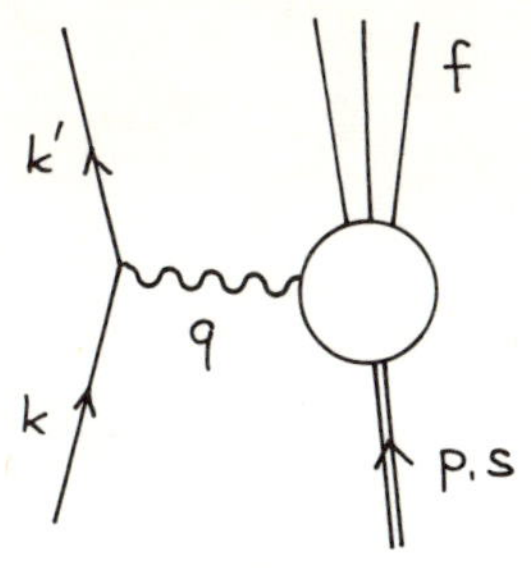

Fig. 1

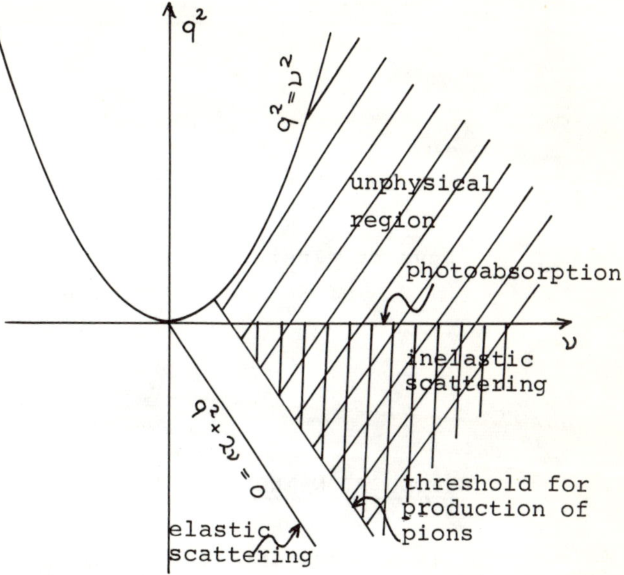

Fig. 2

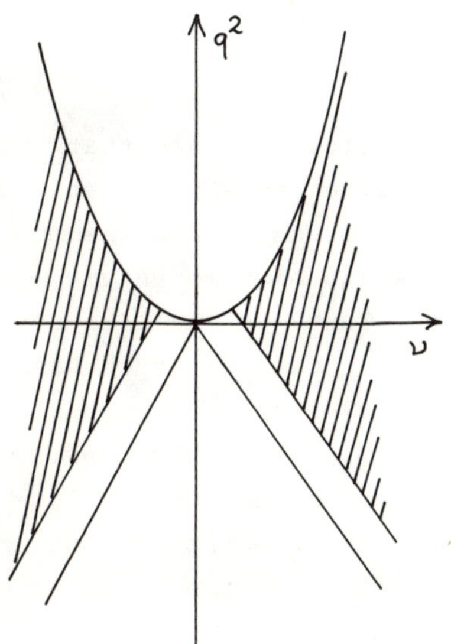

Fig. 3

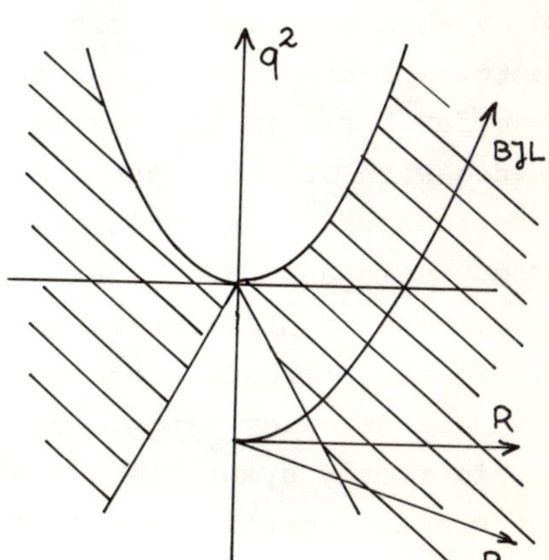

Fig. 4

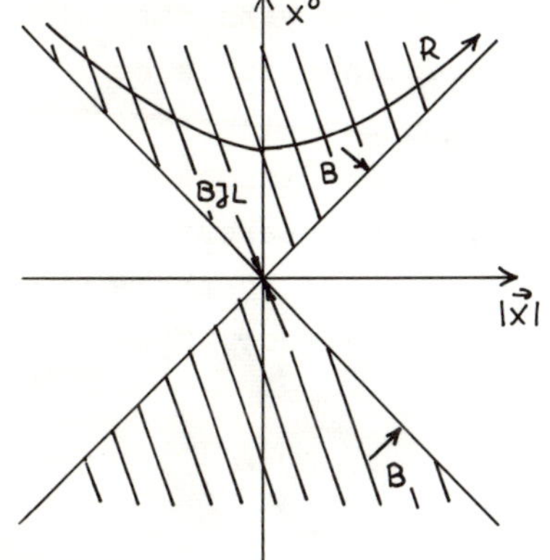

Fig. 5

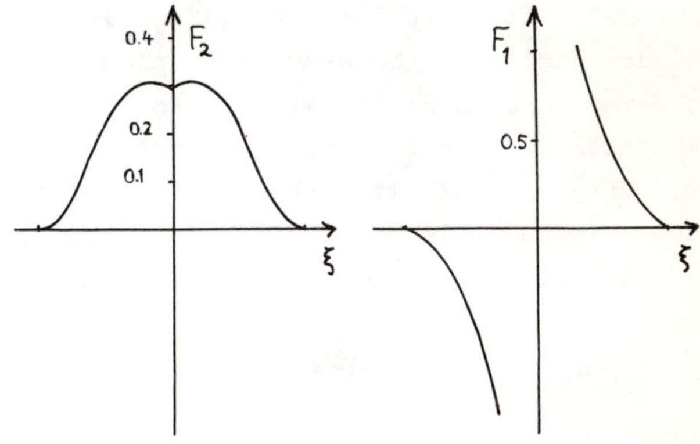

Fig. 6

I.T. Todorov

Joint Institute for Nuclear Research, Dubna, USSR
and
Physical Institute of the Bulgarian Academy
of Science, Sofia, Bulgaria

1. INTRODUCTION AND OUTLOOK

It is a recurrent idea in elementary particle physics that the approximate strong interaction symmetries become exact at short distances when rest masses can be neglected . "Short distances" in the momentum space picture means not only high energies and high momentum transfer but also high external momenta squared. Thus, we shall be concerned with off-mass-shell Green functions. It is known that such Green functions can be associated with observable quantities only through electromagnetic and weak interactions which are treated as perturbations. An example is provided by the off-shell photon line in the deep inelastic electron-proton scattering considered in the approximation of one virtual photon exchange. It is, therefore, wise for the time being to consider only symmetries of the strong interactions and to look at weak and electromagnetic interactions chiefly as means of producing off-mass-shell external legs in the hadronic vertex function.

In these lectures we shall be mainly concerned with conformal invariance[1]. In the example of pion-nucleon interactions it is considered in combination with isotopic spin and γ_5 - invariance. Physically, we expect that under the same asymptotic conditions chiral symmetry will be also exhibited. In fact, the renewed interest in approximate scale and conformal invariance was encouraged by the success of chiral SU(3) x SU(3) and current algebra. For instance, Mack's suggestion[2] that dilatation current is partially conserved (PCDC) was clearly inspired by the idea of PCAC. It is only the smallness of the pion mass which makes PCAC an useful tool for deriving low-energy theorems. Theoretically, it should be expected to become a better symmetry at large momenta when all the masses of the pseudoscalar octet can be neglected (just as in the case of PCDC).

The problem we shall deal with here is a theoretical one: Is it possible to construct an exactly conformal invariant field theory model (for all momenta)? Clearly, if such a model does exist, it may be expected to fit reality for large momenta only. One can also predict that an on-shell S-matrix would not exist in such a model because of infrared problem. In accord with our idea of short distance behaviour we shall only attempt to construct off-shell Green's functions.

The possibility of an affirmative answer to our question was opened by Wilson suggestion[3] that an infinite field-strength renormalization may lead to "anomalous" (i.e. non-canonical) dimensions. Indeed, conformal invariance determines the two-point functions of a field uniquely up to a normalization constant. If dimensions are canonical then we obtain the two-point functions of a free zero-mass field. This implies[4] that the entire theory should be a free one. Only anomalous dimensions which bar the existence of asymptotic free states can be combined with conformal invariance to construct nontrivial Green functions.

The assumption of conformal invariance is attractive theoretically since it allows to determine the "physical" two-and three-point functions[5] . It turns out[6] that for field-dimensions fixed in a certain range the resulting skeleton graph expansion for the many point Green functions is free from ultraviolet divergences. The plan of the present lectures is the following. First, we shall describe briefly the manifestly conformal covariant technique. In Chapter 3 we determine the invariant 2- and 3- point functions by a systematic use of conformal inversion invariance. In Chapter 4 we prove the absence of ultraviolet divergences in the skeleton graph expansion. In conclusion, we will discuss possible relation with Lagrangian field theory. The important problem of studying the renormalized Schwinger-Dyson equations[7,8] (the "bootstrap" equations) is mentioned only briefly.

2.   <u>SYNOPSIS ON THE CONFORMAL GROUP AND THE MANIFESTLY</u>
<u>COVARIANT FORMALISM</u>

A.   <u>The conformal group and its Lie algebra</u>

The conformal group is a 15 parameter group isomorphic to the group of pseudorotation $O(4,2)$ in 6 dimension. It acts in a natural way on a compactification $\overline{M^4}$ of Minkowski space defined as the set of isotropic rays in 6 dimension. In order to exhibit this action we shall first imbed M^4 into the 6-dimensional "light-cone"

$$C_{2,4} = \left\{ \xi = (\xi_0, \xi_1, \xi_2, \xi_3, \xi_5, \xi_6) \mid \xi^2 = g^{ab}\xi_a \xi_b = \xi_0^2 - \xi_1^2 - \xi_2^2 - \xi_3^2 - \xi_5^2 + \xi_6^2 = 0 \right\} \tag{2.1}$$

The space-time coordinates x_μ can be defined as local coordinates on the subset

$$\xi_5 + \xi_6 = \kappa \neq 0 \tag{2.2}$$

of $C_{2,4}$. We can set

$$\xi_\mu = \kappa x_\mu , \ \mu = 0,1,2,3 , \qquad \xi_5 + \xi_6 = \kappa , \ \xi_5 - \xi_6 = \kappa x^2 \tag{2.3}$$

Let $\overset{\circ}{\mathbb{R}}$ be the multiplicative group of similarity transformations $\xi' = \lambda \xi , (\lambda \neq 0)$ of $C_{2,4}$. The above mentioned compactification $\overline{M^4}$ of M^4 is defined as the factor space

$$\overline{M^4} = C_{2,4} / \overset{\circ}{\mathbb{R}} \tag{2.4}$$

which is isomorphic to the set of isotropic rays in 6-dimension. (We can think of $\overline{M^4}$ as obtained from M^4 by adding a "light-cone at infinity": $\xi_5 + \xi_6 = 0 , \ \xi_\mu \xi^\mu = 0$.)

The conformal group $O(4,2)$ acts as a group of pseudorotation on the cone (2.1). Its action obviously commutes with the set of similarity transformations $\xi' = \lambda \xi$. Therefore, it is well defined (and transitive) on the homogeneous space (2.4). The corresponding space-time transformations are compounded of

a) Poincaré transformations $\quad x_\mu' = \Lambda_\mu{}^\nu x_\nu + a_\mu \; , \quad \Lambda \in L^\uparrow, \; \mu = 0,1,2,3$

b) dilations $\quad x_\mu' = \varrho\, x_\mu \; , \quad \varrho > 0$

c) special conformal transformations

$$x_\mu' = (1 - 2cx + c^2 x^2)^{-1} (x_\mu - c_\mu x^2) = (R\, T_c\, R\, x)_\mu \tag{2.5}$$

where $\quad R\,x = - x/x^2 \, , \quad T_c\, x = x + c.$ The mapping (2.5) is not well defined on M^4 : for nonisotropic c the light-cone $(x - \frac{c}{c^2})^2 = 0$ is mapped on the light-cone at infinity, for $c^2 = 0$ the exceptional surface degenerates into a hyperplane.

The generators of special conformal transformations, K_μ , and of dilatations, D , obey the following commutation relations among themselves and with the generators $M_{\mu\nu}$ and P_μ of the Poincaré group:

$$[D, M_{\mu\nu}] = 0 \qquad [D, P_\mu] = -i P_\mu \qquad [D, K_\mu] = i K_\mu$$

$$[K_\mu, K_\nu] = 0 \qquad [K_\mu, M_{\lambda\nu}] = i [g_{\mu\lambda} K_\nu - g_{\mu\nu} K_\lambda],$$

$$[P_\mu, K_\nu] = 2i\, (D\, g_{\mu\nu} + M_{\mu\nu}). \tag{2.6}$$

B. <u>The field transformation law</u>

Let $\psi(x)$ be a Poincaré covariant quantized field, satisfying

$$U(a, \Lambda)\, \psi^\alpha(x)\, U^{-1}(a, \Lambda) = V(\Lambda^{-1})^\alpha{}_\beta\, \psi^\beta(\Lambda x + a) \tag{2.7}$$

where $V(\Lambda)$ is a finite dimensional (in general-double valued) representation of the orthochronous Lorentz group $L^\uparrow$ with generators $S_{\mu\nu}$:

$$[\psi^\alpha(x), M_{\mu\nu}] = [\, i\, (x_\mu \partial_\nu - x_\nu \partial_\mu)\, \delta^\alpha_\beta + (S_{\mu\nu})^\alpha_\beta \,]\, \psi^\beta(x) \tag{2.8}$$

$$(\, S_{\mu\nu} = \frac{i}{4} [\gamma_\mu \gamma_\nu] \quad \text{for a Dirac field}, (S_{\mu\nu})^\kappa_\lambda = i\, (\delta^\kappa_\mu g_{\nu\lambda} - \delta^\kappa_\nu g_{\mu\lambda})$$

for a vector field). We shall say that ψ is a conformal covariant field if it has a definite scale dimension d_ψ (in mass units):

$$U(\varrho)\, \psi(x)\, U^{-1}(\varrho) = \varrho^{d_\psi}\, \psi(\varrho x) \tag{2.9}$$

and if in addition it is covariant under infinitesimal special
conformal transformations:

$$U(\varepsilon)\,\psi(x)\,U^{-1}(\varepsilon) \underset{\varepsilon^\kappa \to 0}{=} \psi(x) - i\,\varepsilon^\kappa \left[\psi(x), \mathcal{K}_\mu\right] \qquad (2.10)$$

where

$$\left[\psi(x), \mathcal{K}_\mu\right] = i\left(2\,d_\psi\,x_\mu + 2\,x_\mu x_\nu \partial^\nu - x^2 \partial_\mu - 2i\,x^\nu s_{\mu\nu}\right)\psi(x) \qquad (2.11)$$

Unlike the global conformal transformations these infinitesimal rules
are well defined for a field in Minkowski space, and do not violate
the causal order of events. The non-existence of global unitary
transformations corresponding to the infinitesimal law (2.10),(2.11)
indicates that the operators $\mathcal{K}_\mu$, though formally hermitian, are
not self-adjoint (cf.[9]).

In what follows we shall be interested in the interaction
of a (pseudo)scalar field $\varphi(x)$, a Dirac spinor field $\psi(x)$
and a conserved vector-current $V^\mu(x)$. They all may
have in addition an internal (say isotopic) index. Henceforth we shall
write $d_\varphi = d$, $d_\psi = d'$. (The canonical values of these
dimensions are $d = 1$, $d' = \frac{3}{2}$.) We shall see that the conserved
current always has canonical dimension $d_V = 3$.

We shall assume that the dilatations and the infinitesimal
special conformal transformations have the vacuum vector invariant,
and therefore that the n-point Wightman and Green functions of our
fields are also conformal invariant in the above sense. Such a
theory is called weakly conformal invariant in the terminology of
ref.[9].

3.   COVARIANT TWO-AND THREE - POINT FUNCTIONS

   A.   Conformal inversion and conservation of dimension. The
         invariant 2-point functions for a spinless field

The general covariant expressions for the 2-point functions
are easily obtained using the manifestly covariant technique[6].
Here we shall give a direct x-space derivation. It is instructive,
because it exhibits the role of the different invariance conditions.

We start with the Wightman 2-point function

$$w_{12}(x_1, x_2) = \langle 0 | \varphi_1(x_1) \varphi_2(x_2) | 0 \rangle$$

of two (pseudo)scalar fields φ_1 and φ_2. Poincaré and scale invariance combined with the spectrum condition give

$$w_{12}(x_1 x_2) = \frac{N}{x_{12}^{d_1 + d_2}}$$

where

$$x_{12} = \left[i 0 (x_1^0 - x_2^0) - (x_1 - x_2)^2 \right]^{1/2} \tag{3.1}$$

It is convenient to consider the analytic continuation of w_{12} to the euclidean region in which x^0 is pure imaginary ($x^0 = i x_4$. Since w_{12} is analytic in the extended tube (defined by $x_{12}^2 = -a < 0$) the euclidean points lie in the domain of analyticity. The advantage of working with euclidean coordinates (and momenta) is twofold. First of all, it allows us to consider global conformal transformations, which, for pseudoeuclidean coordinates, would have destroyed the $i0$ sign convention in expressions like (3.1). Second, the skeleton "Feynman" integrals are absolutely convergent in the euclidean region which allows to justify certain formal manipulations with them[6,8].

The conformal group for euclidean coordinates is $O(5,1)$. It is generated by the euclidean group $E(4)$ (which replaces the Poincaré group) and the inversion R defined in (2.5). In the euclidean case the unproper conformal transformation R assumes the form

$$R x = \frac{x}{|x|^2} \quad , \quad |x|^2 = \underline{x}^2 + x_4^2 \quad \left(= \underline{x}^2 - x_0^2 \right) . \tag{3.2}$$

In the 6-dimensional language of the ξ - variables R corresponds to reflection of the fifth axis: $(\xi_\mu, \xi_5, \xi_6) \to (\xi_\mu, -\xi_5, \xi_6)$. In what follows we shall derive explicit expressions for the 2- and 3-point functions from Poincaré (or $E(4)$), dilatation, and R — invariance.

Let us first turn back to the (pseudo) scalar fields. Under conformal inversion R a spinless "euclidean field" $\varphi(x)$ obeys the transformation law

$$U(R) \varphi(x) U^{-1}(R) = \eta_\varphi \frac{1}{|x|^{2d}} \varphi \left(\frac{x}{|x|^2} \right) \tag{3.3}$$

where $\eta_\varphi = 1$ for a scalar field and $\eta_\varphi = -1$ for a pseudo-scalar field.

A justification of this phase convention will be given later.

Assuming R-invariance for w_{12} and using that

$$| Rx_1 - Rx_2 |^2 = \frac{|x_1 - x_2|^2}{|x_1|^2 |x_2|^2} \tag{3.4}$$

we find

$$\frac{1}{|x_1 - x_2|^{d_1+d_2}} = \frac{\eta_1 \eta_2}{|x_1 - x_2|^{d_1+d_2}} \left| \frac{x_2}{x_1} \right|^{d_1-d_2} .$$

This is only possible for $\eta_1 = \eta_2$ and

$$d_1 = d_2 . \tag{3.5}$$

Thus we obtain, as a consequence of conformal invariance, the "conservation of scale dimension".

Going back to pseudoeuclidean coordinates we can write down the following conformal invariant expression for the 2-point Wightman function of a (pseudo) scalar field

$$\widetilde{\mathcal{F}}_d(x_1 - x_2) = N_\varphi^{-1} \langle 0 | \varphi(x_1) \varphi^*(x_2) | 0 \rangle = \frac{\Gamma(d)}{(4\pi)^2} \left(\frac{2}{x_{12}} \right)^{2d} \tag{3.6}$$

where x_{12} is given by (3.1) and N_φ is a normalization constant. Its Fourier transform is again a homogeneous function (of p):

$$F(p) = \int \mathcal{F}(x) e^{ipx} d^4x = 2\pi \frac{\theta(p_0)}{\Gamma(d-1)} (p^2)_+^{d-2} \tag{3.7}$$

where $\tau_+^\lambda = \theta(\tau)\tau^\lambda$ is defined as a distribution for non-integer negative λ through analytic continuation in λ (cf.[10]). $F(p)$ is a positive distribution for $d \geq 1$ ($for\ d \to 1, F(p) \to 2\pi\,\theta(p_0)\delta(p^2)$). The corresponding invariant expressions for the causal propagator read:

$$\Delta_d^c(x) = \frac{i}{N_\varphi} \langle 0 | T(\varphi(\tfrac{x}{2}) \varphi^*(-\tfrac{x}{2})) | 0 \rangle = i \frac{\Gamma(d)}{(4\pi)^2} \left(i0 - \frac{x^2}{4} \right)^{-d}$$

$$\widetilde{\Delta}_d^c(p) = \frac{\Gamma(2-d)}{(-p^2-i0)^{2-d}} . \tag{3.8}$$

B. <u>The covariant two-point functions for spinor</u>

<u>and vector fields</u>

The transformation law under the inversion R is defined

for an euclidean spinor field $\psi(x)$ of dimension d' by

$$U(R)\,\psi(x)\,U^{-1}(R) = \frac{1}{|x|^{2d'}}\,V(R,x)\,\psi\left(\frac{x}{|x|^2}\right),$$

$$V(R,x) = \hat{x}\,\gamma_5$$

(3.9)

where

$$\hat{x} = \frac{ix}{|x|}\;,\quad \hat{x}^2 = 1\;,\quad (\hat{x}\,\gamma_5)^2 = 1\;.$$

(3.10)

With the above choice of V we have

$$V(R,x)\left(\frac{x}{|x|^2} - \frac{y}{|y|^2}\right) V^{-1}(R,y) = \frac{x-y}{|x||y|}\;.$$

The general Poincaré and scale invariant euclidean 2-point function of a Dirac field and its conjugate is

$$N_\psi\,\mathcal{F}_{\psi\tilde{\psi}}(x_1 - x_2) = \langle 0|\,\psi(x_1)\,\tilde{\psi}(x_2)\,|0\rangle = \langle 0|\,\tilde{\psi}(x_1)\,\psi(x_2)\,|0\rangle =$$

$$= |x_1 - x_2|^{-2d'}(A + B\hat{x}_{12}),\quad \hat{x}_{12} = \frac{i(x_1 - x_2)}{|x_1 - x_2|}\;.$$

(3.11)

(The second in the above chain of equalities is a consequence of TCP-invariance.) It is easily verified that the transformation (3.9), (3.10) leaves only the $B\hat{x}_{12}$ term invariant. Thus, for the conformal covariant 2-point functions of a spinor field we obtain:

$$\mathcal{F}_{\psi\tilde{\psi}}(x_1 - x_2) = \frac{\Gamma(d'+\frac{1}{2})}{(4\pi)^2}\left(\frac{2}{x_{12}}\right)^{2d'}\hat{x}_{12}\;,$$

(3.12)

$$\mathcal{F}_{\psi\tilde{\psi}}(p) = \frac{2\pi}{\Gamma(d'-\frac{3}{2})}\,\slashed{p}\,\theta(p_0)\,(p^2)_+^{d'-\frac{5}{2}}\;;$$

$\mathcal{F}_{\psi\tilde{\psi}}$ is a positive distribution for $d' \geq \frac{3}{2}$.

$$-iS_{d'}^c(x) = \frac{1}{N_\psi}\,\langle 0|\,T(\psi(\tfrac{x}{2})\,\tilde{\psi}(-\tfrac{x}{2}))\,|0\rangle = \frac{\Gamma(d'+\frac{1}{2})}{(4\pi)^2}\,\frac{\hat{x}}{(i0-\frac{x^2}{4})^{d'}}$$

$$\tilde{S}_{d'}^c(p) = \Gamma(\tfrac{5}{2} - d')\,\slashed{p}\,(-p^2 - i0)^{d'-\frac{5}{2}}\;.$$

(3.13)

Remark: Let

$$\varphi_S(x) = A_S : \bar{\psi}(x)\,\psi(x): \quad , \quad \varphi_P(x) = A_P : \bar{\psi}(x)\,\gamma_5\,\psi(x):$$

where

$$: \bar{\psi}(x)\,O\,\psi(y): \; = \lim_{y \to x} \left\{ \bar{\psi}(x)\,O\,\psi(y) - \langle 0|\, \bar{\psi}(x)\,O\,\psi(y)\,|0\rangle \right\} \tag{3.14}$$

(provided that the limit in the **(R)ight** (H) and (S)ide exists in some sense). It is easily seen that for $d = 2d'$ the transformation law (3.3) is a consequence of (3.9). This justifies the choice of the sign in (3.3).

For a vector field $V^\mu(x)$ of dimension d the conformal inversion gives

$$U(R)\,V^\mu(x)\,U^{-1}(R) = \frac{1}{|x|^{2d}} \left(\delta^\mu_\nu + 2\,\frac{x^\mu x_\nu}{|x|^2} \right) V^\nu\left(\frac{x}{|x|^2} \right). \tag{3.14}$$

(We leave to the reader as an exercise to derive (3.14) from (3.9) assuming that $V^\mu(x) = : \bar{\psi}(x)\,\gamma^\mu\,\psi(x):$ and $d = 2d'$.)

The most general Poincaré and scale invariant 2-point function for a vector field $V^\mu(x)$ is

$$\tilde{\mathcal{F}}^{\mu\nu}(x_1 - x_2) = \frac{1}{N_V} \langle 0|\, V^\mu(x_1)\, V^\nu(x_2)\,|0\rangle =$$

$$= \frac{1}{x_{12}^{2d}} \left[A\,g^{\mu\nu} + B\,\frac{(x_1 - x_2)^\mu (x_1 - x_2)^\nu}{x_{12}^2} \right] \tag{3.15}$$

where x_{12} is again given by (3.1). Assuming invariance under the involutive transformation (3.14) we find that $B = 2A$. Thus we can write:

$$\tilde{\mathcal{F}}^{\mu\nu}(x_1 - x_2) = \frac{\Gamma(d+1)}{(4\pi)^2} \left(\frac{2}{x_{12}} \right)^{2d} \left[g^{\mu\nu} + 2\,\frac{(x_1-x_2)^\mu (x_1-x_2)^\nu}{x_{12}^2} \right]$$

$$= \frac{\Gamma(d)}{(4\pi)^2} \left(\frac{g^{\mu\nu}}{2-d}\,\Box + \frac{2}{d-1}\,\partial^\mu \partial^\nu \right) \left(\frac{4}{x_{12}^2} \right)^{d-1} \tag{3.16a}$$

(where $\partial_\mu = \frac{\partial}{\partial x_1^\mu}$, $\Box = \partial_0^2 - \vec{\partial}^2$)

$$\mathcal{F}_d^{\mu\nu}(p) = \frac{2\pi}{\Gamma(d-1)}\,\theta(p_0) \left[(d-1)\,g^{\mu\nu} - 2\,(d-2)\,\frac{p_\mu p_\nu}{p^2} \right] (p^2)_+^{d-2}\,; \tag{3.16b}$$

$$- i\Delta^c_{d\,\mu\nu}(x) = \frac{\Gamma(d)}{(4\pi)^2} \left(\frac{2}{d-1}\,\partial^\mu \partial^\nu - \frac{g^{\mu\nu}}{d-2}\,\Box \right) \left(\frac{4}{i0 - x^2} \right)^{d-1}, \tag{3.17a}$$

$$\tilde{\Delta}^c_{d\,\mu\nu}(p) = \left[(d-1)\,g^{\mu\nu} - 2\,(d-2)\,\frac{p^\mu p^\nu}{p^2} \right] \frac{\Gamma(2-d)}{(-p^2 - i0)^{2-d}}. \tag{3.17b}$$

We note that the expression in square brackets in the first equation (3.16a) is similar to the matrix part of conformal inversion law (3.14). It turns out[11] that the conformal invariant 2-point function for any integer spin can be expressed in terms of a product of such tensor   factors.

It is clear from (3.16) that the current V_μ is conserved if and only if it has the canonical dimension $d = 3$ for a current:

$$\partial^\mu V_\mu = 0 \iff d_V = 3 \;. \tag{3.18}$$

On the other hand we see from (3.17b) that the time-ordered product is not well defined for $d = 3$. In fact, it can be shown (cf.[10]) that no homogeneous Δ^c -distribution exists for $d = 3$. We can, however, define a regularized expression $\Delta^c_{\mu\nu}$ in this case as an "adjoint homogeneous function" which contains a logarithmic term involving an arbitrary constant of dimension of mass squared. Having in mind applications to the isospin current $V^\mu_j (x)$ $(j = 1,2,3)$ we shall write:

$$\delta_{jk} \, \mathcal{F}^{\mu\nu}(x) = \frac{1}{2N_V} \langle 0| V^\mu_j(x) \, V^\nu_k(0) |0\rangle = \frac{\delta_{jk}}{\pi^2} \, R^{\mu\nu}(i\partial) \frac{1}{(i0x_0 - x^2)^2}$$

$$\mathcal{F}^{\mu\nu}(p) = 2\pi \, R^{\mu\nu}(p)\,\theta(p_0)\,\theta(p^2) \;, \quad R^{\mu\nu}(p) = p^\mu p^\nu - g^{\mu\nu} p^2 \;; \tag{3.19}$$

$$\widetilde{\Delta}^c_{\mu\nu} = R_{\mu\nu}(p) \, \log \frac{\tau}{-p^2 - i0} \tag{3.20}$$

C.   <u>The covariant three point functions</u>

Consider first the Wightman function of three scalar fields of scale dimensions d_1 , d_2 and d_3 :

$$\mathcal{W}_3(x_1 \, x_2 \, x_3) = \langle 0| \varphi_1(x_1) \, \varphi_2(x_2) \, \varphi_3(x_3) |0\rangle \tag{3.21}$$

Poincaré and scale invariance allow for any superposition of functions of the type

$$g \, x_{12}^{-\delta_3} \, x_{23}^{-\delta_1} \, x_{13}^{-\delta_2} \tag{3.22}$$

with x_{jk} given by (3.1), provided that $\delta_1 + \delta_2 + \delta_3 = d_1 + d_2 + d_3$.

Under conformal inversion the expression (3.22) goes (for euclidean coordinates) into

$$\frac{|x_1|^{\delta_2 + \delta_3 - 2d_1} \, |x_2|^{\delta_1 + \delta_3 - 2d_2} \, |x_3|^{\delta_1 + \delta_2 - 2d_3}}{|x_1 - x_2|^{\delta_3} \, |x_2 - x_3|^{\delta_1} \, |x_1 - x_3|^{\delta_2}} \; .$$

In order to have R - invariance we have to demand

$$\delta_2 + \delta_3 = 2d_1 \;\; , \quad \delta_1 + \delta_3 = 2d_2 \;\; , \quad \delta_1 + \delta_2 = 2d_3 \; .$$

This gives

$$\delta_1 = d_2 + d_3 - d_1 \;\; , \quad \delta_2 = d_1 + d_3 - d_2 \;\; , \quad \delta_3 = d_1 + d_2 - d_3$$

$$(3.23)$$

Let me write down without derivation the most general covariant expression for the time ordered "pion"-"nucleon" 3-point function. Introducing at this point the isotopic indices we have

$$g_j(x\,;y,z) = \langle 0 | T(\varphi_j(x)\,\psi(y)\,\hat{\psi}(z)) | 0 \rangle =$$

$$= g \; S^c_{\frac{1}{2}d}(y-x)\,\gamma_5\,\tau_j \; S^c_{\frac{1}{2}d}(x-z)\,\Delta^c_{d'-\frac{1}{2}d}(y-z) \;+$$

$$+ g' \; S^c_{d'-\frac{1}{2}d}(y-z)\,\gamma_5\,\tau_j \; \Delta^c_{\frac{1}{2}d}(y-x)\,\Delta^c_{\frac{1}{2}d}(x-z)$$

$$(3.23)$$

where τ_j are the isospin (Pauli) matrices. $[\,\psi(y)$ carries (along with the 4-dimensional Dirac index) a 2-dimensional isotopic index which labels the proton and the neutron states.$]$

If we assume in addition γ_5 -invariance, i.e. invariance under the transformation

$$\psi \to \gamma_5 \psi \;\; , \quad \hat{\psi} \to \hat{\psi}\gamma_5 \;\; , \quad \varphi \to -\varphi \quad (V_\mu \to V_\mu)$$

$$(3.24)$$

then we have to put $g' = 0$ in the RHS of (3.23), so that the 3-point function is determined up to a single constant g .

The 3-point function for a conserved isospin current $V^\mu_j(x)$

and a spinor (and isospinor) field $\psi(y)$ can be written in the form[1]

$$g_j^\mu(x;y,z) = \langle 0| T(V_j^\mu(x)\,\psi(y)\,\widetilde{\psi}(z))|0\rangle =$$

$$= N_\psi\,\Gamma(d'+\tfrac{1}{2})\left\{\frac{1+c}{2}\,S_{3/2}^c(x-y)\,g^\mu\tau_j\,S_{3/2}^c(x-z)\left[i0-\tfrac{1}{4}(y-z)^2\right]^{3/2-d'} + \right.$$

$$\left. + \frac{1-c}{2}\,\frac{y-z}{2}\,\tau_j\left[i0-\tfrac{1}{4}(y-z)^2\right]^{\frac{1}{2}-d'}\left[\Delta_1^c(x-y)\,\overset{\leftrightarrow}{\partial^\mu}\Delta_1^c(x-z)\right]\right\} \quad (3.25)$$

where c is arbitrary constant and $\left[f(x)\,\overset{\leftrightarrow}{\partial^\mu}g(x)\right] =$

$$= f(x)\left(\partial^\mu g(x)\right) - \left(\partial^\mu f(x)\right)g(x)$$

. The constants in the RHS of (3.25) are chosen to ensure the Ward identity

$$\frac{\partial}{\partial x^\mu}\,g_j^\mu(x;y,z) = \langle 0| T(\psi(y)\,\widetilde{\psi}(z)\tau_j)|0\rangle\,\delta(x-z) -$$

$$- \langle 0| T(\tau_j\,\psi(y)\,\widetilde{\psi}(z))|0\rangle\,\delta(x-y) = \quad (3.26)$$

$$= \frac{1}{i}\,N_\psi\,\tau_j\,S_{d'}^c(y-z)\left[\delta(x-z) - \delta(x-y)\right].$$

The identity (3.26) is consistent with the standard current-field equal-time commutation relations

$$\delta(x_0-y_0)\left[V_j^0(x),\,\psi(y)\right] = -\,\delta(x-y)\tau_j\,\psi(y), \quad (3.27)$$

$$\delta(x_0-y_0)\left[V_j^0(x),\,\widetilde{\psi}(y)\right] = \delta(x-y)\,\widetilde{\psi}(y)\tau_j.$$

A consequence of the Ward identity of type (3.26) for a more general vertex function $\langle 0| V^\mu(x)\,\psi_1(y)\,\widetilde{\psi}_2(z)|0\rangle$ is the conservation of scale dimension $(d_1' = d_2')$ also in this case.

One verifies that γ_5-invariance, along with scale invariance and conservation of scale dimension (3.5) is sufficient for obtaining the conformal invariant expressions (3.12), (3.13) for the 2-point functions of a spinor field. We do need R-invariance, however, to fix the 2-point functions (3.16), (3.17) for a vector field.

---

[1] We mention that there is a linear dependence between the three structures written down by Migdal(ref.5) and reproduced in ref.8 (V.Dobrev and D.Stamenov, private communication). The two structures contained in the RHS of (3.25) give the most general conformal and γ_5-invariant expression.

The expression (3.23) and (3.25) can be visualized in terms of an "infraparticle" triangular diagram (Fig.1)

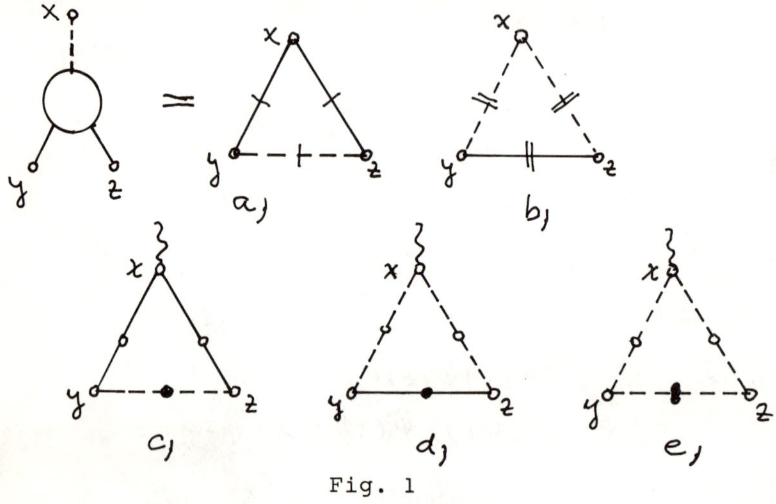

Fig. 1

Infraparticle Feynman diagrams for the invariant vertex functions

a) γ_5 - invariant pion-nucleon Green function;

b) γ_5 - non invariant term in $g_{j\delta}$;

c) d)   contributions in the γ_5 -invariant current-field 3-point functions;

e)   invariant current-pion 3-point Green function.

Analogous expression can be written down for the Green function

$$g^{\mu}_{jk\ell}(x;y,z) = \langle 0|T(V^{\mu}_{j}(x)\, \varphi_{k}(y)\, \varphi_{\ell}(z))|0\rangle \qquad (3.28)$$

where $\varphi_{k}(x)$ is the pion (isovector) field. It is presented graphically on Fig. 1e. The current-field equal-time commutation relations have in this case the form

$$\delta(x_0-y_0)\left[\varphi_{k}(x),\, V^{0}_{j}(y)\right] = i\,\varepsilon_{jk\ell}\,\varphi_{\ell}(x)\,\delta(x-y). \qquad (3.29)$$

The commutation relations (3.27), (3.29) can be supplemented by the standard current algebra relations:

$$\delta(x_0 - y_0)\left[V_j^0(x), V_k^0(y)\right] = i \, \varepsilon_{jk\ell} \, V_\ell^0(x) \, \delta(x - y) \, . \qquad (3.30)$$

According to the general program outlined in the introduction we shall assume that the conserved current V^μ is coupled only electromagnetically (or weakly) to the hadrons and we shall consider matrix elements involving V^μ only in the lowest order in the charge e. [In other words, in the language of Feynman diagrams, no internal wavy lines (corresponding to V) will be considered.]

4. SKELETON GRAPH EXPANSIONS FOR HIGHER GREEN FUNCTIONS AND SELF-CONSISTENCY CONDITIONS

A. Why conformal invariance is not sufficient to fix the n-point functions ($n \geq 4$).

We obtained in the previous chapter explicit expressions for the physical ("dressed") propagators and 3-point Green functions. One cannot proceed further in the same way, since already the 4-point function is not uniquely determined from conformal invariance alone. The reason is that one can construct from 4 points x_1, x_2, x_3, x_4, two conformal invariants given by the two anharmonic ratios

$$\xi = \frac{x_{12}^2 \, x_{34}^2}{x_{13}^2 \, x_{24}^2} \quad , \quad \eta = \frac{x_{14}^2 \, x_{23}^2}{x_{13}^2 \, x_{24}^2} \qquad (4.1)$$

and any function of these two variables will be conformal invariant as well. For instance, the conformal invariant 4-point Wightman function of a (pseudo) scalar field φ of dimension d can be written in the form[12]

$$w_4(x_1 \, x_2 \, x_3 \, x_4) = \langle 0 | \varphi(x_1) \, \varphi(x_2) \, \varphi(x_3) \, \varphi(x_4) | 0 \rangle =$$

$$= \int d\alpha \int d\beta \int d\gamma \; f(\alpha\beta\gamma) \, (x_{12}^2 \, x_{34}^2)^{-\alpha} (x_{13}^2 \, x_{24}^2)^{-\beta} (x_{14}^2 \, x_{23}^2)^{-\gamma} \cdot$$

$$\cdot \; \delta(\alpha + \beta + \gamma - d) \; = \qquad (4.2)$$

$$= \; (x_{13}^2 \, x_{24}^2)^{-d} \iint d\alpha \, d\beta \; f(\alpha, \beta, d - \alpha - \beta) \, \xi^\alpha \, \eta^\beta$$

where X_{jk} are given by (3.1). Locality implies that $f(\alpha,\beta,\gamma)$ should be a symmetric function of its arguments, but we are still left with a large freedom in its choice. This freedom would be reduced if we try to satisfy the positivity requirement which can be obtained as a consequence from a set of generalized unitarity relations (cf.[8] and [13]). It is natural to attempt to fulfil these relation by expanding the n-point Green functions (for $n \geqslant 4$) in skeleton diagrams.

B.   Feynman rules for skeleton diagrams

There are two (equivalent) ways of constructing the skeleton diagrams. First, one can write them down in terms of physical vertex functions (= amputated 3-point Green functions) and propagators (cf.ref.6). Here we shall follow the second way which uses full 3-point Green functions and inverse propagators.

An inverse propagator Δ^{c-1} is defined by the relation

$$\int \Delta^{c-1}(x-y)\, \Delta^c(y-z)\, d^4y = \delta(x-z) .$$

(4.3)

Using the explicit expressions (3.8) and (3.13) for the Fourier transforms of Δ^c and S^c we obtain

$$\Delta^{c-1}_d(x) = (2-d)\,\frac{\sin \pi d}{\pi}\, \Delta^c_{4-d}(x) ,$$

$$S^{c-1}_{d'}(x) = -\frac{1}{\pi}\, \cos \pi d'\, S^c_{4-d'}(x) .$$

(4.4)

(We shall use these formulae for non-integer d and non-half-integer d' only).

For the sake of definiteness we shall speak about the model of conformal and γ_5 - invariant pion-nucleon interaction considered in the previous section. The only non-vanishing 3-point Green function of the pion and nucleon fields is specified by (3.23) with $g' = 0$. We define the n-point Green functions ($n \geqslant 4$) as the sum of all skeleton diagrams, that is, all Feynman diagrams without self-energy part and vertex function corrections (see Fig. 2). It is easily seen that the normalization factors N_ψ and N_φ of (3.8) and (3.13) can be incorporated into g. This is so since the (skeleton) diagrams are invariant under the substitutions indicated in Eq.(4.5) below:

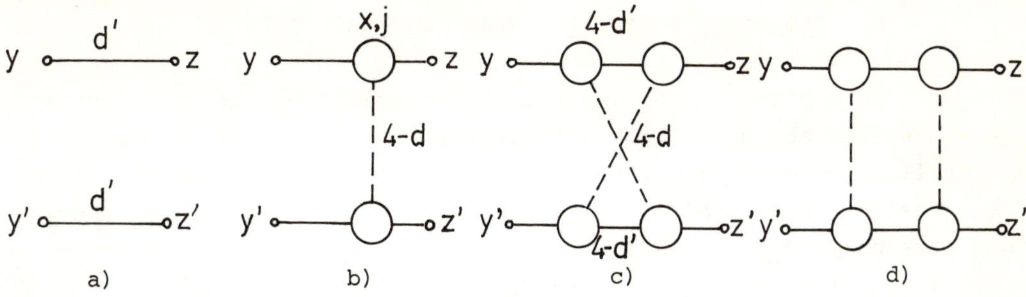

Fig. 2. The set of all nucleon-nucleon scattering skeleton
diagrams up to 4th order in g

$$\longrightarrow \bigcirc \quad 1 \;\longrightarrow\; N_\psi^{-\frac{1}{2}} \qquad\qquad \cdots\!-\!\cdots\bigcirc \quad 1 \;\longrightarrow\; N_\varphi^{-\frac{1}{2}}$$

$$\bigcirc\!-\!\bigcirc \quad S_{d'}^{c^{-1}} \;\longrightarrow\; N_\psi^{-1}\, S_{d'}^{c^{-1}} \qquad\qquad \bigcirc\!\cdots\!\bigcirc \quad \Delta_{d}^{c^{-1}} \;\longrightarrow\; N_\varphi^{-1}\, \Delta_{d}^{c^{-1}}$$

$$g_j \;\longrightarrow\; N_\psi \sqrt{N_\varphi}\; g_j \tag{4.5}$$

Therefore, we shall set in what follows $N_\varphi = N_\psi = 1$.

C. <u>Bootstrap equations. Main results.</u>

In contrast with conventional perturbation theory it turns out [8] that the above mentioned generalized unitarity conditions are not fulfilled identically with respect to the parameters of the theory (g, d and d') by the skeleton graph expansions. They hold true only for those values of the parameters for which the renormalized homogeneous Schwinger-Dyson equations are satisfied. According to [7,6,8] these "bootstrap" equations can be presented graphically as

$$(4.6a)$$

$$(4.6b)$$

$$(4.6c)$$

where and stand for the (two particle) irreducible Bethe-Salpeter kernels of the NN and $\pi\pi$ skeleton diagram expansions.

(The first two diagrams in the expansion of are presented on Fig. 2b,c the graph on Fig. 2d is reducible.) Taking the (4-) divergence of Eqs.(4, 6b, c) and using the Ward identities of type (3.26) we obtain the renormalized equations for the 2-point Green functions.

Our aim in the rest of this chapter will be to give an idea of the proof[6] of absence of ultraviolet and "catastrophic infrared" divergences in the skeleton diagrams as well as in the diagrams involved in the RHS of Eqs.(4.6).

The proof is carried out for euclidean momenta and consists of two steps.

1) We introduce an infrared cutoff m in all (infra) particle propagators and show that the theory thus obtained is convergent, provided that the scale dimensions d and d' lie in the intervals

$$1 < d < 3 \, , \, d \neq 2 \, ; \quad \frac{3}{2} < d' < \frac{5}{2} \qquad (4.7)$$

We note that these intervals are symmetric with respect to transition to the dual (or "shadow") representations ($d \longrightarrow 4 - d$, $d' \longrightarrow 4 - d'$). The corresponding Källen-Lehmann representations for the propagators

$$\widetilde{\Delta}^c_d \, (p) = \frac{1}{\Gamma(d-1)} \int_0^\infty \frac{\tau^{d-2}}{\tau - p^2 - i0} \, d\tau \qquad (4.8a)$$

$$\widetilde{S}^c_{d'} \, (p) = \frac{\not{p}}{\Gamma(d-\frac{3}{2})} \int_0^\infty \frac{\tau^{d'-\frac{5}{2}}}{\tau - p^2 - i0} \, d\tau \qquad (4.8b)$$

need at most one subtraction (in (4.8a) for $d > 2$) in this range of dimensions.

2) One proves that for non-exceptional momenta the limit $m \rightarrow 0$ exists and is conformal invariant. The euclidean momenta $p_1, \cdots p_n$ obeying the conservation law $p_1 + \cdots + p_n = 0$ are called exceptional if some partial sum of them vanishes: $p_{i_1} + \cdots + p_{i_k} = 0 \, (1 \leq k \leq n)$. This step uses essentially Weinberg's power-counting theorem[14].

In the next sections we shall sketch an improved version of the first step of the proof only. The details of the second step can be found in ref.[6].

D.   <u>Feynman rules in momentum space. The convergence</u>
<u>condition</u>

In order to disentangle the problem of ultraviolet divergences
from the low momenta singularities of the Green functions we shall
introduce an infrared cutoff m in the (infra) particle propagators.
Going at the same time to euclidean momenta we will have the following
expressions for the cutoff propagators and of the corresponding
exponents of $(m^2 + |p|^2)^{-1}$ in the case when only vertices of the type
shown in Fig. 1a are present:

$$\tilde{S}^{c-1}_{d'}(p; m) = -\not{p}\,\frac{(m^2 + |p|^2)^{3/2 - d'}}{\Gamma(5/2 - d')} \qquad \left(\lambda = d' - \frac{3}{2}\right)$$

$$\tilde{\Delta}^{c-1}_{d}(p) = \frac{1}{\Gamma(2-d)}\,|p|^{2(2-d)} \qquad (\lambda = d - 2)$$

$$\tilde{S}^{c}_{\frac{1}{2}d}(p) = \Gamma\left(\frac{5-d}{2}\right)\frac{\not{p}}{(m^2 + |p|^2)^{\frac{5-d}{2}}} \qquad \left(\lambda = \frac{5-d}{2}\right)$$

$$\tilde{\Delta}^{c}_{d'-\frac{1}{2}d}(p) = \frac{\Gamma(2 - d' + \frac{1}{2}d)}{(m^2 + |p|^2)^{2 - d' + \frac{1}{2}d}} \qquad \left(\lambda = 2 - d' + \tfrac{1}{2}d\right)$$

$$\tag{4.9}$$

$$\left[\,|p^2| = \underline{p}^2 + p_4^2 \;;\; p_o = i\,p_4\,\right]$$

(We did not introduce an infrared cutoff in $\tilde{\Delta}^{c-1}_{d}$ since it goes
to zero for $p \to 0$ when d is between 1 and 2.)

The intuitive reason why the theory is divergence-free is the
following. The improved behaviour for large momenta of the inverse
propagators more than compensates the stronger singularity of the
g -functions (as compared to canonical perturbation theory).

In order to give a formal proof for the convergence of all
Feynman integrals encountered in our approach we have first to
describe the set of diagrams which appear in the skeleton expansion
and in the bootstrap equations. Each diagram involving marked lines
is obtained from a skeleton graph by substituting the dressed vertices
by the diagram presented on Fig. 1a. We shall use, in particular,
the following properties of the set of diagrams (and subdiagrams) thus

obtained:

(i)   There are no self energy subgraphs.

(ii)   With the exception of the whole uncut graphs in the RHS
of the bootstrap equations (4.6) the only vertex subgraphs are those
presented on Fig. 1 a) c) d) e) .

Obviously, without loss of generality we can restrict our
attention to one-particle irreducible graphs of the above set.

Let λ_h be the exponent of the denominator of the line h
(whose value is fixed by (4.9)). Then, according to the Dyson-
Weinberg power counting theorem[14], (see also[15]) the convergence condition for a graph G is that for any choice of the subgraph $H \subseteq G$

$$\sum_{h \in \mathscr{L}(H)} (\lambda_h - 1) > \left[\frac{\mu_c}{2} \right]$$

(4.10)

Here $\mathscr{L}(H)$ is the set of lines of the subgraph H,
$\mu_c = \mu_c(H)$ is the canonical "superficial degreee of
divergence":

$$\mu_c = 4 + f - n - F - B$$

(4.11a)

or, if there are no 3- boson vertices in H,

$$\mu_c = 4 - B - \frac{3}{2} F$$

(4.11b)

where n is the total number of vertices, f is the number of internal
fermion lines,   B and   F are the numbers of external boson and
fermion lines of H , $[\times]$ stands for the integer part of $\times$.

In order to verify (4.10) we shall express the sum over internal
lines in the L(eft) H(and) S(ide) of (4.10) in terms of a similar sum
over external lines. To do this we observe that for each vertex i,
if $S(i)$ is the set of lines incident to this vertex, we have

$$\sum_{h \in S(i)} (\lambda_h - 1) = 0 \ .$$

(4.12)

This rule could be considered as an extension of the conservation
of dimension (3.5). It holds true also for the elementary vertices
appearing in the triangles of Fig. 1c,d, excluding the upper vertex
(with an external wavy line).

Indeed, we have

$$\tilde{S}^{c}_{3/2}(p) = \frac{\not{p}}{m^2 + |p|^2} \qquad (\lambda = 1)$$

$$\tilde{\Delta}^{c}_{1}(p) = \frac{1}{m^2 + |p|^2} \qquad (\lambda = 1)$$

$$\lambda = \frac{7}{2} - \alpha'$$

$$\lambda = \frac{7}{2} - \alpha'$$ (4.13)

$$\lambda = 3 - \alpha$$

Combining these formulas with the first two equations (4.9) we find that (4.12) holds true for all vertices except the ones appearing in Fig. 1e, where we have instead

$$\sum_{k \in} (\lambda_k - 1) = -1 \tag{4.14}$$

Since the electromagnetic-type vertices (with infrastructure given by (4.13)) appear only once in the bootstrap equations - we shall treat them separately at the end of the convergence proof.

E.   <u>Proof of absence of ultraviolet divergences</u>

Using (4.12) we can rewrite the convergence condition (4.10) in terms of the external parameters only:

$$-\frac{1}{2} \sum_{ext} (\lambda_k - 1) > \left[\frac{\mu}{c}\right] \tag{4.15}$$

where summation is over the external lines attached to subgraph H.

The two diagrams presented on Fig. 3 are the only

Fig. 3
<u>Superficially divergent diagrams</u>

admissible graphs with normal (i.e. uncrossed) external lines for
which $\mu_c \gtrless 0$. In both cases, actually $\mu_c = 0$. It follows from
(4.7) and (4.9) that $\lambda_n - 1 < 0$ for normal external lines so that
the LHS of (4.15) is positive. This implies the validity of (4.15) for
all admissible graphs with normal external lines.

Going to the general case we note that external crossed lines
in a subgraph H can only be obtained by cutting internal lines in a
dressed vertex of the skeleton graph G. Therefore, crossed lines
should always appear in pairs. Let the set E of external lines of a
(sub) diagram H consists of k crossed fermion lines, n crossed
boson lines (n + $\hat{k}$ even) $\mathcal{K}$ normal fermion lines (K + $\hat{k}$ even)
and b normal boson lines. A necessary condition that H is a one-
particle irreducible subdiagram of an admissible graph G is given
by the inequality

$$b + \frac{1}{2}\left(3\,\mathcal{K} + \hat{k}\right) + \min\left(n,k\right) \geqslant 4.\tag{4.16}$$

To see this we note that 1) crossed external boson lines in H can
either be combined with a crossed fermion line to give a normal external fermion line in G or else be connected pairwise between each
other; 2) similarly a pair of crossed fermion lines can be connected
either with each other or in a vertex with a normal boson line, and
3) in a graph with normal external lines only the inequality
$(3/2)\,F + B\ > 4$ should take place. In particular, superficially
divergent subgraphs with crossed external lines (Fig.4)

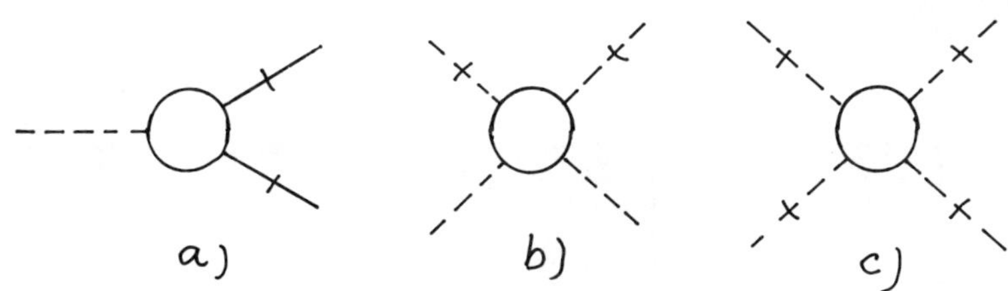

Fig. 4

<u>Superficially divergent subgraphs with crossed external lines</u>

cannot occur. They can only appear as subgraphs of self-energy parts
which are absent by condition (i) stated above. In verifying this
statement we have to use that there are no 3-boson vertex parts, since
any such diagram would involve a fermion loop with an odd number of
lines, and hence, the trace of an odd number of γ -matrices.

Thus condition (4.16) is indeed satisfied.

Let us then have a (sub)graph H with k + n + **x** + b external lines as described above. Taking into account (4.9), we see that the convergence condition (4.15) would be verified if

$$\mu_c + \sum_{int} (\lambda_\ell - 1) = \left(4 - b - m - \tfrac{3}{2}\varkappa - \tfrac{1}{2}k\right) + b(d-3) + \varkappa\left(d' - \tfrac{5}{2}\right)$$

$$+ \left[m - n + \frac{k-n}{2}(1-d)\right] + n\left(\tfrac{3}{2} - d'\right) < 0$$

$$m = \min(k, n) = \frac{k+n}{2} - \frac{|k-n|}{2}. \qquad (4.17)$$

This last inequality is indeed fulfilled since each term in the parentheses is negative because of (4.7) and (4.16). In particular,

$$m - n + \tfrac{1}{2}(k-n)(1-d) = \tfrac{1}{2}\left[(k-n)(2-d) - |k-n|\right] \le \tfrac{1}{2}|k-n|\left(|2-d| - 1\right) \le 0$$

It remains to consider the current-field vertex parts. Using (4.13) we see that the diagram on Fig. 1c) is convergent for $2d' < 5$, which follows from (4.7); the diagrams on figures 1d) and 1e) make sense even under the weaker assumptions $2d' < 6$ and $d < 3$. The only new types of one loop subdiagrams coming from the RHS of the bootstrap equations (4.6b) and (4.6c) are presented on Fig. 5

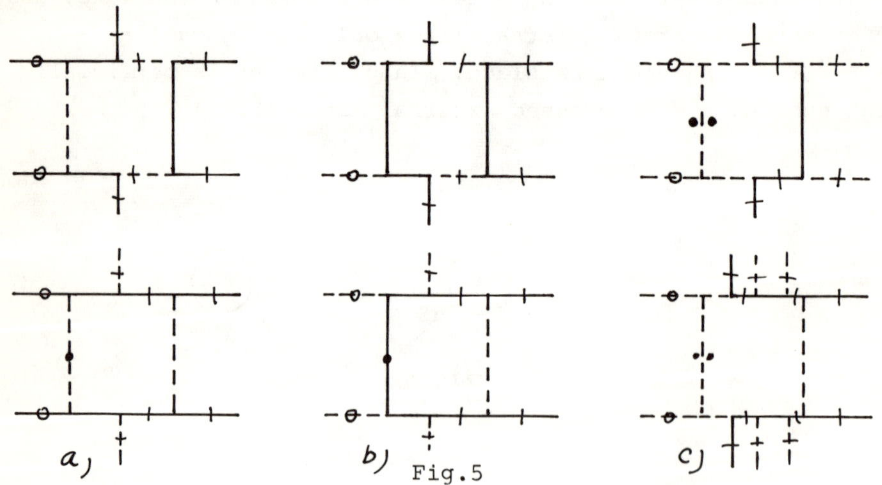

One loop subdiagrams involved in the bootstrap equations for current-field 3-point Green functions

A direct power counting shows that the diagrams on Fig. 5 are convergent if $d > 1$ and $d' > 1$ which is a consequence of (4.7). This completes the proof of absence of ultraviolet divergences.

5. CONCLUDING REMARKS

In these final remarks we shall not attempt to summarize
what was said already, but will discuss briefly some related problems
not treated in the text.

A. Relation to canonical field theory

It is generally believed among experts that conformal invariant
field theory coincides with the Gell-Mann-Low limit in the renormalization group program[16] (the limit corresponding to finite charge renormalization). Stating it more carefully, if there is any relation
between the constructive approach to conformal invariant Green
functions, outlined here, and Lagrangian quantum field theory, it is
only possible through the Gell-Mann-Low limit [17]. This statement is
not at all obvious, since there exists a wider variety of conformal
invariant 3-point functions than local conformal invariant Lagrangians
of third degree. We have found, for instance, an invariant 3-point
function of three scalar fields (Eqs.(3.22),(3.23)) while the $g\,\varphi^3$
coupling is not even scale invariant since g must have dimension
of mass. It is note-worthy that if we apply the power-counting
technique of Sec.4 to the theory of a single scalar field φ with
a vertex function given by (3.22),(3.23) (with $d_1 = d_2 = d_3 = d$) we
find using (4.11a) and (4.14) that the theory is divergence-free only
for $4/3 < d < 8/3$ (so that the canonical value $d = 1$ cannot be approached.)
One might speculate that in this case the bootstrap equations will
have only a trivial solution.

Let us conclude this section with a remark on integer values
of d and half-integer values of d'. We already mentioned that
the canonical $d = 1$ and $d' = 3/2$ correspond to free 0-mass fields.
Consider the theory of single (hermitean) conformal invariant scalar
field of dimension $d = 2$ with a two-point Wightman function normalized
according to (3.6).[18] There is a folklore belief that in this case
$\phi(x)$ should be proportional to the normal product of a free
0 -mass field $\varphi_0(x)$ with itself:

$$\phi(x) = \frac{4\pi}{\sqrt{2}} : \varphi_0^{\,2}(x) : \qquad\qquad (5.1)$$

(The coefficient $4\pi/\sqrt{2}$ is chosen in such a way that to reproduce the 2-point function (3.6) provided that φ_0 is normalized in the canonical way. We are dealing with Wightman functions in this case since the time ordered propagator cannot be defined as a homogeneous function for $d = 2$.)

This conjecture is actually not satisfied without additional assumptions. For example we could define $\phi(x)$ as a generalized free field [19] with a C-number commutator given by

$$\left[\phi\left(\tfrac{x}{2}\right), \phi\left(-\tfrac{x}{2}\right)\right] = w_2\left(\tfrac{x}{2}, -\tfrac{x}{2}\right) - w_2\left(-\tfrac{x}{2}, \tfrac{x}{2}\right) = \frac{2}{\pi i} \varepsilon(x_0)\, \delta'(x^2) \tag{5.2}$$

In this case Eq.(5.1) is obviously wrong. Furthermore, the field ϕ could be reducible in the sense that

$$\phi(x) = \phi_1(x) + \phi_2(x), \quad \left[\phi_1(x), \phi_2(y)\right] = 0 = \langle 0|\phi_1(x)\phi_2(y)|0\rangle \tag{5.3}$$

If we assume that ϕ_1 is a generalized free field with 2-point function

$$\langle 0|\phi_1\left(\tfrac{x}{2}\right)\phi_1\left(-\tfrac{x}{2}\right)|0\rangle = \frac{\alpha}{(4\pi)^2}\, \frac{1}{\left(i 0 x_0 - \tfrac{x^2}{4}\right)^2}\ , \qquad 0 < \alpha < 1$$

and that

$$\phi_2(x) = \sqrt{1-\alpha}\ 2\sqrt{2}\ \pi\ :\varphi_0^2(x):$$

we again arrive at the same conformal invariant 2-point function for ϕ and at a 3-point function of type (3.22):

$$w_3(x_1 x_2 x_3) = \langle 0|\phi(x_1)\phi(x_2)\phi(x_3)|0\rangle = \langle 0|\phi_2(x_1)\phi_2(x_2)\phi_2(x_3)|0\rangle$$

$$= \left(4\sqrt{2}\,\pi\sqrt{1-\alpha}\right)^3 \mathcal{F}_1(x_1-x_2)\,\mathcal{F}(x_2-x_3)\,\mathcal{F}_1(x_1-x_3). \tag{5.4}$$

Reducibility in the sense of (5.3) implies operator reducibility [19], i.e. existence of a bounded operator $B \neq \lambda \mathbb{1}$ weakly commuting with $\phi(x)$. The question arises whether irreducibility and non-vanishing of w_3 would allow us to determine all (conformal invariant) Wightman functions from w_2 . In any case it would be interesting to characterize the set of all conformal invariant theories of a real irreducible scalar field $\phi(x)$ of dimension $d = 2$. The problem becomes even more complex for higher integer (or half-integer) scale dimensions.

B.   <u>Related ideas and possible applications</u>

What was said in these lectures touches only one of the existing approaches to scaling and conformal invariance—let us call it the field theoretic approach. It draws its origin also in the theory of phase transitions (see Polyakov, ref. 5 as well as the recent review [20]). There are at least three other approaches which should be kept in mind.

a)   <u>Phenomenological approach</u>. Two trends should be distinguished in this approach. The first starts with a theoretical analysis of hadronic matrix elements of the current commutator based on the Jost-Lehmann-Dyson representation [21]. It studies the most general conditions under which the Bjorken scaling phenomenon takes place. The second trend consists in making a number of detailed model-type assumptions which are compared with experiment. A representative of this second trend is the work [22] on "automodel behaviour" of form-factors for deep inelastic electromagnetic and weak processes.

b)   <u>Light-cone algebra approach</u> [23]. This is based on Gell-Mann's belief that "Nature reads books on the theory of free fields". In fact, since we saw that conserved currents are bound to have canonical dimensions they might share also other features with the free quark model, in which

$$V_a^\mu (x) = \; : \tilde{q}(x) \, \gamma^\mu \, \lambda_a \, q(x) : \; , \quad d_q = {}^3\!/_2 \; .$$

c)   <u>Operator product expansions</u> (Wilson, ref. 3, see also [24]). The Wilson short-distance expansions find direct application to deep inelastic experiments. Hopefully, they can be derived from (or at least related to) the field theoretic approach described here (see [25, 26]). This appears as one of the most interesting prospects in this general trend from the point of view of physical applications.

ACKNOWLEDGEMENTS

The author would like to thank Professor N.N. Bogolubov
for his kind interest and support of the work.

I have much learned from our collaboration with
Gerhard Mack and take the opportunity to express my sincere
gratitude to him. Stimulating discussions with A.A. Migdal
are also gratefully acknowledged.

It is a pleasure to thank Professor W. Rühl for his kind
hospitality in Kaiserslautern where these lectures were delivered.

REFERENCES

1. The interest in conformal invariance in physics goes back to
Cunningham, Bateman and Dirac. The present trend originates
in the work of J.E. Wess, Nuovo Cim. $\underline{18}$, 1086 (1960),
and H.A. Kastrup, Ann.Physik $\underline{7}$, 388 (1962). For a modern
review and further references see G. Mack and A. Salam,
Ann.Phys. (N.Y.) $\underline{53}$, 174 (1969).

2. G. Mack, Nucl.Phys. $\underline{B5}$, 499 (1968). This suggestion was extended
to conformal symmetry by Mack and Salam, ref.1.

3. K. Wilson, Phys.Rev. $\underline{179}$, 1499 (1969); ibid $\underline{D2}$, 1473 (1970).
Concerning the situation in the Thirring model see also
J.H. Lowenstein, Commun.Math.Phys. $\underline{16}$, 265 (1970; B. Schroer
and J.H. Lowenstein, Phys.Rev. $\underline{D3}$, 1981(1971).

G.F. Dell' Antonio, Y. Frishman and D. Zwanziger, preprint
WIS 71/144, Ph. (1971)

4. P.G. Federbush and K.A. Johnson, Phys.Rev. $\underline{120}$, 1926 (1960).

The result was extended to 0-mass fields by K. Pohlmeyer,
Commun.Math.Phys. 12, 204 (1969).

5. A.M. Polyakov, Zh. ETP Pisma Red. 12, 538 (1970).
    (transl. JETP Lett. 12, 381 (1970));
    A.A. Migdal, Phys. Lett. 37B, 98 (1971);
    G. Parisi and L. Peliti, Lett. Nuovo Cim.2, 627 (1971).
    The invariant vertex function for 3 canonical vector fields
    was written down in E.Schreier, Phys.Rev. D3, 980 (1971).

6. G. Mack and I.T. Todorov, preprint IC/71/139, Trieste (1971).
    This paper contains additional references.

7. E.S. Fradkin, Zh. ETF 29, 121 (1955);
    K. Symanzik, in "Lectures in High Energy Physics",
    ed.B. Jaksič, Zagreb, 1961. (Gordon and Breach, New York,
    1965).
    I. Bialyncki-Birula, Bull.Acad.Pol.Sci. 13, 499 (1965);
    A.A. Migdal, Phys.Lett. 37B, 386 (1971). See also ref.[6].

8. G. Mack and K. Symanzik, preprint DESY 72/20, Hamburg, (1972).

9. M.Hortaçsu, R. Seiler, B. Schroer, preprint NYO-3829-80,
    Univ. of Pittsburgh (1971).

10. I.M.Gel'fand, and G.E. Shilov, Generalized Functions, vol. 1
    (Academic Press, New York, 1964). (See, in particular, Chapt.1,
    Sec. 3 and Chapt. 4).

11. S. Ferrara, R. Gatto, A.F. Grillo and G. Parisi, Lett. Nuovo
    Cimento 4, 115 (1972).

12. This form was suggested to me by A.A. Migdal (private
    communication). It appears as a special case in K. Symanzik
    preprint DESY 72/6, Hamburg (1972).

13. V. Glaser, H. Lehmann, W.Zimmermann, Nuovo Cim. 6, 1122 (1957).

14. S. Weinberg, Phys. Rev. 118, 838 (1960).

15. E. Speer, J. Math. Phys. 9, 1404 (1968);
    Commun.Math.Phys. 23, 23 (1971), ibid. 25, 336 (1972).

16. M.Gell-Mann and F.E. Low, Phys.Rev.95, 1300 (1954).
    About the renormalization group program see N.N. Bogolubov
    and D.V.Shirkov, Introduction to the Theory of Quantized Fields,

Inters. Publ.New York, 1959, I.F. Ginzburg and
D.V. Shirkov, Zh. ETF $\underline{49}$, 335 (1965).
The scale and conformal invariance of the Gell-Mann-Low-
limit is discussed in
K. Wilson, Phys.Rev. $\underline{D3}$, 1818 (1971).
About a possible application of these ideas to the
determination of the fine structure constants in quantum
electrodynamics see the enlightening discussion in the
recent paper by
S. Adler: "Short distance behaviour of quantum electrodynamics
and an eigenvalue condition for α " , preprint,
Institute for Advanced Study, Princeton (1972). It relates,
in particular, the Gell-Mann-Low with the Callan-Symanzik
equations (C.G. Callan, Jr., Phys.Rev. $\underline{D2}$, 1951 (1970);
K. Symanzik, Commun.Math.Phys. $\underline{18}$, 227 (1970)).

17.   Cf.S. Coleman and R. Jackiw, Ann.Phys. (N.Y.) $\underline{67}$, 552 (1971),
        especially the footnote at the end of the paper.

18.   The problem of describing all such fields was raised by
        A.A. Migdal (private communication).

19.   R.F. Streater and A.S. Wightman, "<u>PCT Spin and Statistics and
        All That</u>", (Benjamin, New York, 1964);
        Res Jost, <u>The General Theory of Quantized Fields</u> (American
        Math.Society, Providence, Rhode Island, 1965);
        N.N. Bogolubov, A.A. Logunov and I.T. Todorov, <u>Introduction
        to Axiomatic Quantum Field Theory</u>, (Nauka, Moscow, 1969;
        English translation: Benjamin, New York, to be published).

20.   B. Schroer, Lectures given at IV Simposio Brasileiro De
        Fisica Teorica, preprint Freie Universität, Berlin, (1972).

21.   H. Leutwyler and J. Stern, Nucl.Phys. $\underline{B20}$, 77 (1970);
        R.A. Brandt, Phys.Rev. $\underline{D1}$, 2808 (1970);
        R. Jackiw, R.Van Royen and G.B. West, Phys.Rev.$\underline{D2}$, 2473(1971);
        N.N. Bogolubov,A.N.Tavkhelidze and V.S. Vladimirov,
        preprints P2-6342 and E2-6490, Joint Institute for Nuclear
        Research, Dubna,(1972).

        D. Stichel, Contribution to "Ecole Internationale de la
        Physique des Particules Elémentaire", Basko Polje-Makarska
        (Jugoslavia, 1971).

22.   V.A. Matveev, R.M. Muradyan, A.N. Tavkhelidze, Particles
       and Nucleus, $\underline{2}$, 7 (1971).

23.   H. Fritzsch and M.Gell-Mann, <u>Broken Scale Invariance and the
       Light Cone</u>, Lectures from the Coral Gables Conference on
       Fundamental Interactions at High Energy, Ed.by M. Dal Gin et al.
       (Gordon and Breach, New York, 1971) pp. 1-42. See also:
       P. Carruthers, Physics Reports $\underline{1C}$, 1 (1971) and D.J. Gross and
       S.B. Treiman. Phys.Rev. $\underline{D4}$, 2105 (1971).

24.   W. Zimmermann, <u>Lectures on Elementary Particles and Quantum
       Field Theory</u>, 1970 Brandeis University Summer Institute
       in Theoretical Physics, vol.1, (MIT Press, Cambridge, 1970).
       pp. 395-589. See also lectures presented at this School.

25.   G. Mack, Nucl.Phys. $\underline{B35}$, 592 (1971). See also Lectures
       presented at this School.

26.   B. Schroer, Talk given at the Topical Meeting on "Outlook
       for Broken Conformal Symmetry in Elementary Particle Physics",
       Frascati (1972). See also Lectures presented at this School.

CONFORMAL INVARIANCE AND SHORT DISTANCE BEHAVIOR IN QUANTUM FIELD THEORY

G. Mack

Institut für Theoretische Physik der Universität
BERN, Switzerland

The purpose of these notes will be to give a survey of the bootstrap approach to the construction of a conformal invariant quantum field theory. This approach was started by Migdal[1] and Polyakov[2], and has been further developed by several authors[3-10], in particular Parisi, Peliti and D'Eramo, Todorov, Symanzik and myself. Conformal invariant operator product expansions and related problems were extensively studied by Ferrara, Gatto, Grillo and Parisi, and by Bonora , Sartori and Tonin [12-14].

The bootstrap approach to conformal invariant field theory offers an interesting alternative construction principle to the up to now only available canonical perturbation theory. Its relevance to hadron physics in the real world derives from the fact that one can under certain hypotheses show that the short distance behavior of a finite mass theory is described by a conformal invariant "asymptotic" field theory - the Gell-Mann Low limit theory. This will be discussed in Sec.7.

The material presented here can of course not possibly be exhaustive. The author apologizes for this both to the reader, and to the authors whose results are not discussed in the breadth and detail they would deserve. There is no substitute for reading the original literature, to which these notes are only a guide.

1. <u>CONFORMAL INVARIANCE</u>

Conformal invariance is a space time symmetry. In addition to the Poincaré group one has (c^μ, ρ are 5 additional group parameters):

i) dilatations:

$$x^\mu \to \rho x^\mu, \; \rho > 0 \qquad ds^2 \to \rho^2 ds^2$$

(1)

ii) special conformal transformations:

$$x^\mu \to \sigma(x)^{-1}(x^\mu - c^\mu x^2) \; ; \; ds^2 \to \sigma(x)^{-2} ds^2$$
$$\sigma(x) = 1 - 2cx + c^2 x^2$$

Following Kastrup[15], special conformal transformations may be interpreted as space time dependent dilatations; this can be seen by looking at the transformation law of the line element ds^2.

The most general conformally covariant field transformation law is determined by the theory of induced representations[17]. For the fundamental fields of the theory (i.e. those of lowest dimension, see below)

$$\delta \varphi(x) = \varepsilon (d + x_\nu \partial^\nu) \varphi(x) \tag{2a}$$

for infinitesimal dilatations, and for infinitesimal special conformal transformations

$$\delta \varphi(x) = \varepsilon^\mu (2d x_\mu + 2 x_\mu x_\nu \partial^\nu - x^2 \partial_\mu - 2i x^\nu \Sigma_{\mu\nu}) \varphi(x) \tag{2b}$$

where $\Sigma_{\mu\nu}$ are the spin matrices in the Lorentz transformation law of the field $\varphi(x)$, and d is a new quantum number, called the dimension of mass of the field. If the field $\varphi(x)$ satisfies canonical equal time commutation relations, d=1 resp. $\frac{3}{2}$ for scalar resp. Dirac fields[17]. One knows however, that in a nontrivial exactly conformal invariant quantum field theory canonical equal time commutation relations cannot hold, and as a consequence $d > 1$ resp. $d > \frac{3}{2}$. In this case one speaks of dynamical dimensions. They were proposed by Wilson[19].

The requirement of invariance of the theory under infinitesimal conformal transformations means that the Green functions of the theory (or at least the Wightman functions) satisfy

$$\delta \langle 0 | T \varphi(x_1) \cdots \varphi(x_n) | 0 \rangle = 0$$

under the infinitesimal change of fields Eq.(2) above. Actually there
are some intricacies involved in the use of conformal invariance in
Minkowski space. A correct formulation is as follows: The Green functions are [20] (boundary values of) analytic functions of coordinates x_j.
One requires that they are invariant under (infinitesimal) conformal
transformations _qua_ analytic functions of the x_j. This is the concept
of "weak conformal invariance" of Schroer, Hortacsu and Seiler[21]. It
is most convenient to make the analytic continuation to Euclidean space
(x^o imaginary). The Lorentz group then becomes O (4), and the conformal
group becomes O(5,1). In Euclidean space, even finite conformal transformations present no problem.

Of course an exactly conformal invariant theory is a zero mass
theory. In fact, as has been pointed out by Wilson some time ago[19],
it must necessarily be an "infraparticle theory" so that an S-matrix
does not exist [because the asymptotic condition does not hold due to
infrared effects] . The notion of infraparticle is due to Schroer[22].

Conformal invariant Green functions may also be defined in an
arbitrary, not necessarily integer number of space time dimensions D,
e.g. by using the calculus due to Symanzik which will be described in
Sec. 4 below[6]. (This remark does not apply to pseudoscalar interactions).

2.   <u>CONSTRUCTION PRINCIPLES</u>

The construction of the theory starts from the observation
of Polyakov and Migdal that the integral equations for dressed vertices
(and propagators as well) allow for conformal invariant solutions.
More specifically one proceeds as follows.

    i)   Conformal invariance fixes uniquely the dressed propagator
and 3-point vertex, up to some constants (coupling constant g and field dimension d $)^2$ - see Eqs.(7),(8) below.

    ii) Using these propagators and vertices, the $n \geqq 4$ - point
functions may be constructed by skeleton graph expansion[4,23]. In particular the Bethe-Salpeter kernel

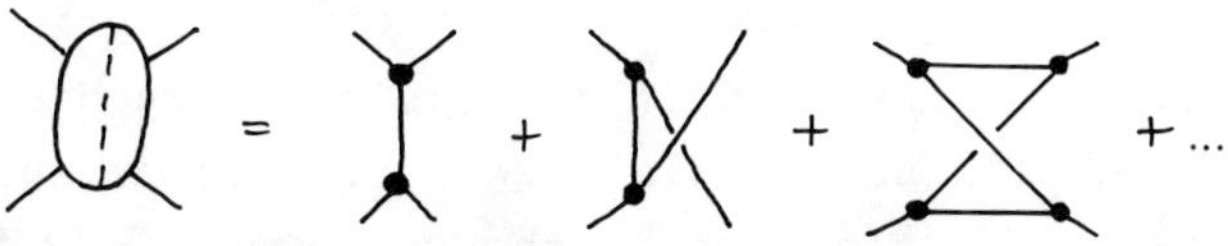

(3)

iii) The renormalized Schwinger Dyson equations (henceforth called bootstraps) for 3-point vertices and propagators are solved by the conformal invariant Ansatz, and one obtains[3] a system of algebraic equations for coupling constants g and field dimensions d. There are as many equations as unknowns, so that one may expect a unique or at most a discrete set of solutions for g, d.

The vertex bootstrap is[1]

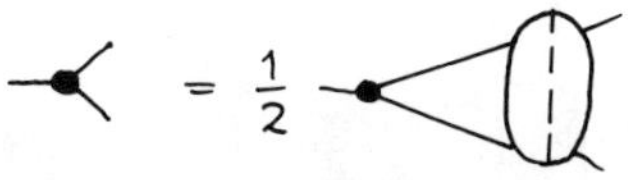

, with B.S.-kernel from Eq.(3) (4)

For propagator bootstrap one can use for instance[7]

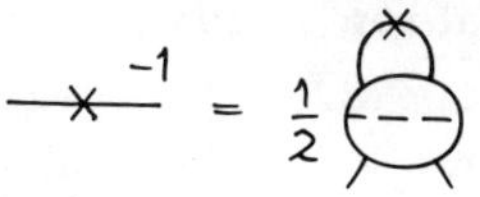

(5)

where 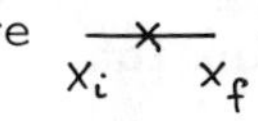 $= Q(x_i^\nu - x_f^\nu)\,G(x_i,x_f)$

and 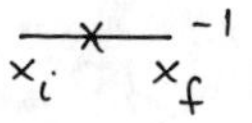 $= Q(x_i^\nu - x_f^\nu)\,G^{-1}(x_i,x_f)$

with G^{-1} the inverse propagator in the convolution sense, and one may choose either Q=internal charge flowing from i to f, or $Q = i\nabla_i^\mu$, $\mu \neq \nu$.

Instead of Eq.(5) one can use generalized 2-point unitarity, as was done in the original paper by Parisi and Peliti[3]. This leads to equivalent results but is less convenient for calculations.

A bare vertex is absent in Eq.(4) since it would not be conformal invariant[1]. This is in agreement with the compositeness condition $Z_1 = Z_3 = 0$.

3. ABSENCE OF DIVERGENCES[4]

In order that the construction explained in the last section makes sense one must show that the expressions so obtained are well defined, i.e. free from divergences, and conformal invariant. (Convergence of the series expansion involved in the skeleton graph expansion is a mute question which will not be discussed here nor anywhere. See however Sec. 8 below.). Verification of the conformal invariance is particularly important for the RHS of the vertex bootstrap, for it is needed there to guarantee that the equation is solved by the conformal invariant Ansatz.

To prove absence of divergences one starts from the observation that every skeleton graph in the theory is represented by a generalized Feynman integral in the sense of Speer[24]. I.e. it is an integral over a product of generalized Feynman propagators

$$2^{\delta}\,\Gamma(\delta)\,(-x^2+i0)^{-\delta} = -i(2\pi)^{-\frac{D}{2}}\int d^D\!p\, e^{-ipx}\, 2^{\lambda}\,\Gamma(\lambda)\,(-p^2-i0)^{-\lambda} \quad (6)$$

$$\lambda = \frac{D}{2} - \delta\ ,$$

where $D = \#$ of space time dimensions. (For spinor lines there is a factor $i\rlap{/}{\partial}$ in front). Such a representation holds in particular for the dressed 3-point vertex and propagator. E.g. in scalar φ^3-theory in D space time dimensions

$$G(x,0) = 2^d\,\Gamma(d)\,(-x^2+i0)^{-d} \qquad ,\ d = \text{field dimension}, \qquad (7)$$

$$V(x_1 x_2 x_3) = \frac{-ig}{(2\pi)^D}\,\Gamma(\delta_1)\,x_{23}^{-2\delta_1}\,\Gamma(\delta_2)\,x_{13}^{-2\delta_2}\,\Gamma(\delta_3)\,x_{12}^{-2\delta_3}\,2^{\Sigma_i \delta_i +1} \qquad (8)$$
$$\cdot\ \Gamma(D-\textstyle\sum_i \delta_i)^{-1}$$

with $x_{ij}^{\alpha} = [-(x_i-x_j)^2 + i0]^{\alpha/2}$, $\ \delta_1 = \delta_2 = \delta_3 = \frac{1}{2}(D-d)$ if all fields have the same dynamical dimension d.

Eq. (8) may be graphically represented as in Fig. 1.

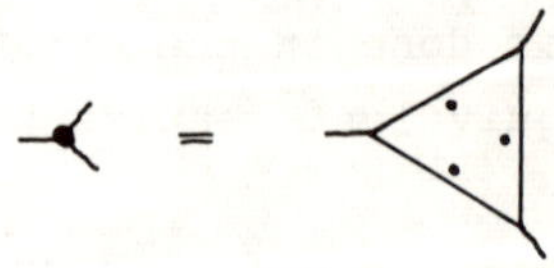

In general there is one generalized Feynman propagator for each line in the Migdal graph G_M which is obtained by substituting Fig.1 for the dressed vertex into a skeleton graph G.

One knows from the work of Speer, or from Dyson's power counting theorem, that these generalized Feynman integrals will be UV-convergent if for every 1-particle irreducible subgraph H of the Migdal graph G_M the following inequality is satisfied:

Fig. 1 Graphical representation of the dressed vertex (8)

$$\sum_{\ell \in \mathcal{L}(H)} (\lambda_\ell - 1) > \tfrac{1}{2}\mu_H \tag{9a}$$

Equivalently,

$$-\sum_{\ell \in \mathcal{L}(H)} \delta_\ell > \tfrac{1}{2}D(n_H - 1) \tag{9b}$$

with $\lambda_\ell = \tfrac{1}{2}D - \delta_\ell$. Summation is over all lines ℓ of the subgraph H, μ_H is the canonical superficial degree of divergence of the subgraph H. We consider φ^3-theory in D space time dimensions. Then

$$\mu_H = (D-2)L_H - D(n_H - 1) \tag{10}$$

where L_H is the number of lines in H, and n_H the number of vertices.

By (7), (8), exponents δ_ℓ in the generalized propagators (6) are δ_ℓ = d resp. $\tfrac{1}{2}$(D-d) for undotted and dotted lines, respectively. As a consequence one has "conservation of dimension" at every vertex v of a Migdal graph,

$$\sum_{S(v)} \delta_\ell = D \tag{11}$$

Summation is over the 3 lines (1 dotted + 2 undotted) incident at vertex v. Eq.(11) allows to express the LHS of Eq.(9a) or (9b) in terms of the configuration of external lines only, for instance

$$\sum_{\ell \in \mathcal{L}(H)} (\lambda_\ell - 1) = \tfrac{1}{2}\mu_H - \tfrac{1}{2}D + \tfrac{1}{2}\sum_{ext.} \delta_\ell \tag{11'}$$

where summation on the RHS is over the external lines of H.

The UV-convergence condition (9) becomes then

$$\tfrac{1}{2}\sum_{ext} \delta_\ell > \tfrac{1}{2}D \tag{9c}$$

Thus the problem is reduced to one of classification of all possible configurations of external lines. For this purpose one uses the fact that dotted external lines of H can only be created by cutting into a dressed vertex of the skeleton graph G. The final result is that condition (9) for UV-convergence is satisfied for some range of values of the field dimensions. For ps.Yukawa theory in 4 space time dimensions one finds the restrictions

$$1 < d_M < 3 \; ; \; \frac{3}{2} < d_B < \frac{5}{2} \qquad (< \text{ is } \leqq \text{ throughout}) \qquad (12)$$

for pseudoscalar and Dirac field, respectively.

One can also prove in a variety of ways absence of infrared divergences of the "catastrophic kind". That means that integrals are well defined for nonexceptional external momenta[26] - i.e. those for which Weinberg's power counting theorem holds. Euclidean external momenta $(p_1 \ldots p_n)$ are nonexceptional except when a partial sum of them vanishes. On the other hand one must of course expect infrared singularities at some exceptional momenta, as in QED. Remember that we are dealing with an infraparticle theory, for which no on-shell S-matrix exists[19].

For later use, I shall finally mention that generalized Feynman integrals can be analytically continued beyond the domain (9a). They are meromorphic functions of all λ_ℓ with poles on hypersurfaces given by (H some 1-particle irreducible subgraph):

$$\sum_{\ell \in \mathcal{L}(H)} (\lambda_\ell - 1) = \tfrac{1}{2} \mu_H - m \; ; \quad m = 0, 1, 2, \ldots \qquad (13)$$

In general there are additional "infrared type" singularity surfaces[25], but not inside the domain of values of $\{\lambda_\ell\}$ which interests us, and never for conformal invariant integrals, i.e. when Eq. (11) holds.

4. <u>MANIFESTLY CONFORMAL INVARIANT GENERALIZED FEYNMAN RULES</u>[27]

One must also show conformal invariance of the expressions constructed according to the rules in Sec.2. This can be done most conveniently be rewriting the generalized Feynman rules given there in a manifestly conformal invariant form. They specify a corresponding integral to every given Migdal graph G_M. These conformal invariant rules are also very helpful for dealing with spin. I shall give them for pseudoscalar Yukawa theory in D=4 space time dimensions.

One introduces projective coordinates ξ^A, $A = 0 \ldots 3, 5, 6$ related to Minkowski coordinates x^μ by

$$\xi^\mu = \kappa x^\mu \; (\mu = 0, \ldots 3), \quad \kappa = \xi^6 - \xi^5 \; ; \; \xi^2 = g_{AB} \, \xi^A \xi^B = 0$$
$$g_{AB} = \text{diag}(+---,-+) \qquad (14)$$

One makes use also of the 6-dimensional Clifford algebra of 8x8 matrices β_A with defining property

$$\{\beta_A, \beta_B\}_+ = 2g_{AB} \tag{15}$$

Out of them one can construct a pseudoscalar β_7 and an 8-dimensional representation of $O(2,4)$-algebra by matrices s_{AB},

$$\beta_7 = -\beta_0\beta_1\beta_2\beta_3\beta_5\beta_6 \quad , \quad s_{AB} = \frac{i}{4}[\beta_A, \beta_B]_- \tag{16}$$

In addition one introduces a measure

$$d\mu_\eta(\xi) = 2d^6\xi \, \delta(\xi^2) \, \delta(\xi\eta - 1) \tag{17}$$

where η is a 6-vector with $\eta^2 = 0$. Choosing it as $\eta^A = (\underline{0};1,1)$ gives

$$\int d\mu_\eta(\xi) f(\xi) = \int d^4x \, f(x, \kappa = 1)$$

A pseudoscalar field over ξ-space may be introduced by

$$\varphi(\xi) = \kappa^{-d_M} \varphi(x)$$

in terms of the ps. field $\varphi(x)$ over Minkowski space with field dimension d_M. Evidently $\varphi(\xi)$ is a homogeneous function of ξ. As a consequence of the multivaluedness of the factor κ^{-d_M} the proper domain of definition of $\varphi(\xi)$ is an (in general)∞-sheeted covering of the cone (14).

One also introduces an 8-spinor field $\chi(\xi)$. It is homogeneous in ξ of degree $-d_B + \frac{1}{2}$ and satisfies a subsidiary condition

$$\xi \cdot \beta \, \chi(\xi) = 0 \tag{18}$$

It is related to the Dirac field in x-space by

$$\psi(x) = \kappa^{d_B - \frac{1}{2}} B(x) \chi(\xi)$$

where $B(x)$ is the translational boost matrix

$$B(x) = \exp\{-i(s_{5\mu} + s_{6\mu})x^\mu\}$$

The Dirac field $\psi(x)$ so defined has only 4 nonvanishing components as a consequence of the subsidiary condition (18). The Dirac adjoint of $\chi(\xi)$ is defined by $\tilde{\chi} = \chi^* A$ in terms of the 8x8 matrix A with defining property $A\beta_C = -\beta_C A$.

These new fields transform under conformal $O(2,4)$ transformations in a manifestly covariant way, e.g.

$$\delta\chi(\xi) = -\tfrac{i}{2}\,\varepsilon^{AB}\,(L_{AB} + S_{AB})\chi(\xi)\,;\quad L_{AB} = i\left(\xi_A\frac{\partial}{\partial\xi^B} - \xi_B\frac{\partial}{\partial\xi^A}\right)$$

This is equivalent to transformation law (2) for $\Psi(x)$ by virtue of homogeneity and subsidiary condition for χ .

Given the correspondence of fields, one can also translate Green functions into the ξ-space form. Given any Migdal graph G_M the rules of Sec.2 read then as follows (for ps. Yukawa theory):

> For every vertex v of G_M include a factor $\xi_v\cdot\beta$, and an extra factor β_7 for vertices with incident undotted meson line.
>
> For every line ℓ with initial resp. final vertex i_ℓ and f_ℓ write down a stripped propagator $\Gamma(\delta^\ell)(\xi_{i_\ell}\cdot\xi_{f_\ell}+i0)^{-\delta_\ell}$ where $\delta_\ell = d_M$ resp. $d_B+\tfrac{1}{2}$ for undotted meson resp. nucleon lines, and $\delta_\ell = \tfrac{1}{2}d_M - d_B + 2$ resp. $-\tfrac{1}{2}d_M + \tfrac{5}{2}$ for dotted ones.
>
> Integrate over variables ξ associated with internal vertices, with measure $d\mu(\xi)$ given by Eq.(17).
>
> For every dressed vertex subgraph shown in Fig.1 include a factor $-ig/\Gamma(d_B + \tfrac{1}{2}d_M - 2)$.

The factors $\xi\cdot\beta$ have to be arranged in the same order as vertices are arranged along fermion lines, and traces are to be taken over internal fermion loops. Such traces can be performed by familiar procedure using anti-commutation relations (15).

These rules may be used also for the LHS of the vertex bootstrap (4), i.e. the dressed vertex itself.

By these rules, the amplitudes are given by integrals whose integrands are manifestly $O(2,4)$ invariant since they depend on scalar products of 6-vectors only. It only remains to be shown that the integrals are also independent of the 6-vector η in the definition (17) of the measure $d\mu(\xi)$. In accordance with the discussion in Sec.1 one performs analytic continuation $\xi^0 \to i\xi^0 = \xi^4$ [i.e. $x^0 \to ix^0$, see (14)] so that the metric becomes $g_{AB} = \mathrm{diag}(----,-+)$ and integration is effectively over Euclidean space $\{x^1 \ldots x^4\}$. Using homogeneity of the integrand in all integration variables ξ_v one may then apply the fundamental

Covariance Lemma: Let $f(\xi)$ defined on the forward cone $\xi^2 = 0$, $\xi^6 > 0$ and such that $I = \int d\mu_\eta(\xi) f(\xi)$ exists (for some η). If $f(\rho\xi) = \rho^4 f(\xi)$ for all $\rho > 0$ then I is independent of the positive lightlike 6-vector η in the measure $d\mu_\eta(\xi) = 2d^6\xi\, \delta(\xi^2)\, \delta(\xi\cdot\eta - 1)$.

5. LOCALITY, SPECTRUM CONDITION AND GENERALIZED UNITARITY[6,7]

The conformal invariant skeleton graph expansions satisfy locality and spectrum condition term by term. These properties are manifest in the Symanzik representation[6] of a general Migdal graph. It reads

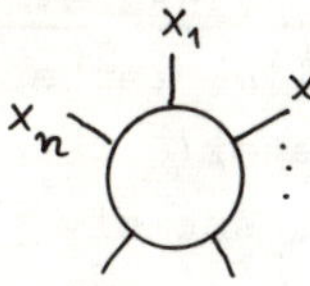

$$= \int_{\delta^0 + i\Gamma} \omega\, K(\delta)$$

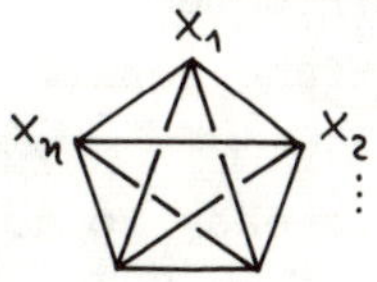

(19)

where each line stands for a factor $2^{\delta_{ij}}\, \Gamma(\delta_{ij})\, x_{ij}^{-2\delta_{ij}}$, with $x_{ij}^{-2\delta} = [-(x_i - x_j)^2 + i0]^{-\delta}$ as before. Integration is over the imaginary part of exponents δ_{ij} subject to the restriction that (for Green functions)

$$\delta^0 + i\Gamma : \qquad \sum_j \delta_{ij} = d_i \tag{20}$$

d_i = dimension of field that couples to external vertex x_i. ω is the differential form of the surface Γ defined by (20).

More explicitly, one may parametrize arguments δ_{ij} satisfying (20) as

$$\delta_{ij} = \delta_{ij}^0 + \sum_k c_{ij}{}^k s_k \qquad (i \neq j)$$

where δ^0_{ij} is a particular solution of (20) which should be chosen such that $0 < \mathrm{Re}\,\delta^0_{ij} < \frac{1}{2}D$ for all $i \neq j$. Matrices c^k, $k = 1 \ldots \frac{1}{2}n(n-3)$, are real symmetric, with zeroes on the diagonal, and $\sum_j c_{ij}{}^k = 0$ for all i,k.

With this parametrization

$$\omega = (2\pi i)^{-\frac{1}{2}n(n-3)} \, ds_1 \cdots \, ds_{\frac{1}{2}n(n-3)}$$

and integration is over $s_k = -i\infty \ldots +i\infty$. $K(\delta)$ is a kernel depending on the δ_{ij}. It depends of course on the graph. For n=3 there is no integration left after (20) is imposed, Eq.(19) then reduces to the infraparticle representation Fig.1.

The representation (19) was derived by Symanzik by noting that it holds for the elementary n-star shown in Fig.2, with $K(\delta) = 1$. This was established for n=3 by Parisi, Peliti and d'Eramo[9], and generalized to all n by Symanzik[6]. Using the n-star formula one can successively integrate out all internal vertices, until one arrives at the form (19). The kernel $K(\delta)$ comes out as an integral over a product of B-functions.

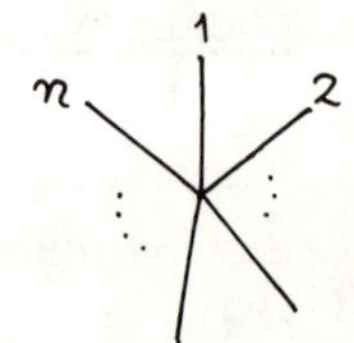

Fig.2 <u>The n-star vertex.</u> Every line represents a propagator(6).Exponents δ_i must satisfy $\sum_i \delta_i = D$

In Eq.(19) it is easy to go over to the Wightman function. All one has to do is change the $i\varepsilon$ -prescription to

$$x_{ij}^{-2\delta} = \left[-(x_i - x_j)^2 + i\varepsilon(x_i^0 - x_j^0) \right]^{-\delta}$$

Locality is then manifest: Wightman functions corresponding to different orderings of fields are known to be boundary values of the same analytic function[20]. They are seen to agree for interchange of fields with relatively space-like arguments since

$$\left[-x^2 + i\varepsilon x^0 \right]^{-\delta} = \left[-x^2 - i\varepsilon x^0 \right]^{-\delta} \text{ for } x^2 < 0$$

The spectrum condition is also manifest since

$$\left[-x^2 + i\varepsilon x^0 \right]^{-\delta} \propto \int_{p^2 > 0} d^D p \, e^{-ipx} \, \Theta(p^0) \, (p^2)_+^{-\frac{1}{2}D + \delta}$$

The Symanzik representation of a conformal invariant n-point function has great theoretical interest quite independent from a theory based on skeleton graph expansion. This is because it provides a new type of Ansatz which satisfies locality and spectrum condition, and which is different from the perturbation theoretical one.

While locality and spectrum condition hold term by term in the skeleton graph expansion, generalized unitarity will be satisfied

only when vertex and propagator bootstrap conditions are fulfilled, i.e.
for a presumably discrete or even unique set of values for coupling
constant g and dynamical field dimension d. This has been proven by
Symanzik and myself[7]. It is instructive to compare this result with
canonical perturbation theory[28]. There locality and spectrum condition
are also satisfied term by term in the expansion, and unitarity holds
in the sense of formal power series in g, i.e. in a sense identically
in the coupling constant g.

I will not go into the details of the proof of generalized
unitarity here since it is a somewhat technical affair. Instead I will
be content with writing down the generalized unitarity of the propagator in the form suitable for an infraparticle theory. As far as I
know it is originally due to Symanzik and Veltman[29] and was also discussed by Polyakov[5]. It reads

$$\text{disc} \ {\underline{}}^{-1} = \sum_{\substack{n \geq 2 \\ (\Gamma_1, \Gamma_2)}} \ \boxed{\Gamma_1}\ \vdots\ \vdots\ \boxed{\Gamma_2} \ , \quad p^0 > 0 \qquad \text{(21)}$$

$$n \ \text{cut lines}$$

A cut line stands for the positive frequency absorptive part of a
<u>dressed</u> propagator. Summation is over all pairs of skeleton graphs
(Γ_1, Γ_2) such that the combined graph does not contain a self energy
subgraph. Complex conjugation is understood on the shaded side of the
cut.

Eq.(21) has been first used as a propagator bootstrap by
Parisi and Peliti as discussed in Sec.2.

6. <u>CURRENTS AND STRESS TENSOR</u>[7]

To construct matrix elements of currents associated with
(exact) internal symmetries one starts again with the unique conformal
invariant Ansatz for its 3-point function[1]. For coupling to a Dirac
field in 4 space time dimensions the Ansatz depends[12] on 2 constants
a_S, a_V,

$$\langle T\{ j^\mu(x_3)\Psi(x_1)\bar\Psi(x_2)\}\rangle = \sum_{i=S,V} a_i \Omega_i \ \hat{x}_{23} \ x_{23}^{-4} \ y^\mu \ \hat{x}_{31} \ x_{31}^{-4} \ \Omega_i \ 2^{d+\frac{1}{2}} \ \Gamma(d+\tfrac{1}{2}) x_{12}^{1-2d}$$

$$\times \ \text{Tr} \left(\Omega_i \hat{x}_{12} \Omega_i \hat{x}_{12} \right)$$

where

$$\hat{x}_{ij} = (x_i - x_j)\gamma \; , \; x_{ij}^{\alpha} = (-(x_i - x_j)^2 + i\varepsilon)^{\alpha/2}, \; \Omega_i = (1, \gamma_\alpha)$$

The vertex function $\Gamma^\mu(x_3; x_1 x_2)$ is obtained from this expression by full-propagator amputation on variables x_1, x_2; it will be represented in the following by the symbol shown on the LHS of Eq.(23) below. The Ward identity reads

$$\nabla_y^\mu \Gamma_\mu (y; x_1 x_2) = -Q\left[\delta(y - x_1) - \delta(y - x_2)\right] G^{-1}(x_1 x_2) \qquad (22a)$$

where Q is the charge of the field Ψ . It is satisfied by the Ansatz if

$$a_s + 4 a_v = Q/8\pi^2 \qquad (22b)$$

where Q is the charge (e.m. charge for e.m. current etc.).

The remaining a_v is not a free constant, however, but is determined by the dynamics. In order that the current be a local operator it must satisfy the appropriate Bethe-Salpeter equation. If there are no charged fundamental Bose fields it reads

$$\mu \bullet\!\!\bigcirc \qquad = \tfrac{1}{2}\mu \; \langle\!\!\bigcirc\!\!=\!\!\bigcirc\!\rangle \qquad\qquad (23)$$

This equation is conformal invariant and free from divergences. It will therefore be solved by the most general conformal invariant Ansatz. As a result, one obtains a system of <u>homogeneous</u> linear equations for the coefficients a_s, a_v. In order that it have a non-trivial solution satisfying (22b) it is necessary that the RHS of (23) reproduces also validity of Ward identity (22a). Since conformal invariance fixes the momentum dependence of the vertex function it suffices to check at zero momentum transfer, i.e. W.I. obtained from (22a) by integrating over $\int dx\, x^\nu$. Let us define a cross operation as in Eq.(5), with Q = internal charge. Inserting the $\int dx\, x^\nu$ - integrated form of (22a) on both sides of Eq.(23) we obtain an equation identical with the propagator bootstrap (5). It is thus fulfilled by hypothesis.

With Ward identity (22b) consistently imposed, B.S. Eq.(23) reduces to an inhomogeneous linear (system of) equation(s) for the coefficient(s) a_v (,...) that is (are) then still free. Barring unexpected degeneracies, it will determine the coefficient(s) uniquely.

With the current's 3-point functions known, the higher
n+1 -point functions can be determined by skeleton graph expansion.

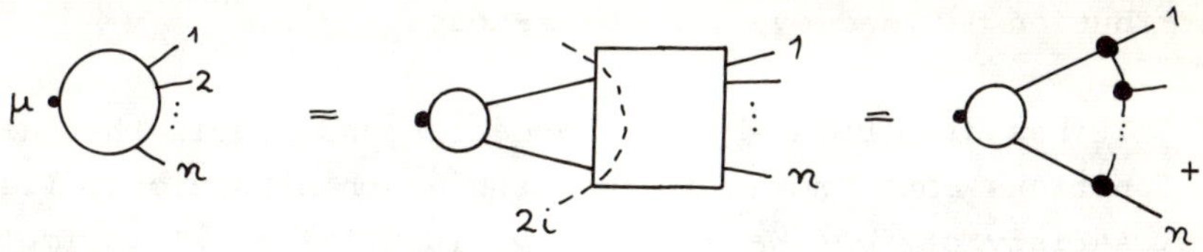

$$\mu \bullet \bigcirc \!\!\!\begin{smallmatrix} 1 \\ 2 \\ \vdots \\ n \end{smallmatrix} \quad = \quad \bullet \bigcirc \!\!-\!\!\Box\!\!\begin{smallmatrix} 1 \\ \vdots \\ n \end{smallmatrix} \quad = \quad \bullet \bigcirc \!\!\triangleleft \!\!\begin{smallmatrix} 1 \\ \\ n \end{smallmatrix} \ + \ \dots \tag{24}$$

The box stands for all n+2 - point skeleton graphs such that
the whole graph is connected, 1-particle irreducible and satisfies the
skeleton condition (i.e. no vertex correction or self energy subgraphs
exist). The last requirement is ensured by the indicated 2-particle
irreducibility requirement -----2i. It means that one should sum only
over those skeleton graphs inside the box which cannot be cut into two
pieces along the dashed line without cutting more than 2 internal lines.

The discussion of the stress energy tensor runs exactly parallel to that of an internal symmetry current, apart from one point:
The conventional choice of seagulls adopted e.g. by Callan, Coleman
and Jackiw[30] is no good for our purpose. In order to avoid divergences
it is necessary to choose seagulls such that the time ordered Green
functions are traceless.

One can also construct amplitudes involving two or more
currents. For instance $\langle 0|T(j^\rho (x) j^\sigma (0)) | 0 \rangle$ is given by

$$\rho \!-\!\!\times\!\!-\! \sigma \ = \ \rho \bullet \bigcirc\!\!\!\overset{\times}{\bigcirc}\!\!\!\bigcirc \bullet \sigma \ + \ \rho \bullet \bigcirc\!\!\!\fbox{$*$}\!\!\!\bigcirc \bullet \sigma \tag{25}$$

The cross operation herein is defined by

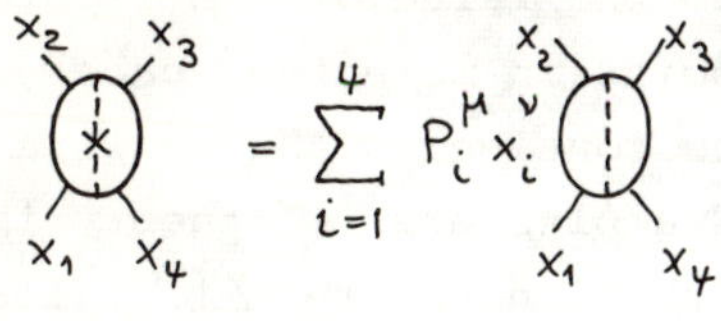

$$\begin{smallmatrix} x_2 & & x_3 \end{smallmatrix} \atop \bigcirc\!\!\!\fbox{$*$} \atop \begin{smallmatrix} x_1 & & x_4 \end{smallmatrix} \quad = \quad \sum_{i=1}^{4} P_i^\mu x_i^\nu \ \begin{smallmatrix} x_2 & & x_3 \end{smallmatrix} \atop \bigcirc \atop \begin{smallmatrix} x_1 & & x_4 \end{smallmatrix} \qquad , \mu \neq \nu, \ P_i^\mu = \text{momentum flowing} \atop \text{into leg } i \tag{26}$$

The crossed propagator was already defined in (5), with $Q = P^\mu$.

The Eq.(25) is well defined and free from divergences for
non-coinciding external coordinates $x \neq 0$. There is an ambiguity at
$x = 0$ for even number of space time dimensions, which may however be
traced to the ambiguous definition of the T^* - product. In other

words, one can show that the 2-point Wightman function, which may be
obtained from the Green function in the standard way by changing the
$i\varepsilon$ -prescription, possesses a unique extension to a dilatation invariant distribution defined over all of x-space.

It has also been shown by Symanzik and myself that all the
n-point functions constructed by skeleton graph expansions, i.e. Eq.(24),
(25) etc. satisfy correct Ward-Takahashi identities. This implies in
particular that the internal symmetry currents satisfy correct equal
time commutation relations

$$\left[j_o^a(x), \; j_\mu^b(0) \right]_{x^o=0} = iC^{abc} \, j^c(0) \; \delta(\underset{\sim}{x}) + S.T. \tag{27}$$

where C^{abc} are the structure constants of the symmetry Lie algebra.

7.   <u>RELEVANCE TO THE REAL WORLD</u>[26,31]

The exactly conformal invariant theory allows to make statements relevant to the real world if it is Gell-Mann Low large momentum
asymptote of a realistic finite mass theory[32].

Let us consider n-point vertex functions $\Gamma(p_1 \ldots p_n)$, i.e.
one-particle irreducible, full propagator amputated Green functions.
We know from the work of Callan[33] and Symanzik[26] that all the n-point
vertex functions of a perturbation theoretically renormalizable massive
theory satisfy differential equations

$$\left[m^2 \frac{\partial}{\partial m^2} + \beta(\bar{g}) \frac{\partial}{\partial \bar{g}} - n\,\gamma(\bar{g}) \right] \Gamma(\lambda p_1, \ldots \lambda p_n ; m^2, \bar{g}) = \Delta\Gamma(\lambda p_1 \ldots \lambda p_n ; m^2, \bar{g}) \tag{28}$$

where $\bar{g}$ is the physical coupling constant (i.e. the value of the 3-point
vertex of the <u>massive</u> theory at a particular point in momentum space),
and m is the rest mass. For simplicity let us consider a theory with
only one dimensionless coupling constant $\bar{g}$. (Examples are φ^3-theory in
D=6 or 6+ ε space time dimensions). In perturbation theory $\Delta\Gamma$ falls
off at large λ faster than Γ itself, by a power of λ . One assumes
that this remains true after summing up the perturbation series. Follow-
ing Symanzik one may then define the vertex functions $\Gamma_{as}(p_1 \ldots p_n ; m^2, \bar{g})$
of a "pre-asymptotic" zero mass theory which are solutions of the
<u>homogeneous</u> Callan Symanzik Eq. (28) and approximate Γ at large λ .

They depend effectively on one parameter $\bar{g}$ since a change of renormalization momentum, i.e. of m^2, may be compensated by a change of coupling constant $\bar{g}$ and a change of normalization - this is precisely the feature expressed by validity of a homogeneous CS-equation (28). In contrast with the Gell-Mann Low limit theory, to be defined below, the pre-asymptotic theory may still be constructed by canonical perturbation theory, i.e. as a formal power series in $\bar{g}$ (which may or may not converge and thereby define a function of $\bar{g}$).

To solve the homogeneous equation (28) one introduces in place of $\bar{g}$ a new variable ρ defined by $\rho(\bar{g}) = {}_0\!\int^{\bar{g}} dg' \beta(g')^{-1}$ so that $\beta\frac{\partial}{\partial g} = \frac{\partial}{\partial \rho}$.

The most general solution is then easily found to satisfy

$$\Gamma_{as}(p_1 \cdots p_n; \lambda^{-2} m^2, \bar{g}) = a(\lambda)^{-n} \Gamma_{as}(p_1 \cdots p_n; m^2, g(\lambda)) \tag{29a}$$

where the "effective coupling constant" $g(\lambda) = \rho^{-1}(\ln \lambda^2 + \rho(\bar{g}))$. I.e. it is implicitly defined by

$$\ln \lambda^2 = \int_{\bar{g}}^{g(\lambda)} dg' \, \beta(g')^{-1} \tag{29b}$$

Furthermore,

$$a(\lambda) = \int_{\bar{g}}^{g(\lambda)} dg' \, \beta(g')^{-1} \, \gamma(g')$$

From Eq.(29a) one gets a statement about momentum dependence by virtue of the relation

$$\Gamma(\lambda p_1 \cdots \lambda p_n; m^2, \bar{g}) = \lambda^{D - \frac{1}{2}n(D-2)} \Gamma(p_1 \cdots p_n; \lambda^{-2} m^2, \bar{g}) \tag{29c}$$

which follows from ordinary dimensional arguments if $\bar{g}$ is chosen dimensionless, and holds also for Γ_{as}.

If the function $\beta(\bar{g})$ has a nontrivial simple zero at $\bar{g} = g_\infty \neq 0$, viz.

$$\beta(g_\infty) = 0$$

and Γ_{as} exists as a function of $\bar{g}$ and is continuous at $\bar{g} = g_\infty$, then one may define the n-point vertex functions of the Gell-Mann Low limit theory by

$$\Gamma_{GML}(p_1 \cdots p_n) = \lim_{\bar{g} \to g_\infty} \Gamma_{as}(p_1 \cdots p_n; m^2, \bar{g})$$

They depend only trivially on m^2, i.e. through an overall factor of a fractional power of m^2 (see below).

Let us now assume that the slope $\beta'(g_\infty) < 0$. Then it follows from (29b) that $g(\lambda) \to g_\infty$ as $\lambda \to \infty$. Therefore by (29a,c,d) Γ_{GML} describes the asymptotic behavior of Γ_{as}, and therefore also of vertex functions $\Gamma(\lambda p \ldots)$ of the massive theory, in the limit $\lambda \to \infty$. A more precise statement, valid if $_{as}$ is also differentiable at $\bar{g} = g_\infty$, is as follows: The vertex functions of the (realistic) massive theory behave in the large momentum limit as

$$\Gamma(\lambda p_1, \ldots \lambda p_n; m^2, g) = \lambda^{D-nd} r(g)^{-n} \left\{ \Gamma_{GML}(p_1 \cdots p_n) + c(\bar{g})(\bar{g} - g_\infty) \lambda^{2\beta'} \times \right.$$
$$\left. \times \Gamma^{(1)}(p_1 \cdots p_n) + s(\bar{g}) \lambda^{-2+\sigma} \ell n^m \lambda \, \Delta\Gamma_{as}(p_1 \cdots p_n) + \cdots \right\} \tag{30}$$

provided $\beta' = (d/d\bar{g}) \beta(\bar{g})\big|_{\bar{g}=g_\infty} < 0$. Here m=0 or 1; r, c, s some $\bar{g}$-dependent constants, and

$$d = \tfrac{1}{2}(D-2) + 2\gamma(g_\infty)$$

is the dynamical dimension of the field. For φ^3-theory one can show[7] that σ is the anomalous part of the dimension of the conformal invariant field φ^2, viz. $\sigma = 2 - d$ (see Sec.8A). $\Delta\Gamma_{as}$ is the GML asymptote of the inhomogeneous term $\Delta\Gamma$ in (28), and $\Gamma^{(1)}\genfrac{}{}{0pt}{}{as}{(p)} = \left\{-n\gamma'(g_\infty)/\beta' + \frac{\partial}{\partial\bar{g}}\right\}\Gamma_{as}(\ldots)\big|_{\bar{g}=\dot{g}_\infty}$.

Obviously $\beta'(g_\infty) < 0$ is valid if g_∞ is the first non-trivial zero > 0 and $\beta'(0) > 0$. (The reader may draw for himself a picture plotting β against g). This last condition can be checked by low order perturbation theory. Eq.(30) holds also for Γ_{as}, but then the $\Delta\underline{\Gamma_{as}}$-term on the RHS is absent.

If on the other hand $\beta'(g_\infty) > 0$ then by (29b) $g(\lambda) \to g_\infty$ when $\lambda \to 0$. Therefore the GML-limit now describes the <u>infrared</u> behavior of the preasymptotic zero mass theory. Eq.(30) holds again for Γ_{as} (without the $\Delta\underline{\Gamma_{as}}$-term on the RHS) but now for $\lambda \to 0$. This case is of interest for critical phenomena in statistical mechanics, where one is interested in the large-distance behavior of correlation functions (at the critical point where the system can be described by a kind of zero mass theory) [26,32,36].

Since the λ-dependence of every term in Eq.(30) is explicit, this equation shows in what sense and to what accuracy the Gell-Mann Low limit theory is the large momentum asymptote of the finite mass theory. In particular, the leading term is independent of the coupling constant $\bar{g}$, apart from an overall factor, and is always given by the Gell-Mann Low limit theory. Then there are two correction terms which scale with smaller powers of λ , which powers are also independent of the physical coupling constant $\bar{g}$. These correction terms stem from the coupling constant and mass dependence of the massive theory respectively.

Note that the first of the correction terms vanishes when $\bar{g} = g_\infty$. Since presumably it falls off more slowly with λ than the second correction term $\Delta\Gamma_{\underline{as}}$ (this is true if β' and σ are close to their canonical values 0), we see that "asymptopia" will be approached fastest if the physical coupling constant $\bar{g}$ takes the particular value

$$\bar{g} = g_\infty \, . \tag{31}$$

It is tempting to conjecture[4] on the basis of the SLAC-results on electro-production that this is the real situation in nature. This would allow in principle to compute the physical coupling constant $\bar{g}$ of the massive theory by determining the zero g_∞ of $\beta(\bar{g})$. (A similar suggestion has been made by Adler for QED[34], the physics of it is however very different there, since in QED $\beta' = 0$). The dots in Eq.(30) stand for terms that are smaller than the bigger one of the two explicit correction terms.

Eq.(30) holds in particular for the inverse propagator, so that (a,b are some constants)

$$\tilde{G}^{-1}(p) \propto -i(-p^2)^{\frac{1}{2}D-d}\left\{1+a(-p^2)^{\beta'}+b(-p^2)^{-1+\frac{1}{2}\sigma}\ell n^m(-p^2)+...\right\} \tag{32}$$

Eq.(30) shows that the Gell-Mann Low limit theory is dilatation invariant. Schroer has demonstrated[36] that it must also be conformal invariant. Basically his argument amounts to showing that the trace of the stress energy tensor in the GML limit theory is zero, as one expects on the basis of dilatation symmetry. It follows then that also the conformal currents are conserved

$$\tag{33}$$

$$\theta_\mu^\mu(x) = 0 \implies K_{\nu\mu}(x) = 2x^\rho x_\mu \theta_{\nu\rho}(x) - x^2 \theta_{\nu\mu}(x) \qquad \text{conserved}$$

Now the Green functions of the GML-theory much as those of the massive
field theory must also satisfy the renormalized Schwinger-Dyson equations for dressed propagator and 3-point vertex. Thus the Migdal-Polyakov
bootstrap approach emerges as a natural way to construct the GML-theory
which, as we saw, is the large momentum asymptote of the massive theory.

It is amusing that the scaling behavior of the symmetry breaking correction terms in (30), (32) can in principle be determined from
an eigenvalue equation involving only quantities that can be constructed in the frame work of the Gell-Mann Low Migdal theory.
This was discussed in Ref.31 by Symanzik and the author. The result is
not unexpected: According to Wilson's spurion analysis[19], the first
order correction terms should be suppressed by a power λ^{-D+d_u}
where d_u is the dimension of some scalar field u(x) in the Lagrangean
density. In the φ^3-model, the relevant fields are φ^2 and φ^3 (there
could also be a linear term in φ). Comparing with Eq.(30) and following discussion we see that

$$-2 + \sigma = -D + \dim \varphi^2 \quad , \quad 2\beta' = -D + \dim \varphi^3 < 0 \qquad (34)$$

The computation of the dimension of φ^2 will be exemplified in Sec.8A.
It is interesting to note that a conformal invariant large momentum
limit requires that the interaction Lagrangean density $\mathcal{L}_I = \varphi^3$ must
have an anomalous dimension strictly less than D.

8.A   SOLUTION OF BOOTSTRAP EQUATIONS IN 6+ε SPACE TIME

DIMENSIONS [8]

To obtain values for the coupling constant(s) g and dynamical
dimension(s) d of fields one needs to solve the algebraic equations
obtained from the bootstrap conditions(4),(5) by factoring out the
common p-dependence. They are of the form

$$g = g^3 f_1(d) + g^5 f_2(d) + ...$$
$$1 = g^2 h_1(d) + g^4 h_2(d) + ... \qquad (35)$$

for the vertex bootstrap and propagator bootstrap respectively. We thus
have a system of algebraic equations whose RHS is given by an infinite
series, and whose solutions would consist of numerical values for g,d.

If looking at it this way one is frustrated with a shapeless problem.

However, one gets a lot more structure to play with if one starts by looking at a more general problem. To this end we consider the (not necessarily integer) number D of space time dimensions as an extra variable. This trick was inspired by Wilson's recent work on statistical mechanics in 3.99 dimensions[37]. It allows to put to use the analyticity properties in dimensions discussed in Sec.3. Solutions of bootstraps (35) will form 1-dimensional hypersurfaces in (D,d,g^2)-space. For simplicity we consider φ^3-theory, which is perturbation theoretically renormalizable in D=6 space time dimensions. The aim will however be to construct a solution of (35) in D=7 dimensions or so rather than D=6. (This corresponds to a conformal invariant solution of a nonrenormalizable theory). We shall show that there exists a solution manifold of bootstraps (35) which passes through the point $(D,d,g^2) = (6,2,0)$, and we will determine the germ of this surface. Note that for D=6 the theory is free (g=o).

We know from the results of Sec.3 that the functions $f_k(d)$ resp. $h_k(d)$ are meromorphic functions of d. They have poles at

$$d = d_c^V = \tfrac{1}{3}D \qquad \text{resp. at} \qquad d = d_c^P = \tfrac{1}{2}D-1 \qquad (36)$$

For D=6 space time dimensions both expressions coincide, and $d_c^V=d_c^P=2$ is the canonical field dimension. Expressions (36) are the only singularity surfaces passing through a neighborhood of the point $(D,d)=(6,2)$, provided the dressed vertex is normalized as in (8).

Because of meromorphy, we may write down convergent Laurent expansions, viz.

$$f_k(d) = \sum_{m=-1}^{\infty} \Delta'^{m} f_{km} \quad , \qquad \Delta' = d - d_c^V \qquad (a)$$

$$h_k(d) = \sum_{m=-1}^{\infty} \Delta^{m} h_{km} \quad , \qquad \Delta = d - d_c^P \qquad (\ ) \tag{37}$$

The coefficients h_{km}, f_{km} will of course still depend on D.

To determine the germ of the solution manifold mentioned above, we expand g^2, Δ, Δ' in power series in $\varepsilon = D - 6$:

$$g^2 = G_1(\tfrac{\varepsilon}{6}) + G_2(\tfrac{\varepsilon}{6})^2 + \dots \tag{38}$$

cont.

$$\Delta = E_1 \left(\frac{\varepsilon}{6}\right) + E_2 \left(\frac{\varepsilon}{6}\right)^2 + \ldots$$

$$\Delta' = (E_1+1)\frac{\varepsilon}{6} + E_2 \left(\frac{\varepsilon}{6}\right)^2 + \ldots$$

We used that $\Delta' = \Delta + \frac{1}{6}\varepsilon$ according to Eq.(36). We substitute expansions (38) into (37) and insert the result into bootstraps (35). We demand that these bootstrap equations are fulfilled identically in ε. The coefficients G_i, E_i are thereby determined, viz.

$$G_1 = (f_{1,-1} - h_{1,-1})^{-1} \quad , \qquad E_1 = h_{1,-1}(f_{1,-1} - h_{1,-1})^{-1} \text{ etc.} \qquad (39a)$$

where $f_{1,-1}$, $h_{1,-1}$ are to be computed at D=6.

The first few coefficients can be worked out in closed form by computing generalized Feynman integrals in the appropriate approximation. Let us sketch the lowest order calculation, for scalar singlet φ^3-theory.

Consider first the simplest vertex graph, Fig.3a, which defines $f_1(d)$. To get $f_{1,-1}$ we need to evaluate it to order $(d-d_c^V)^{-1}$, for D=6. Associated with it is a generalized Feynman integral involving 12 propagators, each one represents a factor $\Gamma(\lambda_\ell)(-\frac{1}{2}q_\ell^2-io)^{-\lambda_\ell}$ with $\lambda_\ell = \frac{1}{2}D-d$ resp. $\frac{1}{2}d$ for undotted and dotted lines respectively (s.Sec.3). When D=6 and $d \to 2$, all $\lambda_\ell \to 1$. The trick is now to treat the λ_ℓ as independent and use analyticity in the λ_ℓ. According to the result of Sec.3, the generalized Feynman integral will admit of a Laurent expansion around $\{\lambda_\ell=1\}$. For D=6 one has $\mu_{G_M} = 0$, therefore the expansion will be of the form

$$\frac{1}{\sum_\ell (\lambda_\ell-1)}\left\{a_{-1} + \sum_\ell a_o^\ell (\lambda_\ell-1) + \ldots \right\} \qquad \text{(sum over all lines)} \qquad (39b)$$

To determine $f_{1,-1}$ we only need to know the singular term, viz. a_{-1}. This allows us to let $\lambda_\ell \to 1$ independently, keeping however(cf.Eq.(11'))

$$\sum_\ell (\lambda_\ell-1) = -\frac{3}{2}(2-d) \qquad (39c)$$

as is true when λ_ℓ have the above mentioned values prescribed by conformal symmetry. This trick allows us to put $\lambda_\ell=1$ for dotted lines first. In this way, expression (8) for the dressed vertex becomes, for D=6

$$V(x_1 x_2 x_3) \longrightarrow ig\,\delta(x_1-x_3)\,\delta(x_2-x_3) \qquad \text{as } \lambda_i = 2-\delta_i \longrightarrow 1 \qquad (39d)$$

This is proven by writing down the Feynman parametric representation of (8) in momentum space. For the remaining (undotted) lines, we put $\lambda_\ell = \tfrac{1}{2}d$ so that Eq.(39c) is satisfied. To the desired approximation, the graph of Fig. 3a then gives, in x-space language(cf.Eq.(6))

$$-ig^3 \, \Gamma(\delta) \, x_{23}^{-2\delta} \, \Gamma(\delta) \, x_{13}^{-2\delta} \, \Gamma(\delta) \, x_{12}^{-2\delta} \, 2^{3\delta} \; ; \; \delta = \tfrac{1}{2}(D-d) \qquad (39e)$$

But this is just $\tfrac{1}{2}g^2 \, \Gamma(\tfrac{3}{2}[\,d-2\,])\,(2\pi)^D$ times the LHS of the vertex bootstrap (4), i.e. the dressed vertex (8). Therefore, by (35), (37) we have $f_{1,-1}=\tfrac{1}{3}(2\pi)^D$. The simplest propagator graph, Fig.3b, is evaluated by the same method, using(39d).

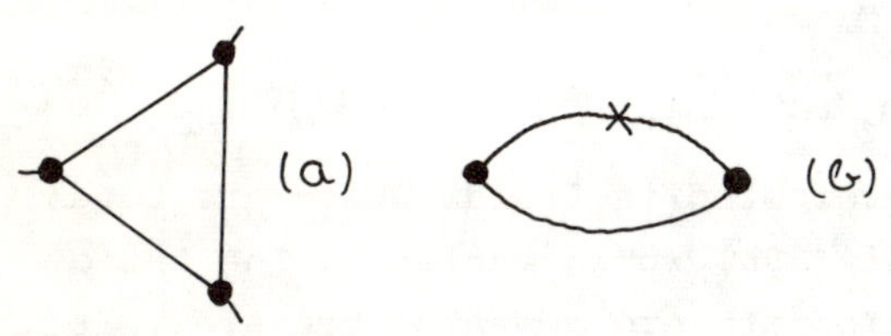

Fig.3. The lowest order vertex-(a) and propagator graph (b) contributing to bootstrap Eqs.(4), (5)

The reader is advised to work it out for himself. The result is $h_{1,-1}=\tfrac{1}{12}(2\pi)^D$. Hence finally we obtain from Eqs.(39a) and (38) the coupling constant and anomalous part of the field dimension as

$$\Delta = d - (\tfrac{1}{2}D - 1) = \tfrac{1}{18}\varepsilon \; + \ldots \; ; \; g^2 = \tfrac{2}{3}(2\pi)^D \varepsilon + \ldots \qquad (39f)$$

A physically acceptable solution must satisfy $g^2 > 0$, and $\Delta > 0$ because of the Lehmann representation of the propagator. This requires $h_{1,-1} > 0$, which can be shown to be always true. On the other hand, in the present φ^3-model, $g^2 > 0$ also requires that $\varepsilon > 0$ so that we must restrict ourselves to considering $D > 6$.

The apparent smallness of g^2 is due only to our unconventional normalization of propagator (7). Wit normalization such that in the canonical limit (39d) is valid and $\widetilde{G}^{-1}(p) \longrightarrow -i(-p^2-i0)^{-1}$, one has instead $g^2 = (4\pi)^{D/2}\tfrac{2}{3}\varepsilon + \ldots$. This is a typical strong interaction coupling constant, compare e.g. the physical πN-coupling constant in the real world (D=4), viz. $g^2_N = 1.2(4\pi)^{D/2}$.

It is of theoretical interest in connection with Bjorken scaling (see Sec.9) to compute also the anomalous part σ_s of the field dimensions d_s of (w.l.o.g. traceless) symmetrical tensor fields $O_{\mu_1 \ldots \mu_s}(x)$,

$$d_s \equiv D-2+s + \sigma_s \quad .$$

In a free field theory, these tensor fields would be of the form $:\varphi \overset{\leftrightarrow}{D}_{\mu_1\cdots\mu_s}\varphi:$, where $\overset{\leftrightarrow}{D}_{\cdots}$ is some differential operator. Let us define

$$G_{\mu_1\cdots\mu_s}(y;x_1 x_2) = \langle T(O_{\mu_1\cdots\mu_s}(y)\,\varphi(x_1)\varphi(x_2))\rangle \tag{39g}$$

This 3-point function must satisfy the BS-equation (23), with $\mu=(\mu_1\cdots\mu_s)$. It is most convenient to go to zero momentum transfer, i.e. integrate over y. Once can verify by inspection of the unique conformal invariant expression for $G_{\mu_1\cdots\mu_s}$ that this does not lead to an infrared divergence. For reasons of Lorentz and dilatation symmetry,

$$\int d^D y\, G_{\mu_1\cdots\mu_s}(y;x\,0) = (-i)^{s-1}\{x_{\mu_1}\nabla_{\mu_2}\cdots\nabla_{\mu_s} - traces\}\Gamma(d+\tfrac{1}{2}\sigma_s)\cdot(-\tfrac{1}{2}x^2+i0)^{-d-\frac{1}{2}\sigma_s} \tag{39h}$$

up to normalization.(We are not interested in fields which are total derivatives and for which the LHS of (39h) would vanish.) The LHS of (23) involves then the vertex function at zero momentum transfer. It is obtained from expression (39h) by amputation on arguments x_1, x_2, i.e. multiplication with inverse propagators which are easily obtained from Eqs.(6),(7). The computation to order ε then proceeds exactly as described above, using (39d). Inserting (39f), the result is

$$\tfrac{1}{2}\sigma_s = \left[\tfrac{1}{18} - \frac{2}{3(s+2)(s+1)}\right]\varepsilon + \ldots \tag{39i}$$

Hence in particular the stress tensor $O_{\mu\nu}$ has canonical dimension $d_2=D$. This was clear beforehand,since the propagator bootstrap (5) is __identical__ with the BS-equation for expression (39h) with s=2. The s=0 field φ^2 has dimension $d_o= D-d$ (cp.(39f)), i.e. it is the "shadow-operator" of φ in the nomenclature of Ferrara et al.[46] It is interesting to note that the anomalous parts of the dimension of the "constituent fields" φ of $O_{\mu_1\cdots\mu_s}$ become additive when $s\to\infty$,viz $\sigma_s \to 2\Delta$ as $s \to \infty$.

The result (39i) implies quite a complicated light cone structure of products of 2 fields (or currents when the model is extended to fields with e.g. isospin 2). For instance, according to Wilson's dimensional rules [19]

$$\varphi(x)\,\varphi(0) \underset{x^2\to 0}{=} \sum_{s=0,2,\ldots}^{\infty} x^{\mu_1}\cdots x^{\mu_s}\,(-x^2+i\varepsilon x_o)^{-\Delta+\frac{1}{2}\sigma_s}\,O_{\mu_1\cdots\mu_s}(0) +\ldots$$

where the dots stand for less singular terms, and terms involving derivatives of fields (cp.Sec.9).

A closer examination of the steps leading to (39a) reveals that the Ansatz (38) works because singularity surfaces (36) of the propagator and vertex graphs cross in the point $(D,d) = (6,2)$. Since there are many other such crossings of singularity surfaces elsewhere, the method has great prospect of generalization. By a further study of the geometry of singularity surfaces - some of which are related to light cone singularities exhibited by operator product expansions - one may hope to approach the goal of a qualitative classification of solution manifolds of bootstraps (35). This is because the germs of such manifolds which are attached to various crossings of singularity surfaces must somehow connect to form global 1-dimensional surfaces in (D,d,g^2)-space.

8.B   <u>FEYNMAN GRAPH EXPANSION OF DYNAMICAL DIMENSIONS</u>

There exists another method for computing dynamical dimensions of fields for the conformal invariant φ^3-theory in $6+\varepsilon$ space time dimensions. This is Wilson's celebrated Feynman graph expansion for critical exponents, which has been used with great success in the description of phase transitions[37]. In presenting his theory, Wilson made essential use of a nontrivial renormalization group argument. For pedagogical reasons I shall present here a modified version of Wilson's approach which was shown to me by K. Symanzik[38] and is based on use of the Callan-Symanzik equations. The pedagogical reasons are that the CS-equations will be discussed in considerable detail at this school, while the (equivalent) renormalization group arguments will probably not. As we shall see, the Feynman graph expansions give identical results with the conformal bootstrap approach described before.

Let us consider a prae-asymptotic zero mass theory (see Sec.7) in $D=6+\varepsilon$ space time dimensions. It depends on a renormalization momentum U^2 (the argument m^2 in Γ_{as} of Sec.7) and a coupling constant. The coupling constant has dimension $(mass)^{-\varepsilon/2}$, we write it as $VU^{-\varepsilon/2}$ such that V is a dimensionless parameter. The prae-asymptotic theory can be constructed by standard perturbation theory, i.e. Feynman graph expansions. The Feynman propagator is $-i(-p^2-i0)^{-1}$, and integration is over loop momenta $d^D q$. After going to Euclidean space, $\vec{q} = (q^1, \dots q^4 = iq^0)$ real, and inserting $(\vec{q}^2)^{-1} = \int_0^\infty d\alpha \; \exp\{-\alpha \vec{q}^2\}$ one only needs the integral $\int d^D\vec{q} \; \exp\{-A\vec{q}^2\} = (\pi/A)^{D/2}$.

Polynomial subtractions have to be made on self-energy and vertex graphs such that the appropriate normalization conditions are fulfilled (for <u>infinitesimal</u> ε the theory is renormalizable!):

$$\Gamma(0,0\,;\,U^2,V) = 0 \qquad (a)$$

$$\Gamma(p,-p\,;\,U^2,V)_{p^2=-U^2} = -iU^2 \qquad (b)$$

$$\text{where } (\hat{p}_i\hat{p}_j) = -\tfrac{1}{2}(3\delta_{ij}-1)U^2$$

$$\Gamma(\hat{p}_1\hat{p}_2\hat{p}_3\,;\,U^2,V) = -iVU^{-\varepsilon/2} \qquad (c)$$

$$(40)$$

For dimensional reasons, the vertex functions Γ have homogeneity properties

$$\Gamma(\lambda p_1\dots\,\lambda p_n\,;\,\lambda^2 U^2,V) = \lambda^{D-\tfrac{1}{2}n(D-2)}\,\Gamma(p_1\dots p_n\,;\,U^2,V)$$

$$= \begin{cases} \lambda^2\,\Gamma(p,-p\,;\,U^2,V) & (n=2) \\[1ex] \lambda^{-\varepsilon/2}\,\Gamma(p_1 p_2 p_3\,;\,U^2,V) & (n=3) \end{cases} \qquad (41)$$

As remarked in Sec.7, the vertex functions of the praeasymptotic theory satisfy a homogeneous Callan-Symanzik equation. This implies (with new β,γ)

$$\left[U^2\frac{\partial}{\partial U^2} + \beta(V,\varepsilon)\frac{\partial}{\partial V} - n\gamma(V,\varepsilon)\right]\Gamma(p_1\dots p_n\,;\,U^2,V) = 0 \qquad (42)$$

We have indicated that β and γ will of course also depend on the space time dimension D, i.e. on ε. We are looking for a zero V_∞ of $\beta(V,\varepsilon)$ considered as a function of V. $\Gamma(p_1\dots p_n, U^2, V_\infty)$ are then vertex functions of a conformal invariant theory (they depend on U^2 in a trivial way), with field dimension

$$d = \tfrac{1}{2}(D-2) + 2\gamma(V_\infty) \qquad (43)$$

We shall show that for small enough ε there exists a zero V_∞ of β which can be found by power series expansion in ε :

$$V_\infty^2 = a_0\varepsilon + \dots \qquad (44)$$

Consequently there exists a conformal invariant theory with small coupling constant V_∞^2 (of order ε); its field dimension will be given by a power series in ε. It is obtained by substituting (44) into the expansion of γ in powers of V. Since we are interested in a zero at small value of V, we can restrict ourselves to compute β and γ to the lowest orders in V. Higher order corrections will allow to compute the field dimension d to higher order in ε .

Let us consider CS-equation (42) for n=2 at $p^2=-U^2$. Using homogeneity (41) it can be rewritten as

$$0 = \left[-p^2\frac{\partial}{\partial p^2} + \beta(V)\frac{\partial}{\partial V} - (2\gamma(V)-1)\right]\ \Gamma(p,-p;U^2,V)_{p^2=-U^2}$$

The last two terms are determined by the normalization condition (40b), the term $\beta\frac{\partial}{\partial V}$ vanishes since the RHS of Eq.(40b) does not depend on V. As a consequence

$$1 - 2\gamma(V) = -i\frac{\partial}{\partial p^2}\ \Gamma(p,-p)_{p^2=-U^2} \tag{45}$$

Next we consider CS-equation (42) for n=3 with p_i at the symmetry point p_i given by (40). Using homogeneity (41) again we rewrite it as

$$0 = \left[-\frac{\varepsilon}{4} - \lambda^2\frac{\partial}{\partial\lambda^2} + \beta(V)\frac{\partial}{\partial V} - 3\gamma(V)\right]\Gamma(\lambda\hat{p}_1\ldots\lambda\hat{p}_3;U^2,V)_{\lambda=1}$$

Evaluating all but the second term with the help of the normalization condition (40c) gives

$$\beta(V,\varepsilon) = V\left[\frac{\varepsilon}{4}+3\gamma(V)\right] + iU^{\varepsilon/2}\lambda^2\frac{\partial}{\partial\lambda^2}\Gamma(\lambda\hat{p}_1\ldots\lambda\hat{p}_3;U^2,V)_{\lambda=1} \tag{46}$$

It remains to compute the derivatives of vertex functions which occur on the RHS's of (45) and (46). They are of order V^2 resp. V^3, and are of course finite at $\varepsilon = 0$, and also differentiable there. Consequently

$$\gamma(V) = c_o V^2 + O(V^2\varepsilon) + O(V^3)$$

$$\beta(V) = \frac{\varepsilon}{4}V + b_o V^3 + O(V^3\varepsilon) + O(V^4)$$

where c_o and b_o are constants. They can therefore be determined by evaluating the lowest order Feynman graphs contribution to $-i$ $(\partial/\partial p^2)\,\Gamma(p,-p)$ resp. $\lambda^2(\partial/\partial\lambda^2)\,\Gamma(\lambda p_1\ldots\lambda p_3)$ at $\varepsilon = 0$, i.e. in D=6 space time dimensions.

Consider for instance $-i(\partial/\partial p^2)\,\Gamma(p,-p)$. It is given by an integral which needs a subtraction. One considers therefore first the second derivative $(\partial/\partial p^2)^2\, \Gamma(p,-p)$ which is finite without subtraction. To order v^2

$$(\partial/\partial p^2)^2\, \Gamma(p,-p) = \tfrac{1}{2}(\partial/\partial p^2)^2 \bigcirc\!\!\!\!\!\bigcirc + \dots$$

For p Euclidean (i.e. $p^2 < 0$) one derives the Feynman parameter representation

$$= \tfrac{i}{2}\, v^2 (4\pi)^{-3} \int_0^\infty\!\!\int_0^\infty d\alpha_1 d\alpha_2\, \alpha_1^2 \alpha_2^2 (\alpha_1+\alpha_2)^{-5} \exp \frac{\alpha_1\alpha_2}{\alpha_1+\alpha_2} p^2 = \tfrac{i}{12} v^2 (4\pi)^{-3} (-p^2)^{-1}$$

Using the normalization conditions (40a,b) to fix constants of integration we recover

$$\Gamma(p,-p) = i p^2 - i\, v^2 (4\pi)^{-3} \tfrac{1}{12} p^2 \ln(-p^2/U^2) \qquad (D=6)$$

Therefore by (45)

$$2\,\gamma(V) = \frac{1}{12}(4\pi)^{-3} v^2 \qquad (D=6)$$

viz.

$$c_o = \frac{1}{24}(4\pi)^{-3}$$

The vertex function on the RHS of (46) is even easier to compute; b_o is given by an integral that converges without subtraction. The result is $b_o = -\tfrac{3}{8}(4\pi)^{-3}$. Thus

$$\beta(V) = \frac{\varepsilon}{4} V - \tfrac{3}{8}(4\pi)^{-3} V^3 + \dots$$
$$2\,\gamma(V) = \frac{1}{12}(4\pi)^{-3} v^2 + \dots$$

We see that $\beta(V)$ has a zero at $V_\infty^2 = \tfrac{2}{3}(4\pi)^3 \varepsilon + \dots$. The anomalous part of the field dimension is given by Eq.(43), viz

$$2\,\gamma(V_\infty) = \frac{1}{18}\varepsilon + \dots \tag{47}$$

This agrees with our previous result of Sec. 8 A.

Since the square of the coupling constant V_∞^2 must be positive, we must have

$$\varepsilon > 0,$$

i.e. we get an acceptable theory only in D>6 dimensions. The slope

$$\frac{\partial}{\partial V}\,\beta(V)\Big|_{V=V_\infty} = -\tfrac{1}{2}\varepsilon + \ldots < 0 \tag{48}$$

so that the condition mentioned after Eq.(30) is satisfied, and the theory may be considered as Gell-Mann Low large momentum asymptote of a massive or prae-asymptotic massless theory. (For φ^4-theory in 3.99 dimensions the slope $\beta'(V_\infty) > 0$. That is basically why that model is relevant for critical phenomena in statistical mechanics, i.e. <u>large distance</u> behavior, and not for short distance behavior as interests us here - see the discussion in Sec.7).

9.   <u>THE FERRARA-GRILLO-PARISI-GATTO THEOREM AND NON-BJORKEN</u>
        <u>SCALING</u>

It is known[43] that a canonical Bjorken scaling law of electroproduction requires an infinite set of (w.l.o.g. traceless) symmetrical tensor fields $O_{\mu_1\cdots\mu_s}$, s =2,4,6, ... with canonical dimension d_s,

$$d_s = s + 2 \qquad , \; s=2,4,6,\ldots \;(D=4) \tag{49}$$

Gatto, Ferrara and Grillo have made extensive studies of conformal invariant operator product expansions[12]. They have in particular shown that the fields $O_{\mu_1\cdots\mu_s}$ relevant for scaling must transform according to Eq.(2b) in the conformal invariant Gell-Mann Low limit. Fields with this transformation law ($\varkappa_\mu = 0$ in the notation of Ref.17) are said to have conformal weight σ .

There is furthermore a remarkable theorem due to Ferrara, Grillo, Parisi and Gatto[13]. It says that in a conformal invariant theory a traceless symmetrical tensor field $O_{\mu_1\cdots\mu_s}(x)$ of conformal weight 0 and with canonical dimension (49) is necessarily conserved

$$\nabla^{\mu_1}O_{\mu_1\cdots\mu_s}(x) = 0 \quad , \tag{50}$$

assuming there are no vacuum troubles. This implies in particular that a canonical Bjorken scaling law would require an infinite set of <u>conserved</u> tensor fields $O_{\mu_1\cdots\mu_s}(x)$.

The proof of this theorem is very simple. Essentially it

amounts to showing that the most general conformal invariant expression for the 2-point function $\langle 0| O_{\mu_1 \ldots \mu_s}(x) \; O_{\nu_1 \ldots \nu_s}(y)|0\rangle$ will necessarily satisfy

$$\langle 0| \nabla_x^{\mu_1} O_{\mu_1 \ldots \mu_s}(x) \nabla_y^{\nu_1} O_{\nu_1 \ldots \nu_s}(y)|0\rangle = 0 \qquad (51)$$

Eq.(50) follows then e.g. as a consequence of positivity of probability.

Strictly speaking it has not yet been established that the hypotheses of this theorem are fulfilled in a conformal invariant theory constructed according to the bootstrap approach. There could be vacuum or domain-problems, especially since the generators K_μ of conformal transformations are not selfadjoint[21]. More precisely one does not know yet whether the 2-point functions $\langle O_{\mu_1 \ldots \mu_s}(x) \; O_{\nu_1 \ldots \nu_s}(0)\rangle$ will come out conformal invariant, for they have to be computed from equations like (25) which are _not_ manifestly conformal invariant because of the crosses. (Note that this has nothing to do with the ambiguous definition of the $\overset{*}{T}$-product mentioned after Eq.(25). We are here considering the Wightman functions.)

Assuming the theorem holds and there is no vacuum trouble, one may be able to show that a canonical Bjorken scaling law cannot possibly hold (if the GML- limit theory is not free). Arguments to this effect have been advanced by Parisi[40].

For simplicity let us consider φ^3-theory in D=6 dimensions. To understand the problem let us consider for instance the conserved charge associated with a conserved 4-th rank tensor field with dimension d_s =s+D-2 = 8. It is formally defined by

$$Q_{\mu_1 \ldots \mu_3} = \int d\sigma^{\mu_4} O_{\mu_1 \ldots \mu_4}(x)$$

We assume that it annihilates the vacuum. Let us then consider the commutator $[\varphi(x), Q_{\mu_1 \ldots \mu_3}]$. It must represent a traceless 3rd rank tensor field of dimension $d_\varphi +3$. The only candidate is $-i\partial_{\mu_1} \ldots \partial_{\mu_3}\varphi(x)-$ -traces. Hence for suitable normalization of Q

$$[\varphi(x), Q_{\mu_1 \ldots \mu_3}] = -i\partial_{\mu_1} \ldots \partial_{\mu_3}\varphi(x)$$

The proportionality factor is not zero as may be verified by explicit computation using the known expression for $\langle 0|\varphi(x)\varphi(y)O_{\mu_1 \ldots \mu_3}(z)|0\rangle$

which is fixed by conformal invariance.

The conservation of the new charge would have to be respected by all Green functions. In particular the 3-point function $W(x_1 x_2 x_3) = \langle \varphi(x_1) \varphi(x_2) \varphi(x_3) \rangle$ should satisfy

$$\delta W(x_1 x_2 x_3) = -i \sum_{i=1}^{3} \nabla_i^{\mu_1} \nabla_i^{\mu_2} \nabla_i^{\mu_3} W(x_1 x_2 x_3) - \text{traces} = 0 \qquad (52)$$

Now the 3-point function $W(x_1 x_2 x_3)$ is explicitly known from conformal invariance. One can therefore check whether (52) is fulfilled. One finds that it is not. So there cannot be any nontrivial good conserved tensor charge which annihilates the vacuum.

To rule out all possibilities of a canonical Bjorken scaling one would also have to dispose of the possibility of some kind of a spontaneously broken symmetry. However, we shall not further pursue this line of thought here[40].

Let us ask instead what positive information is available at present. We assume of course that the massive theory possesses a (conformal invariant) Gell-Mann Low limit theory as discussed in Sec.7. This guarantees[44] an asymptotically dilatation invariant operator product expansion around the tip of the light cone for $J_\mu^{e.m.}(x) J_\nu^{e.m.}(0)$. This will probably be discussed in the lectures of Profs.Schroer and Zimmermann.

According to Leutwyler and Stern[48], electroproduction in the Bjorken limit amounts to considering the imaginary part of the virtual forward Compton scattering amplitude $T_{\mu\nu}(p,q)$ in the limit

$$q = k + En \quad ; \quad E \longrightarrow \infty \qquad (53)$$

where n is a __lightlike__ vector , and k is a fixed spacelike one. As usual, p is the target nucleon's momentum and q the photon momentum. In the following $\nu = pq$ and $\omega = -q^2/\nu$ (target mass = 1). Because $n^2 = 0$ one is faced with a problem involving exceptional __Minkowski__ momenta in the nomenclature of Symanzik[26] and this is a difficult problem.

However, something can nevertheless be said because one is only interested in the imaginary part of the amplitude. The trick is to use dispersion relations[43] and thereby reduce the problem to one involving Euclidean exceptional momenta, which can be mastered with the help of Wilson expansions[26].

More specifically one considers the virtual Compton amplitude

$$T_{\mu\nu}(p,\lambda\hat{q})\ ,\ \lambda\to\infty\ ;\ \hat{q}^2<0 \quad \text{(Euclidean)} \tag{54}$$

i.e. essentially the "old" Bj limit.(In this limit the amplitude becomes real for spectral reasons if q^μ is real). The asymptotic expansion in λ is determined by the Wilson expansions of $J_\mu^{e.m.}(x)J_\nu^{e.m.}(0)$ around the tip of the light cone. Making an O(3) expansion in p one can project out the contribution from tensor fields $O_{\mu_1\ldots\mu_s}(x)$ with any given rank s in that expansion.

The final step is to use dispersion relations in ν which relate the asymptotic behavior of (54) to the large $-q^2$-behavior of the moments $\int\omega^{s-2}\nu W_2(q^2,\omega)d\omega$ etc. of the electroproduction structure functions. Let us assume that $T_2(q^2,\nu)$ satisfies unsubtracted dispersion relations in ν^2 for $q^2<0$. The result is then [43]

$$\int_0^2 d\omega\,\omega^{s-2}\,\nu W_2(q^2,\omega)\sim(-q^2)^{-\frac{1}{2}\sigma_s}\ ,\ q^2\to-\infty\ , \tag{55}$$

$$s=2,4,6,\ldots$$

where

$$\sigma_s=d_s-s-2$$

is the anomalous part of the dimension of the spin s field $O_{\mu_1\ldots\mu_s}$ with lowest dimension that occurs in the Wilson expansion of the product of two e.m. currents. Because of positivity of W_2 ,

$$\sigma_{s+2}\geqq\sigma_s\ , \tag{56}$$

The Callan-Gross Integral (s=2 in (55)) scales canonically if $O_{\mu_1\mu_2}$ is the stress tensor[43].

Stronger statements appear possible only on the basis of extra assumptions which we have been unable to verify in the frame work of the bootstrap approach to conformal invariant field theory.

However, experience with field theory in $6+\varepsilon$ dimensions suggests that the anomalous part of field dimensions is usually very small (compare Eq.(39i) of Sec.8A). If so, an anomalous scaling law (55) might not be inconsistent with the data.

10. <u>CONFORMAL INVARIANT OPERATOR PRODUCT EXPANSIONS</u>

Operator expansions will be discussed in some detail in the lectures of Profs. Schroer and Zimmermann. I shall therefore be content with mentioning one more result due to Ferrara, Gatto and Grillo which relies specifically on conformal invariance[12].

Let us assume that the exactly conformal invariant theory also possesses exactly conformal invariant operator product expansions. (If one is dealing with a Gell-Mann Low limit theory (see Sec.7) it should be possible and would be desirable to prove this, e.g. by generalizing arguments of Schroer[36]. One could also rely on the conformal analog of the Callan Symanzik equations[45] which were derived by Parisi, and Callan and Gross).

Suppose that $C(x)$ is a scalar field operator with conformal transformation law (2b). Suppose further that it occurs in the operator product expansion of $A(x)B(0)$. Then all its derivatives $\nabla_{\mu_1}\cdots\nabla_{\mu_s}C(x)$ will also occur therein. Ferrara, Gatto and Grillo have shown that conformal invariance uniquely fixes the coefficients of all these derivatives. For instance [12]

$$A(x)B(0) \underset{x^2\to 0}{\sim} (-x^2)^{-\frac{1}{2}(d_A+d_B-d_C)}\,{}_1F_1\left(\tfrac{1}{2}[d_A+d_B-d_C],d_C; x\cdot\nabla\right)C(0)+\ldots \tag{57}$$

if A,B are scalar fields with dimension d_A, d_B. Expanding the confluent hypergeometric function in a power series gives an explicit formula for the above mentioned coefficients.

The great beauty of formula (57) is this: If one commutes $A(x)B(0)$ with another local field $O(y)$ the result should vanish whenever $(x-y)^2 < 0$ and $y^2 < 0$, but not in some larger region. Ferrara et <u>al</u>. have shown that the summation (57) over derivatives is just what is needed to restore this property - which would not be true for the individual terms on the RHS of the ordinary Wilson expansion[19].

Eq. (57) has been widely generalized and also forms the starting point of the "shadow operator formalism" of Ferrara et al.[46], and Bonora, Sartori and Tonin[14]. These developments are however rather involved technically, and we will therefore at this point leave the reader, advising him to consult the original literature for further information[46,47].

REFERENCES

1.   A.A. Migdal, Phys.Letters 37B, 98, 386 (1971).

2.   A.M. Polyakov, JETP Letters 12, 381 (1970).

3.   G. Parisi and L. Peliti, Lett. Nuovo Cimento 2, 627 (1971).

4.   G. Mack and I. Todorov, IC/71/139, Trieste, Oct.1971
      (submitted to Phys.Rev.).

5.   A.M. Polyakov, JETP Letters, 32, 296 (1971).

6.   K. Symanzik, Lett. Nuovo Cimento 3, 734 (1972).

7.   G. Mack and K. Symanzik, DESY 72/20 Hamburg, Apr.1972
      (to appear in Commun.Math.Phys.).

8.   G.Mack, Expansions around canonical dimensions in conformal
      invariant quantum field theory, Bern, Apr.1972(unpublished).

9.   M. d'Eramo, G. Parisi and L. Peliti, Lett.Nuovo Cimento 2,
      878 (1971).

10.   G. Parisi, LNF-72/13, Frascati, Feb.1972; LNF - 72/35
      (Apr.1972).

11.   R. Gatto, Riv. Nuovo Cimento 1, 514 (1969).

12.   S. Ferrara, R. Gatto and A.F. Grillo, Springer Tracts (to
      appear), and ref. 39.

13.   S. Ferrara, A.E. Grillo, G. Parisi and R. Gatto, Phys.Letters
      38B, 333 (1972).

14.   L. Bonora, L. Sartori, M. Tonin, Padova preprint IFPTH-6/71
      (1971).

15.   H.A. Kastrup, Nucl. Phys. 58, 561 (1964) and references
      cited in 17.

16.   J.E. Wess, Nuovo Cimento 18, 1086 (1960).

17.   G. Mack and Abdus Salam, Ann.Phys. (N.Y.) 53, 174 (1969).

18.   G. Mack, Nucl.Phys. B5, 499 (1969).

19.   K. Wilson, Phys. Rev. 179, 1499 (1969).

20.   R.F. Streater and A.S. Wightman, "PCT, Spin and Statistics,
      and all that", W.A. Benjamin, New York 1964.

21.   M. Hortaçsu, R. Seiler, B. Schroer, Phys.Rev. D5, 2518(1972).

22.   B. Schroer, Fortschritte der Physik 11, 1 (1963).

23.   J.D. Bjorken and S.D. Drell, "Relativistic Quantum Fields",
McGraw-Hill, New York 1965.

24.   E.R. Speer, "Generalized Feynman Amplitudes", Princeton
University press, Princeton 1969.

25.   G. Mack (unpublished).

26.   K. Symanzik, Commun.Math. Phys. $\underline{23}$, 49 (1971).

27.   Ref.4, Appendix C.(Sec.IIIC of revised version, to be
published in Phys.Rev.).

28.   H. Epstein, V. Glaser, CERN Th 1156, Geneva, May 1970.
O. Steinmann, "Perturbation expansions in Axiomatic Field
Theory", Lecture notes in Physics Vol. $\underline{11}$, Berlin:Springer
1971.

29.   M. Veltman, Physica $\underline{29}$, 186 (1963);
K. Symanzik, J. Math. Phys. $\underline{1}$, 249(1960) esp.Appendix A.

30.   C.G. Callan, S. Coleman, R. Jackiw, Ann.Phys. $\underline{59}$, 42(1970).

31.   Ref.7, Appendix C.

32.   M. Gell-Mann, F.E. Low, Phys. Rev.$\underline{95}$, 1300 (1954),
K.G. Wilson, Phys. Rev. $\underline{D3}$, 1818 (1971).

33.   C.G. Callan, Phys. Rev. $\underline{D2}$, 1451 (1970).

34.   S.L. Adler, Phys. Rev. $\underline{D5}$, 3021 (1972); see also Ref.35.

35.   M. Baker, K. Johnson, Phys.Rev. $\underline{183}$, 1292 (1969), $\underline{D3}$, 2516
(1971).
S.L. Adler, W.A. Bardeen, Phys.Rev. $\underline{D4}$, 3045 (1971).

36.   B. Schroer, Lett. Nuovo Cimento $\underline{2}$, 867 (1971).

37.   K. Wilson, Phys. Rev. Letters $\underline{28}$, 548 (1972).
K. Wilson and M.E. Fisher, Phys.Rev.Letters $\underline{28}$, 240 (1972).

38.   K. Symanzik, private communication.

39.   S. Ferrara, R. Gatto and A.F. Grillo, Phys.Letters $\underline{36B}$,
124 (1971), Nucl.Phys. $\underline{B34}$, 349 (1971),Lett.Nuovo Cimento $\underline{2}$,
1335 (1971).

40.   G. Parisi, Phys. Letters (to appear).

41.   B. Schroer, lectures presented at this summer school, and
references given there to work of Schroer and Lowenstein.

42.   W. Zimmermann, lectures presented at this summer school.

43.   G. Mack, Nucl. Phys. $\underline{B35}$,(1971) 592, Sec.3

44.   N. Christ, B. Hasslacher, A. Mueller, preprint CO-3067(2)-9
        (Columbia, 1972)

45.   G. Parisi, Phys. Letters $\underline{39B}$, 643(1972),
        C.G. Callan, D. Gross, "Broken conformal invariance"
        preprint Princeton 1972.

46.   S. Ferrara, G. Parisi, Nucl. Phys. $\underline{B42}$, 281 (1972).
        S. Ferrara, A.F. Grillo, G. Parisi and R. Gatto, Lett.Nuovo
        Cimento $\underline{4}$, 115 (1972).

47.   S. Ferrara, A.F. Grillo and G. Parisi, Frascati preprint
        LNF-72/38 (1972).

48.   H. Leutwyler and J. Stern, Nucl.Phys. $\underline{B20}$, (1970) 77.

KINEMATICAL ASPECTS OF CONFORMAL INVARIANCE

A.F. Grillo

Laboratori Nazionali del CNEN, Frascati, Italy

The actual relevance of the study of Conformal Invariance
(C.I.) in the framework of Quantum Field Theory (Q.F.T.)[1] does not need
to be stressed: it is a growing up field of research, as it has been
pointed up in the lectures given by Profs. Mack and Todorov at this
School.

In this seminar I want to present a few relevant results
that follow from the requirement of C.I. from the very beginning: I
refer to work done in collaboration with S. Ferrara, G. Gatto and
G. Parisi[2].

Since I do not want to use the informations of C.I. to build
up bootstrap equations[3] (in the sense explained in their lectures by
Mack and Todorov) I will call our approach "kinematical". It is however to be stressed that this is not to be intended in any trivial
sense: it took a lot of sophisticated and deep theoretical investigations only to recognize the role of C.I. in the renormalized Q.F.T.[4].

The content of the present seminar is roughly divided into
two parts: in the first one I will discuss consequences of C.I. on
vacuum expectations values (V.E.V.'s) of local operators; the second
one will be devoted to some remarks concerning field theoretical models
in two dimensional space-time.

For my purposes it will be useful to characterize conformal
transformations in a way slightly different from that used by Prof.To-
dorov: to this will be devoted the next section.

CONFORMAL TRANSFORMATIONS[+]

In a D-dimensional pseudo-euclidean space-time (D-1 space dimensions), conformal transformations are defined as those leaving the flat metric invariant, apart from multiplication by a scalar function of the coordinates:

$$g'_{\mu\nu}(x') = \frac{\partial x^{\lambda}}{\partial x'^{\mu}} \frac{\partial x^{\rho}}{\partial x'^{\nu}} g_{\lambda\rho} = \Lambda(x) g_{\mu\nu} \tag{1}$$

where the first equality is simply the definition of the transformed metric tensor. Their name is due to the fact that they leave the cosine of infinitesimal angles

$$\cos\phi = \frac{dx^{\mu} dy_{\mu}}{(dx^2 dy^2)^{\frac{1}{2}}}$$

invariant.

Eq. 1 gives, for infinitesimal changes

$$x'_{\mu} \simeq x_{\mu} + \delta x_{\mu} , \quad \Lambda(x) \simeq 1 - \lambda(x)$$

the differential equation

$$\partial_{\mu}\delta x_{\nu} + \partial_{\nu}\delta x_{\mu} = \lambda(x) g_{\mu\nu} \tag{2}$$

From a different point of view, it is easy to show that Eq. 2. follows from the requirement of the conservation of the Noether current associated with a transformation

$$\partial^{\mu}(\Theta_{\mu\nu}\delta x^{\nu}) = \Theta_{\mu\nu}\partial^{\mu}\delta x^{\nu} = 0 \tag{3}$$

if and only if this current is generated by a symmetric and traceless energy-momentum tensor $\Theta^{\mu\nu}$. This can be easily seen by decomposing $\partial_{\mu}\delta x_{\nu}$ into irreducible components.

The method for obtaining the solution of Eq. 2.is clearly stated in Ref.[5]; here I want only to write down the equation that follows for the generating function $\lambda(x)$

$$[(D-2)\partial_{\mu}\partial_{\nu} + g_{\mu\nu}\Box] \lambda(x) = 0 \tag{4}$$

+) The method here outlined follows very closely the one contained in
   in Ref.[5].

which in turn implies, if $\underline{D \neq 2}$,

$$\partial_\mu \partial_\nu \lambda(x) = 0 \tag{5}$$

This means that $\lambda(x)$ is a linear function of x, so that Eq.2 implies that δx_μ can be at most a quadratic function, so giving:

$$\delta x_\mu = \delta a_\mu + \delta \omega_{\mu\nu} x^\nu + x_\mu \delta \rho + (x^2 \delta c_\mu - 2 x_\mu \delta c \cdot x) \tag{6}$$

Conformal transformations are so seen to form a group which depends on $(D+2)(D+1)/2$ parameters, which contains the Poincaré group as a subgroup, formed by the vector δa_μ and the antisymmetric tensor $\delta \omega_{\mu\nu}$, the dilatation subgroup ($\delta \rho$), and the subgroup of special conformal transformation (δc_μ).

From Eq. 4 it is evident that the case D = 2 is a special one, implying only

$$\Box \lambda(x) = 0 \tag{7}$$

We shall see later the consequences of this well known fact.

As a last point, note that in the definition Eq. 1 is contained also the finite transformation called coordinate inversion:

$$R : x_\mu \rightarrow - \frac{x_\mu}{x^2} \tag{8}$$

CONFORMAL COVARIANT VACUUM EXPECTATION VALUES

The action of conformal group on fields and operators has already been defined in the lectures by Prof. Todorov, as well as the theoretical problems connected with the definition of conformal transformations in a field theoretical sense[6]; here I want only to give some definitions.

A tensor operator will be said to be conformal irreducible if it is a Lorentz irreducible tensor (i.e. symmetric and traceless), of definite dimension ℓ_n, and moreover its commutator with the special conformal transformation generator vanishes at the point x = 0

$$[O_{\alpha_1 \ldots \alpha_n}(0), K_\lambda] = 0 \qquad (9)$$

It is said to have <u>canonical dimensions</u> if the dimension and the order of the tensor are related by:

$$\ell_n = 2+n$$

Note again that the divergence of a conformal irreducible tensor is conformal irreducible only if the tensor has canonical dimension

$$[\partial^{\alpha_1} O_{\alpha_1 \ldots \alpha_n}(0), K_\lambda] \sim (\ell_n - 2 - n) \, O_{\lambda \alpha_2 \ldots \alpha_n}(0) \qquad (10)$$

Application of C.I. on n-point V.E.V.'s implies that every conformal covariant V.E.V. depends on an arbitrary function of $N = (n-3)n/2$ conformal invariant variables (the harmonic ratios, or crossed ratios defined by Prof.Todorov). We are interested in the 2-point Wightman function: in this case $N = -1$, i.e. the function is overdetermined. This is more clearly stated by the following selection rule:

The 2-point Wightman function of conformal irreducible tensors

$$\langle 0| O_{\alpha_1 \ldots \alpha_n}(x) O_{\beta_1 \ldots \beta_m}(y)|0\rangle = M_{\alpha_1 \ldots \alpha_n, \, \beta_1 \ldots \beta_m}(x-y) \qquad (11)$$

vanishes, unless $\ell_n = \ell_m$ and $n = m$.

The proof of the selection rule is very simple in the simplest cases; in general it requires the use of the 6-dimensional formalism or of the coordinate inversion, and will not be given here.

From the selection rule a theorem follows:

<u>"Any irreducible conformal tensor of canonical dimension is conserved"</u>

The proof is very simple[7]: Eq.10 and the selection rule imply that

$$\langle 0| \partial_x^{\alpha_1} O_{\alpha_1 \ldots \alpha_n}(x) \, O_{\alpha_1 \ldots \alpha_n}(y)|0\rangle = 0 \qquad (12)$$

This can be alternatively proved by explicitly computing the more general conformal covariant two-point function and noting that it is conserved if $\ell_n = 2+n$.

Eq. 12. in turn implies

$$\| \partial^{\alpha_1} O_{\alpha_1 \dots \alpha_n}(x) |0\rangle \| = 0 \tag{13}$$

so that operator conservation follows if the Hilbert space has positive definite metric (positive definitness could be relaxed by using again the selection rule and conformal covariant operator product expansions).

This is a very strong result: in fact a theory with infinitely many conserved charges is possibly only a free theory.

What is the conclusion that follows from the previous discussion? From the existing experimental facts about canonical scaling in deep inelastic electroproduction one is led to assume that such a string of canonical dimension operators does indeed exist: the additional requirement of conformal invariance then implies that they are conserved.

In my opinion this conclusion can be avoided in three ways: the first one is that conformal invariance is not really relevant in the deep inelastic limit. From a physical point of view the other two alternatives seem to be more interesting: that the observed scaling is not really canonical (remind that a little amount of non-canonical dimensions is compatible with the experimental data and that bootstrap calculations seem to suggest that physical dimensions are non canonical, but __almost__ canonical), and that the very requirement of the existence of infinitely many conformal irreducible tensors is a too strong one, so fixing the theory to be free from the beginning; also this point has to be proved (or disproved) in the framework of the bootstrap theory.

CONFORMAL INVARIANCE IN TWO DIMENSIONAL SPACE-TIME[8]

As we have already seen, the case D = 2 is a special one: in fact in this case the requirements on infinitesimal transformations are simply

$$\Box \lambda(x) = 0 \quad \text{and} \quad \Box \delta x_\mu = 0 \tag{14}$$

The solution of equation 2. in this case is simply given by

$$\delta x_\mu = m_\mu g(u) + n_\mu h(v) \tag{15}$$

where

$$u = n_\mu x^\mu \ , \quad n_\mu \equiv (1, 1)$$
$$v = m_\mu x^\mu \ , \quad m_\mu \equiv (1, -1) \tag{16}$$

and h(u) and g(v) are arbitrary functions of u and v respectively: the group of conformal transformations turns out to be infinitely dimensional. To every infinitesimal transformation an infinitesimal generator can be associated

$$T(f) = \int d\sigma_\mu \Theta^{\mu\nu} \delta x_\nu$$

which can be decomposed in the sum of two terms depending only on u and v respectively.

Let me consider only the u part of the group: in order to study the commutation relations of the generators one has to give a model for the commutator of the energy-momentum tensor: this can be taken, for example, from the free Thirring model. I do not want to enter into details: the relevant result is the following commutator

$$[T^\mu(h), T^\mu(h')] = T^\mu([h, h']) + \Delta(h, h') \tag{17}$$

(and a similar relation for the v part)

where

$$[h, h'](u) = h(u) \frac{d}{du} h'(u) - \left(\frac{d}{du} h(u)\right) h'(u) \tag{18}$$

and $\Delta(h, h')$ is a c-number functional of h,h'.

It is interesting to note that, by choosing the following basis

$$h(u) = u^{1-n}, \quad h'(u) = u^{1-m}$$

$$[T_n, T_m] = (m-n) T_{n+m} + c\text{-number} \tag{19}$$

one obtains an algebra formally identical with the gauge algebra of the dual models[9].

As a last point, it is possible to show that the algebra Eq. 17. can be obtained also by quantizing a 4-dimensional field theory on a light-like plane[10].

<u>ACKNOWLEDGEMENTS</u>

I thank Profs. H. Leutwyler, G. Mack, B. Schroer and I.T. Todorov for many interesting discussions.

<u>REFERENCES</u>

1.   For earlier references see, for example:
G. Mack and A. Salam, Ann. Phys.(N.Y.) <u>53</u>, 174 (1969).

2.   S. Ferrara, R. Gatto, A.F. Grillo, Frascati Preprint
LNF 71/79 (1971), to appear in Springer Tracts in
Modern Physics.

S. Ferrara, R. Gatto, A.F. Grillo and G. Parisi, Lett.Nuovo
Cimento <u>4</u>, 115 (1972).

For application to operator product expansions, see also:

S. Ferrara, R. Gatto and A.F. Grillo, Phys. Letters <u>36B</u>,
124 (1971)

L. Bonora, G. Sartori and M. Tonin, to appear in Nuovo
Cimento.

3.   For a review, see:

G. Mack, extended version of the talk given at the Informal

Meeting on Broken Conformal Symmetry, Frascati (May 1972)

and, for the original papers:

A.A. Migdal, Phys.Letters $\underline{31B}$, 386 (1971)

A.M. Polyakov, JETP Letters $\underline{12}$, 381 (1970)

G. Parisi and L. Peliti, Lett.Nuovo Cimento $\underline{2}$, 627 (1971)

M.D'Eramo, G. Parisi and L. Peliti, Lett.Nuovo Cimento $\underline{2}$, 878 (1971)

A.M. Polyakov, JETP Letters $\underline{32}$, 296 (1971)
G. Mack and I.T. Todorov, Trieste Preprint IC/71/139 (1971)
G. Mack and K. Symanzik, DESY Report 72/19 (1972)

4.   C.G. Callan jr., Phys. Rev. $\underline{D2}$, 1541 (1970)
        K. Symanzik, Comm.Math.Phys. $\underline{18}$, 227 (1970)
        B. Schroer, Lett. Nuovo Cimento $\underline{2}$, 887 (1971)
        G. Parisi, Frascati Preprint L.N.F.-72/13 (1972)

        B. Schroer, Extended Version of the Lectures given at
        the IV Simposio Brasileiro de Fisica Teorica,
        Rio de Janeiro (1972)

5.   D.G. Boulware, L.S. Brown and R. Peccei,Phys.Rev. $\underline{D2}$, 293 (1970)

6.   M. Hortaçsu, R. Seiler and B. Schroer, University of Pittsburgh Preprint NYO-3829-80, to appear in Phys.Rev.

7.   S. Ferrara, R. Gatto, A.F. Grillo and G. Parisi, Phys.Letters $\underline{38B}$, 333 (1972)

8.   For a wider discussion, see:
        S. Ferrara, R. Gatto and A.F. Grillo, Rome Preprint 385, (1972), to appear in Nuovo Cimento

9.   S. Fubini, E. Del Giudice and P.Di Vecchia, M.I.T.Preprint 209(1971), to appear in Annals of Physics

10.   H. Leutwyler, Lectures given at this School

OPERATOR PRODUCT EXPANSIONS

Wolfhart Zimmermann

Physics Department

New York University

and

Physik-Department

Technische Universität

München

1.   GENERAL THEORY OF THE SHORT DISTANCE EXPANSION

The purpose of this series of lectures is to review some recent work on operator product expansions. We will be concerned with the behaviour of the operator product

$$A_1(x_1)\, A_2(x_2) \tag{1.1}$$

of two field operators near the origin ($x_1 \to x_2$) and near the light cone ($(x_1 - x_2)^2 \to 0$).

Information on the behaviour of the operator product (1.1) near the origin is provided by Wilson's short distance expansion. We begin with a derivation of the short distance expansion in the framework of general field theory[2]. The behaviour of matrix elements

$$\left(\Phi ,\, A_1(x_1) A_2(x_2)\, \Psi \right) \tag{1.2}$$

at short distances in general depends on the direction of the four-vector $x_1 - x_2$ when the origin $x_1 - x_2$ is approached. We therefore keep the direction fixed for the beginning and set

$$x_1 = x + \xi , \quad x_2 = x - \xi , \quad \xi = \varrho\, \eta \tag{1.3}$$

The four-vector η will be fixed at a value off the light cone $\eta^2 \neq 0$. g is a positive parameter: $g > 0$. For simplicity we consider the case of one scalar field $A_1 = A_2 = A$. We then study the g-dependence of the matrix elements

$$(\Phi, A(x+g\eta) A(x-g\eta) \Psi) \tag{1.4}$$

Among all these matrix elements we choose a particular one

$$f_1(g) = (\hat{\Phi}, A(\hat{x}+g\hat{\eta})A(\hat{x}-g\hat{\eta}) \hat{\Psi}) \tag{1.5}$$

which for $g \to 0$ behaves stronger or equally strong as compared to any other matrix element. This means that the limit

$$\lim_{g \to 0} \frac{(\Phi, A(x+g\eta)A(x-g\eta) \Psi)}{f_1(g)} \tag{1.6}$$

is finite for any matrix element. Then the composite operator C_1 defined by

$$C_1(x,\eta) = \lim_{g \to 0} \frac{A(x+g\eta) A(x-g\eta)}{f_1(g)} \tag{1.7}$$

has finite matrix elements. C_1 is not identically zero since

$$(\hat{\Phi}, C_1(x,\hat{\eta}) \hat{\Psi}) = 1$$

As a consequence of the causality condition

$$[A(x), A(y)] = 0 \quad \text{if } (x-y)^2 < 0$$

we further have

$$[C_1(x,\eta), A(y)] = 0 , \quad (x-\eta)^2 < 0 \tag{1.8}$$

Hence C_1 is local for any value of η and may therefore be interpreted as a composite operator. It may, however, happen that C_1 is a multiple of the identity.

In order to get some information on the leading singularity of (1.4) we introduce an operator P_2 by

$$A(x+g\eta)A(x-g\eta) = f_1(g)C_1(x,\eta) + P_2(x,g,\eta) \tag{1.9}$$

It then follows

345

$$\lim_{g \to 0} \frac{P_2(x, g, \eta)}{f_1(g)} = \lim_{g \to 0} \frac{A(x+g\eta)A(x-g\eta)}{f_1(g)} - C_1(x, \eta) = 0$$

The term $f_1 C_1$ therefore gives the leading singularity for $g \to 0$. It has the form of a c-number function f_1 which becomes singular for $g \to 0$ times a field operator C_1 which is local relative to the center of mass point x. In addition C_1 depends on the four-vector η .

In order to get higher order terms we apply the same procedure to the remainder P_2. Again one chooses a matrix element f_2 of P_2 with a maximal behaviour for $g \to 0$ and defines

$$C_2(x, \eta) = \lim_{g \to 0} \frac{P_2(x, g, \eta)}{f_2(g)} \tag{1.10}$$

Inserting

$$P_2 = f_2 C_2 + P_3$$

into (1.9) we find

$$A(x+g\eta)A(x-g\eta) = f_1 C_1 + f_2 C_2 + P_3 \tag{1.11}$$

with

$$\lim_{g \to 0} \frac{f_2}{f_1} = 0 \,, \qquad \lim_{g \to 0} \frac{P_3}{f_1} = 0$$

Repeating this procedure we finally get

$$A(x+g\eta)A(x-g\eta) = \sum_{i=1}^{k} f_i(g) C_i(x\eta) + P_{k+1}(x, g, \eta) \tag{1.12}$$

by induction. The functions f_j are ordered according to decreasing behaviour

$$\lim_{g \to 0} \frac{f_{i+1}}{f_i} \tag{1.13}$$

and the remainder satisfies

$$\lim_{g \to 0} \frac{P_{k+1}}{f_k} \tag{1.14}$$

Symbolically we write

$$A(x+g\eta)A(x-g\eta) = \sum_{i=1}^{\infty} f_i(g) C_i(x, \eta) \tag{1.15}$$

with the understanding that the series is asymptotic in the sense of (1.14 - 15).

For every value of η the operators C_j are local relative to x and may be interpreted as composite operators. This definition is a generalization of the subtraction technique first employed by Dirac and Heisenberg[3] which for free fields leads to the Wick product as an appropriate definition of composite operators[4].

The main assumption involved in this derivation is that at each step k there be a matrix element of P_k which behaves stronger or equally strong for $g \to 0$ as compared to any other matrix element of P_k. The rigorous formulation of this assumption used in ref. 2 will be given now. First the operator product is smeared out in x and with a suitable test function t (x, η). The operator

$$P(t, g) = \int dx\, d\eta\, t(x, \eta)\, A(x+g\eta)\, A(x-g\eta) \tag{1.16}$$

is well-defined for every $g > 0$ on the usual domain D_o of the field operator. The assumptions needed in addition to Wightman's **postulates** can be formulated in the following way. One considers the set of all functions ($\mathcal{H}$ denotes the Hilbert space)

$$\varphi(g) = \sum_{k=1}^{A} \left(\Phi_k,\, P(t_k, g)\, \Psi_k \right), \qquad \Psi \in D_o,\ \Phi_k \in \mathcal{H} \tag{1.17}$$

which satisfy

$$\lim_{g \to 0} \varphi(g) \neq 0 \quad \text{or} \quad = \infty \tag{1.18}$$

This set is denoted as class $\mathcal{A}$. First assume that any ratio

$$\varphi_1 / \varphi_2 \qquad (\varphi_1, \varphi_2 \in \mathcal{A})$$

has only one accumulation point. Accordingly

$$\lim_{g \to 0} \varphi_1 / \varphi_2 \quad \text{exists} \quad \text{or} \quad = \infty$$

This permits introducing an equivalence relation

$$\varphi_1 \sim \varphi_2 \quad \text{if} \quad \lim_{g \to 0} \varphi_1 / \varphi_2 \neq 0, \infty \tag{1.19}$$

We now make the hypothesis that there is only a finite number of equivalence classes in $\mathcal{A}$. Then the **ex**pansion

$$P(t, g) = \sum_{i=1}^{k} f_i(g)\, C_i(t) + P_{k+1}(t, g) \tag{1.20}$$

follows with

$$\lim_{\S \to 0} \frac{f_{i+1}(\S)}{f_i(\S)} = 0 \quad , \quad \lim_{\S \to 0} f_k(\S) \neq 0 \tag{1.21}$$

$$\lim_{\S \to 0} \left(\Phi , P_{k+1}(t,\S) \Psi \right) = 0 \quad , \quad \Phi \in \mathcal{H} , \Psi \in D_0 \tag{1.22}$$

The operators $C_k(t)$ are defined on D_0 and are distributions with respect to t. Moreover, the stated hypothesis is necessary and sufficient for (1.20 - 22).

An asymptotic series of the short distance expansion is obtained if we make the stronger assumption that any set $\mathcal{A}_N$ of all functions (1.17) with

$$\lim_{\S \to 0} \varphi(\S) / \S^N \neq 0 \tag{1.23}$$

has only a finite number of equivalence classes. It then follows

$$\lim_{\S \to 0} P_{k+1} / \S^N = 0 \tag{1.24}$$

provided k is chosen larger than a suitable number $K(N)$. With (1.24) the expansion is asymptotic in the sense that the remainder decreases faster than any power of $\S$.

The η_5 -dependence of the composite operators can be made rather explicit[5]. The condition

$$C_i(x, -\eta) = C_i(x, \eta) \quad \text{if} \quad \eta^2 < 0 \tag{1.25}$$

follows recursively from

$$A(x+\S\eta) A(x-\S\eta) = A(x-\S\eta) A(x+\S\eta) \quad \text{if} \quad \eta^2 < 0$$

The spectrum conditions imply (for details see ref. 5)

$$\tilde{C}_i(x, u) = 0 \quad \text{unless} \quad u^2 \geq 0 , u_0 \geq 0 \tag{1.26}$$

where $\tilde{C}_j$ denotes the Fourier transform of C_j with respect to η . By application of a theorem of Bogoliubov and Vladimirov (1.25) and (1.26) imply that C_j must be polynomial in η for η^2 given. Moreover, the η^2 -dependence of the C_j only involves powers of η^2 and $\lg \eta^2$. For $\eta^2 = -1$ we have

$$C_i(x, \eta) = \sum_{m=1}^{M(i)} \eta_{\mu_1} \cdots \eta_{\mu_m} B_i^{\mu_1 \cdots \mu_m}(x) \tag{1.27}$$

The $B_j^{\mu_1 \cdots \mu_m}$ are local operators transforming as tensors

under Lorentz transformations. (1.27) can further be simplified by decom--
posing $B_j^{\mu_1\cdots\mu_m}$ into irreducible tensors and separating of all $g^{\mu\nu}$ -
factors. For instance, if the decomposition of $B^{\mu_1\cdots\mu_4}$ contains a term
of the form

$$g^{\mu_1\mu_2} F^{\mu_3\mu_4}$$

its contribution

$$\eta_{\mu_1}\cdots\eta_{\mu_4}\, g^{\mu_1\mu_2} F^{\mu_3\mu_4} = -\eta_{\mu_3}\,\eta_{\mu_4}\, F^{\mu_3\mu_4}$$

may be combined with the term n = 2 of the sum (1.27). We assume that
all factors $g^{\mu\nu}$ have been eliminated in this way. The splitting of the
vector ξ into $\rho\eta$ was only a technical device to handle the directional
dependence. We may now return to the original variable ξ by setting

$$\rho = \sqrt{-\xi^2} \quad , \quad \eta = \xi / \sqrt{-\xi^2} \tag{1.28}$$

With the notation

$$g_{j\mu_1\cdots\mu_m} = f_j\left(\sqrt{-\xi^2}\right) \frac{\xi_{\mu_1}\cdots\xi_{\mu_m}}{\left(\sqrt{-\xi^2}\right)^m} \tag{1.29}$$

we finally get

$$A_1(x+\xi)\, A_2(x-\xi) = \sum_j g_j(\xi)\, B_j(x) = \sum_{j=1}^{\infty}\sum_{m=1}^{r(j)} g_{j\mu_1\cdots\mu_m}(\xi)\, B_j^{\mu_1\cdots\mu_m} \tag{1.30}$$

for the general case of two different fields A_1, A_2. The coefficients
$g_{j\mu_1\cdots\mu_m}$ may carry additional spinor or tensor indices corresponding
to the transformation properties of A_1 and A_2. It can always be arranged
that the composite operators C_1 occurring in (1.30) are linearly independent.

From the derivation of the Wilson expansion it is obvious that there is
considerable arbitrariness in the choice of the functions f_j as well as
the operators C_j. The extent of arbitrariness is given by the following
theorem[6]. Suppose

$$A_1(x_1)\, A_2(x_2) = \sum_{j=1}^{k} f_j(\rho)\, C_j(x,\eta) + R(x,\rho,\eta)$$

$$A_1(x_1)\, A_2(x_2) = \sum_{j=1}^{k'} f_j'(\rho)\, C_j'(x,\eta) + R'(x,\rho,\eta) \tag{1.31}$$

are two forms of the principal part which both satisfy the conditions

$$\lim_{\rho\to 0} f_{j+1}/f_j = 0 \quad , \quad \lim_{\rho\to 0} f_k \neq 0 \quad , \quad \lim_{\rho\to 0} R = 0 \tag{1.32}$$

and similarly for the primed quantities. The operators C_j and C_j' are
then related by a triangular linear transformation

$$C' = \Delta C \tag{1.33}$$

with

$$\Delta = \begin{pmatrix} a_{11} & 0 & \cdots & 0 \\ a_{21} & a_{22} & & \vdots \\ \vdots & & & \vdots \\ a_{\ell 1} & a_{\ell 2} & \cdots & a_{\ell\ell} \end{pmatrix}, \quad C = \begin{pmatrix} C_1 \\ \vdots \\ \vdots \\ C_\ell \end{pmatrix}, \quad C' = \begin{pmatrix} C_1' \\ \vdots \\ \vdots \\ C_\ell' \end{pmatrix}$$

$$\ell = \ell' \, , \quad a_{ii} \neq 0 \, ,$$

$$f' = \tilde{\Delta}^{-1} + h \, , \quad \lim_{g \to 0} h_i = 0$$

Wilson has shown that in a scale invariant theory the short distance
behaviour of the coefficients f_j is determined by a simple dimensional
rule. Wilson's rule states that

$$d(A_1) + d(A_2) = d(g_j) + d(B_j) \tag{1.34}$$

holds for each term of the short distance expansion. The dimension
of an operator 0 is defined by the transformation law under scale
transformations

$$U(s)\, O(x)\, U(s)^{-1} = s^{d(0)}\, O(sx) \tag{1.35}$$

The dimension of the coefficient f_j is defined by the transformation
law

$$g_j(\xi) = s^{d(g_j)}\, g_j(s\xi) \tag{1.36}$$

(1.36) and the rule (1.34) are obtained by applying the transformation
(1.35) to both sides of the expansion (1.30), expanding $A_1(sx_1)A_2(sx_2)$
on the left hand side and comparing the coefficients.

As an example we give the short distance expansion for the
product of two free fields. We have

$$A_1(x+\xi)\, A_2(x-\xi) = \langle A_1(\xi)A_2(-\xi)\rangle \mathbb{1} \; + \; : A_1(x+\xi)A_2(x-\xi): \tag{1.37}$$

After smearing out in x the Wick product : $A_1(x + \xi$ $) A_2(x - \xi$):
is finite at the origin and on the light cone and can be expanded into
a Taylor series

$$: A_1(x+\xi)\, A_2(x-\xi): \ =$$

$$= \sum \frac{1}{m!}\ \xi^{\mu_1}\cdots\xi^{\mu_m}\ \partial_{\mu_1}^{\xi}\cdots\partial_{\mu_m}^{\xi}: A_1(x+\xi)A_2(x-\xi):\Big|_{\xi=0} \tag{1.38}$$

The dimensional rule (1.34) is certainly correct for (1.37 - 38).

Dell'Antonio[7] observed recently that one step in this argument is not conclusive. We consider the model of a single scalar field A. The composite operators of the system were defined in terms of A by the short distance expansion. Accordingly one should postulate scale invariance for the field A only and derive it for composite operators rather than assuming it.

In order to find the scaling law of the composite operators we apply the scale transformation

$$U(s)\, A(x)\, U(s)^{-1} = s^d\, A(sx)$$

to the short distance expansion of $A(x + \xi\eta)A(x - \xi\eta)$

$$s^{2d}\, A(sx+s\xi\eta)\, A(sx-s\xi\eta) =$$

$$= \sum g_j(s\xi)\, s^{2d}\, C_j(x,\eta) + P_{a+1}(sx,s\xi,\eta) \tag{1.39}$$

This must be identical to

$$U(s)\, A(x+\xi\eta)\, A(x-\xi\eta)\, U(s)^{-1} =$$

$$= \sum_j g_j(\xi)\, U(s)\, C_j(x,\eta)\, U(s)^{-1} + U(s)\, P_{a+1}(x,\xi,\eta)\, U(s)^{-1} \tag{1.40}$$

By the equivalence theorem (1.33) the composite operators of (1.39) and (1.40) must be related by a triangular transformation

$$U(s)\, C_j(x\eta)\, U(s)^{-1} =$$

$$= \sum_{j'=1}^{j} a_{jj'}(s)\, s^{2d}\, C_{j'}(sx,\eta) \tag{1.41}$$

This implies a similar relation for the operators B_j:

$$U(s)\, B(x)\, U(s)^{-1} = b(s)\, B(sx)$$

$$\tag{1.42}$$

Applying one scale transformation after another we find the multi-plication law

$$b(s_1)\, b(s_2) = b(s_1 s_2) \tag{1.43}$$

Hence the transformation matrices form a representation of the multi-plicative group of the real numbers. With a suitable basis of the operators B_j the matrices b(s) can be brought into the normal form

$$b(s) = \begin{pmatrix} b_1 & 0 & \cdot & - & \cdots & 0 \\ 0 & b_2 & & & & \\ \vdots & & \ddots & & & \\ 0 & & & & & \end{pmatrix} \tag{1.44}$$

$$b_j(s) = s^{C_j} \begin{pmatrix} 1 & 0 & 0 & & & 0 \\ \lg s & 1 & 0 & & & \cdot \\ \frac{1}{2}\lg^2 s & \lg s & 1 & & & \vdots \\ \vdots & \vdots & \vdots & \ddots & & \\ & & & & 1 & 0 \\ \dfrac{\lg^{N_j-1} s}{(N_j-1)!} & & & & \lg s & 1 \end{pmatrix} \tag{1.45}$$

One thus finds the possibility of scale transformations involving logs of the scaling parameter. It is not known whether this possibility occurs in realistic models. This question is being investigated by Brandt and Ng.[8]

2. <u>DEFINITION OF COMPOSITE OPERATORS IN PERTURBATION THEORY</u>[9,10]

We next turn to a discussion of the Wilson expansion in pertur-bation theory. We begin with the definition of the composite operators. The renormalized Green's functions will be defined by the Gell-Mann Low formula

$$\langle T\, A(x_1) \ldots A(x_n)\rangle = $$
$$= F_{in} \langle T\, e^{i\int \mathcal{L}_{int}^0\, dz}\, A_0(x_1) \ldots A_0(x_n)\rangle_0 \tag{2.1}$$

with the interaction Lagrangian

$$\mathcal{L}_{int} = -\lambda\left(\frac{A^4}{4!} - \frac{1}{2}a A^2 - b \mathcal{L}_0\right)$$

$$\mathcal{L}_0 = \frac{1}{2}\partial_\mu A \partial^\mu A - \frac{1}{2}m^2 A^2 \tag{2.2}$$

The symbol "Fin" means that the finite part is formed in the sense of Bogoliubov, with subtractions made around external momenta $p_j = 0$ in momentum space[11]. m is the finite mass, λ the renormalized coupling constant. a and b are power series in λ with finite coefficients which are recursively determined such that the propagator $\tilde{\Delta}_F'$ (p) has the pole at $p^2 = m^2$ with residue i. Bogoliubov's method of renormalization can also be used for defining composite operators. Suppose we want to define A^n through the Green's functions

$$\langle T\, A^n(x)\, A(y_1) \cdots A(y_n)\rangle$$

We then form the Gell-Mann Low expansion which would formally follow in the interaction representation

$$\langle T\, N[A^n(x)]\, A(y_1) \cdots A(y_n)\rangle =$$

$$= \langle T\, e^{i\int \mathcal{L}_I^0\, dz}\, A_0(x)^n\, A_0(y_1) \cdots A_0(y_n)\rangle_0 \tag{2.3}$$

and take the finite part of each Feynman integral. The Green's functions thus constructed define an operator which we denote by $N[A^n(x)]$.

Apart from the usual subtractions additional subtractions have to be made for the new vertex parts containing x, corresponding to the superficial degree of divergence. For instance, for defining

$$\langle T\, N[A^2(x)]\, A(y_1) A(y_2)\rangle \tag{2.4}$$

we first form the unrenormalized integrand

$$\int dk \cdots\; I(k, p-k, \cdots q_1, q_2)$$

By power counting the superficial divergence of the whole diagram is 0. Accordingly we define the renormalized contribution by

$$\int dk \cdots (1 - t^0_{pq_1q_2})(\cdots)\, I(k, p-k, \cdots q_1, q_2) \tag{2.5}$$

$t^n_{pq_1q_2}$ applied to a function of p, q_1, q_2 denotes the Taylor series with respect to these variables up to and including terms of order n. Hence $t^0_{pq_1q_2}$ gives the value of the function at p, q_1, $q_2 = 0$. The bracket $(\cdots)$ refers to subtractions corresponding to subdiagrams. As an example we give the first

order contribution to the Fourier transform of (2.4) (apart from
trivial factors)

$$\int dk \, (1 - t^o_p) \, \frac{1}{k^2 - m^2 + i\varepsilon} \, \frac{1}{(p-k)^2 - m^2 + i\varepsilon} \tag{2.6}$$

Definition (2.3) can be extended to any monomial in A and its
derivatives. The normal product $N\left[M\{A(x)\}\right]$ is constructed by

$$\langle T \, N\left[M\{A(x)\}\right] A(y_1) \dots A(y_n) \rangle =$$

$$= \langle T \, e^{i\int \mathcal{L}^o_I dz} \, N\left[M\{A_0(x)\}\right] A_0(y_1) \dots A_0(y_n) \rangle_0 \tag{2.7}$$

It is often useful to construct normal products which are
formed with additional subtractions. These products are denoted by
$N_a\left[M\{A(x)\}\right]$ where

$$a = d + c \, , \quad c = 0, 1, 2 \dots \tag{2.8}$$

a is called the degree of the normal product. d is the dimension
of M in the naive sense, i.e. the dimension 1 is assigned to
the field A and the differential operator ∂_μ . c is the number
of additional subtractions. For instance, in order to define the
first order contribution to $\langle T N_4\left[A(x)^2\right] A(y_1)(y_2) \rangle$ in momentum
space one forms

$$\int dk \, (1 - t^2_p) \, \frac{1}{k^2 - m^2 + i\varepsilon} \, \frac{1}{(p-k)^2 - m^2 + i\varepsilon} \tag{2.9}$$

which contains two subtractions more than (2.6). More general

$$\int dk \dots (1 - t^2_p)(\dots) \, I(k, p-k, \dots q_1, q_2) \tag{2.10}$$

with the function I of (2.5), is the contribution to the Fourier
transform of $\langle T N_4\left[A(x)^2\right] A(y_1) A(y_2) \rangle$ from an arbitrary diagram.
It should be noted that also the number of subtractions for subdiagrams
containing x must be increased.
The normal product with the smallest number of subtractions (c = 0) is
called minimal and also denoted without subscript

$$N_d\left[M\{A(x)\}\right] = N\left[M\{A(x)\}\right]$$

There are linear relations between composite operators of different
degree. In particular, every normal product can be written as a
linear combination of minimal ones

$$N_a[M] = \sum C_a(M, M') \, N(M') \tag{2.11}$$

The sum extends over all monomials M' with dimension $d(M') \leqq$ a.
For instance, the expansion of $N_4 [A^2]$ reads

$$N_4[A^2] = a\, N[A^2] + b\, N[\partial_\mu A \partial^\mu A] +$$
$$+ c\, N[A \square A] + d\, N[A^4]$$

(2.12)

In (2.12) it was used that only products with an even number of
operators A occur and that the coefficients must be invariant quantities. A term involving the identity is missing since the vacuum expectation values of all other composite operators vanish. The following differentiation rule was shown by Lowenstein[12]

$$\partial_\mu N_a [M\{A(x)\}] = N_{a+1} [\partial_\mu M\{A(x)\}]$$

(2.13)

A corresponding rule holds for the Green's functions

$$\partial_\mu \langle T N_a [M\{A(x)\}] A(y_1) \dots A(y_n) \rangle =$$
$$= \langle T N_{a+1} [\partial_\mu M\{A(x)\}] A(y_1) \dots A(y_n) \rangle$$

(2.14)

The normal product formalism has found many useful applications, we mention Lowenstein's construction of the energy-momentum
tensor[12], a concise derivation of the Callan-Symanzik equation and
the Gell-Mann Low differential equation of the renormalization group[13],
a treatment of the Adler anomalies without use of a cutoff[14], finally
the discussion of gauge invariance in vector meson theories[14]. These
applications will be discussed in the lectures of Schroer. As a simple
illustration of the usefulness of the normal product formalism we will
now derive a finite form of the field equation.
We will find the following identity

$$-(\square + m^2)A = \frac{\lambda}{3!}\, N[A^3] - a A + b\,(\square + m^2)A$$

(2.15)

It is remarkable that (2.15) is just the equation of motion which
would formally follow from the effective Lagrangian

$$\mathcal{L}_{EFF} = \mathcal{L}_0 - \frac{\lambda}{4!} A^4 + \frac{1}{2} a A^2 + b\,\mathcal{L}_0$$

$$\mathcal{L}_0 = \frac{1}{2}\partial_\mu A \partial^\mu A - \frac{1}{2} m^2 A^2$$

(2.16)

which we used for setting up renormalized perturbation theory (see (2.2).
Only the normal product has to be taken for the interaction term A^3 .

We first prove a corresponding field equation for the Green's function $\langle T\,A(x)A(y_1)\,\ldots\,A(y_n)\rangle$ by applying the Klein-Gordon operator $-\,(\Box_x + m^2)$ to each Feynman integral of the perturbative expansion. There are three types of Feynman diagrams involved (see Fig. 1-3). For the sum of all diagrams of a given type we obtain

$$-(\Box_x + m^2)\langle \mathrm{I}\rangle = \frac{\lambda}{3!}\,\langle T\,N\,[A(x)^3]\,A(y_1)\ldots A(y_n)\rangle$$

$$-(\Box_x + m^2)\langle \mathrm{II}\rangle = -a + b\,(\Box_x + m^2)\langle T\,A(x)A(y_1)\ldots A(y_n)\rangle \quad (2.17)$$

$$-(\Box_x + m^2)\langle \mathrm{III}\rangle = i\,\delta(x - y_j)\langle T\,A(y_1)\ldots A(y_{j-1})A(y_{j+1})\ldots A(y_n)\rangle$$

We thus find as field equation for the Green's function

$$\left\langle T\,N\left[\frac{\partial \mathcal{L}_{EFF}}{\partial A} - \partial_\mu \frac{\partial \mathcal{L}_{EFF}}{\partial \partial_\mu A}\right]A(y_1)\ldots A(y_n)\right\rangle =$$

$$= i\sum_{j=1}^{n}\delta(x - y_i)\langle T\,A(y_1)\ldots A(y_{j-1})A(y_{j+1})\ldots A(y_n)\rangle \quad (2.18)$$

with the normal products formed at the point x. By applying the reduction formulae to $y_1,\,\ldots\,y_n$ we get the stated field equation (2.15). The related formula

$$\left\langle T\,N\left[\left(\frac{\partial \mathcal{L}_{EFF}}{\partial A} - \partial_\mu \frac{\partial \mathcal{L}_{EFF}}{\partial \partial_\mu A}\right)\partial_\nu A\right]A(y_1)\ldots A(y_n)\right\rangle =$$

$$= i\sum_{i=1}^{n}\delta(x - y_i)\langle T\,\partial_\nu^x A(x)A(y_1)\ldots A(y_{j-1})A(y_{j+1})\ldots A(y_n)\rangle \quad (2.19)$$

was proved by Lowenstein and used for deriving Ward identities of the energy-momentum tensor[12].

3. <u>WILSON'S SHORT DISTANCE EXPANSION IN PERTURBATION THEORY</u>[9,15]

In this section we will discuss the verification of the Wilson expansion in perturbation theory. To this end we introduce a normal product $N_a\left[A(x_1)A(x_2)\right]$ at different arguments which approaches $N_a\left[A(x^2)\right]$ in the limit $x_j \to x$. The definition of the corresponding Green's functions is very simple. If

$$e^{-ipx}\int dk\ldots F(p-k,\,k,\,\ldots,\,q_1,\,\ldots\,q_n) \quad (3.1)$$

is the contribution to the Fourier transform $\langle T N_a\left[A(x)^2\right]\tilde{A}(q_1)\ldots\tilde{A}(q_n)\rangle$ of $\langle T N_a\left[A(x)^2\right]A(y_1)\ldots A(y_n)\rangle$ with respect to the y_j we define

$$\int dk\ldots e^{i(p-k)x_1 + ikx_2}F(p-k,\,k,\,\ldots,\,q_1,\,\ldots\,q_n) \quad (3.2)$$

as the corresponding contribution to $\langle TN_a \left[A(x_1)A(x_2) \right] \, \tilde{A}(q_1)\ldots\tilde{A}(q_n) \rangle$. Similarly, one defines

$$N \left[\partial_{\mu_1} \ldots A(x_1) \, \partial_{\mu_{21}} \ldots A(x_2) \right]$$

The following differentiation rules can be proved

$$\partial_\xi^\mu \, N_a \left[A_1(x_1) A_2(x_2) \right] = N_a \, \partial_\xi^\mu \left[A_1(x_1) A_2(x_2) \right] \tag{3.3}$$

$$\partial_x^\mu \, N_a \left[A_1(x_1) A_2(x_2) \right] = N_{a+1} \, \partial_x^\mu \left[A_1(x_1) A_2(x_2) \right] \tag{3.4}$$

$$\xi = x_1 - x_2 \qquad x = \frac{x_1 + x_2}{2}$$

A_1, A_2 derivatives of A

Rule (3.3) shows that more and more derivatives with respect to ξ exist provided the degree a is chosen large enough. For

$$\partial_\xi^{\mu_1} \ldots \partial_\xi^{\mu_m} N_a \left[A(x_1) A(x_2) \right] =$$

$$= N_a \left[\partial_\xi^{\mu_1} \ldots \partial_\xi^{\mu_m} A(x_1) A(x_2) \right] \tag{3.5}$$

exists if $x_j \rightarrow x$ provided $a \geq 2 + m$.

The main theorem concerning normal products of different arguments will now be quoted (for the proof see ref.[9]). Equivalent to the given definition of the normal product at different arguments is the formula

$$N_a \left[A(x_1) A(x_2) \right] =$$

$$= T A(x_1) A(x_2) - \sum_{j=1}^{N} G_j(\xi) \, B_j(x) \tag{3.6}$$

The sum extends over all minimal composite operators B_j of degree $d \leq a$. j includes Lorentz indices. The operators B_j are taken from the basis $N \left[\partial_{\mu_1} \ldots A(x) \partial_{\mu_{21}} \ldots A(x) \ldots \partial_{\mu_{r_1}} \ldots A(x) \right]$. The coefficients G_j are Green's functions which can be expressed by vacuum expectation values of time-ordered products in closed form. Instead of the general formula the explicit form of (3.6) is given for $a = 2$

$$N_2 \left[A(x_1) A(x_2) \right] = T A(x_1) A(x_2) - \langle T A(\tfrac{\xi}{2}) A(-\tfrac{\xi}{2}) \rangle 1 - \tfrac{1}{2} G_2(\xi) N \left[A^2(x) \right]$$

$$G_2(\xi) = \frac{\langle T A(\tfrac{\xi}{2}) A(-\tfrac{\xi}{2}) \tilde{A}(0) \tilde{A}(0) \rangle^{CONN}}{\tilde{\Delta}_F'(0)^2} - 2 \tag{3.7}$$

The identiy (3.6) already has the form of the Wilson expansion if written like

$$T\, A(x_1)\, A(x_2) = \sum_{j=1}^{N} G_j(\xi)\, B_j(x) + N_a\left[A(x_1)A(x_2)\right] \tag{3.8}$$

However, the remainder $N_a\left[A(x_1)A(x_2)\right]$ approaches $N_a\left[A(x)^2\right]$ for $\xi \to 0$ which is finite but does not vanish in general. For the series to be asymptotic the remainder should vanish faster than any power. To remedy this defect we use the fact that more and more derivatives of the normal product N_a with respect to ξ exist in the limit. We form

$$M_a\left[A(x_1)A(x_2)\right] =$$
$$= (1 - t_\xi^{a-2})\, N_a\left[A(x_1)A(x_2)\right] = N_a\left[(1 - t_\xi^{a-2})\, A(x_1)A(x_2)\right] \tag{3.9}$$

t_ξ^{a-2} applied to a function of ξ denotes the Taylor series with all terms up to and including the order $a - 2$. The used derivatives all exist in the limit $\xi \to 0$ and (3.9) vanishes like g^{a-2} for $\xi = g\eta$ in the limit $g \to 0$. Hence the product (3.9) qualifies as remainder of the short distance expansion. There remains the technical problem of rewriting (3.8) with the new remainder (for details see ref. 9,16).

$$T\, A(x_1)\, A(x_2) = \sum H_j(\xi)\, B_j(x) + M_a\left[A(x_1)A(x_2)\right] \tag{3.10}$$
$$d(B_j) \leqq a$$

where now
$$\lim_{g \to 0} \frac{M_a\left[A(x+g\eta)A(x-g\eta)\right]}{g^{a-2}} \tag{3.11}$$

(3.10 - 11) represents the Wilson expansion in perturbation theory.

4.   LIGHT CONE EXPANSIONS[16]

The operator product $A(x_1)A(x_2)$ is not only singular at the origin but also along the light cone $(x_1 - x_2)^2$. We have seen that the Wilson expansion provides information on the singularities near the origin, but it does not immediately give information on the light cone singularities. The reason is that the Wilson expansion is asymptotic near the origin but not near the light cone. Let for instance two coefficients f_j in (1.30) be of the

358

form $\lg \zeta^2$ and $\zeta^{\mu_1} \cdots \zeta^{\mu_r} \lg \zeta^2$.
Near the origin the first coefficient behaves like $\lg \rho$ while the
second one decreases like $\rho^r \lg \rho$. Near the light cone both terms
have the same logarithmic behaviour since $\zeta^{\mu_1} \cdots \zeta^{\mu_r}$ is in general
different from zero.

The most direct way of obtaining information on the light
cone singularities is the method of Brandt and Preparata[17]. For its
derivation we make the drastic assumption that the Wilson expansion
is absolutely convergent everywhere for $\zeta^2 \neq 0$ provided matrix elements are taken between states of sharp energy-momentum. Each term
of the expansion (1.30) has the form

$$g_{j\mu_1 \cdots \mu_r}(\zeta) \, B_j^{\mu_1 \cdots \mu_r}(x) = \ell_j(\zeta^2) \, \zeta_{\mu_1} \cdots \zeta_{\mu_r} \, B_j^{\mu_1 \cdots \mu_r}(x) \tag{4.1}$$

We assume that factors g in the decomposition of the operators B_j
have already been eliminated. The light cone behaviour of the term
(4.1) is then given by the singularity of the function l_j at $\zeta^2 = 0$.
The coefficient may of course vanish for special matrix elements. We
now reorder the terms of the expansion according to decreasing singularity near the light cone. If some l_j are equivalent near the light
cone, i.e.

$$\lim_{\zeta^2 \to 0} \frac{\ell_j}{\ell_i} \neq 0, \infty \tag{4.2}$$

we will replace them by a single function. Combining such terms we
eventually obtain

$$A(x+\zeta) \, A(x-\zeta) = \sum_{j=1}^{\infty} h_j(\zeta^2) \sum \zeta_{\mu_1} \cdots \zeta_{\mu_r} \, B_j^{\mu_1 \cdots \mu_r} \tag{4.3}$$

with

$$\lim_{\zeta^2 \to 0} \frac{h_{j+1}}{h_j} = 0 \tag{4.4}$$

This is the light cone expansion of Brandt and Preparata[17]. The
advantage of this method is that it is straightforward and gives explicit rules which immediately follow from the short distance expansion. On the other hand absolute convergence, or even convergence,
seems to be a very strong assumption. We will now try to justify the
result without convergence assumptions using the Wilson expansion as
asymptotic series only. First we assume that there exists a function
$h(\zeta^2)$ which represents the leading singularity near the light
cone in the sense that

$$\lim_{\xi \to \zeta} \frac{:A(x+\xi)A(x-\xi):}{h(\xi^2)} = E(x,\zeta) \qquad \zeta^2 = 0 \qquad (4.5)$$

$$:A(x_1)A(x_2): = A(x_1)A(x_2) - \langle A(x_1)A(x_2)\rangle$$

$E(x,\zeta)$ is supposed to be finite everywhere, but not identically zero.
More precisely, E may vanish along a lower dimensional subset of the
light cone. This is a rather weak assumption which is correct in
perturbation theory. If the Wilson expansion should converge we may
set $h = h_1$ and have

$$E(x,\zeta) = \sum \zeta_{\mu_1} \cdots \zeta_{\mu_r} B_1^{\mu_1 \cdots \mu_r} \qquad (4.6)$$

The function h can be replaced by any equivalent function h' which
satisfies

$$\lim_{\xi^2 \to 0} \frac{h}{h'} \neq 0, \infty \qquad (4.7)$$

We will now relate the behaviour near the light cone to the short
distance expansion by using the asymptotic form

$$:A(x+\xi)A(x-\xi): = \sum_{j=1}^{n} f_j(\xi) B_j(x) + R_n(x,\xi)$$
$$R_n \sim \rho^{N+1} \text{ for } \rho \to 0 \quad \text{with } \xi = \rho\eta \qquad (4.8)$$

Again, we rewrite the finite sum ordering the terms according to decreasing singularity near the light cone

$$:A(x+\xi)A(x-\xi): = \sum_{j=1}^{m} h_j(\xi^2) \sum \xi_{\mu_1} \cdots \xi_{\mu_r} B_j^{\mu_1 \cdots \mu_r}(x)$$
$$+ R_n(x,\xi) \qquad (4.9)$$

$$\lim_{\xi^2 \to 0} \frac{h_{j+1}}{h_j} = 0$$

We first consider the case that

$$\lim_{\xi^2 \to 0} \frac{h(\xi^2)}{h_1(\xi^2)} \neq \infty \qquad (4.10)$$

This implies that

$$\lim_{\xi \to \zeta} = \frac{R_n}{h_1} \qquad (\zeta^2 = 0) \qquad (4.11)$$

is finite. We then set

$$R_n = \sum h_j(\xi^2) S_n^{(i)}(x,\xi) \qquad (4.12)$$

such that

   (i)   the $S_n^{(j)}$ are finite on the light cone,
   (ii) behave like ρ^{N+1} near the origin,
   (iii) form bilocal operators in the sense

$$\left[S_n^{(j)}(x,\xi), A(z) \right] = 0 \text{ if } (x \pm \xi - z)^2 < 0 \text{ and } (x-z)^2 < 0 \qquad (4.13)$$

360

A representation (4.12) can always be found. If $S_2^{(i)}, \ldots S_n^{(j)}$
are chosen arbitrarily such that the conditions (i) - (iii) are
satisfied the operator $S_1^{(j)}$ automatically fulfills these conditions
also. Finally we obtain

$$: A(x+\zeta) A(x-\zeta): = \sum_{j=1}^{m} h_j(\zeta^2)\, E_j(x,\zeta) \tag{4.14}$$

with

$$E_j(x,\zeta) = \sum \zeta_{\mu_1} \ldots \zeta_{\mu_r}\, B_j^{\mu_1 \cdots \mu_r} + S_n^{(j)} \tag{4.15}$$

In the remaining case

$$\lim_{\zeta^2 \to 0} \frac{h}{h_1} = \infty \tag{4.16}$$

we get a similar result

$$: A(x+\zeta) A(x-\zeta): \sum_{j=0}^{m} h_j(\zeta^2)\, E_j(x,\zeta) \tag{4.17}$$

by setting

$$R_n = h\, S_n^{(0)} + \sum_{j=1}^{m} h_j\, S_n^{(j)} \tag{4.18}$$

$$h_0 = h$$

The operators E_j occurring in (4.14), (4.17) satisfy the three conditions

    (i)   the E_j are finite on the light cone,

    (ii)   may be differentiated N-times with respect to ζ at the
           origin

    (iii) form bilocal operators in the sense

$$\left[E(x,\zeta), A(z)\right] = 0 \quad \text{if } (x \pm \zeta - z)^2 < 0 \ \text{ and } \ (x-z)^2 < 0 \tag{4.19}$$

We thus arrive at the concept of bilocal operators proposed by
Frishman[18] and recently used by Fritzsch and Gell-Mann[19].
The result (4.14) with the properties (i) - (iii) may be interpreted
as an asymtotic double series. For each E_j we have a Taylor expansion
in ζ up to and including terms of order N with a remainder vanishing
like ζ^{N+1} . If we increase the number N of required derivatives we
will expect more and more terms in the sum. If the limit $N \to \infty$
could be taken we would get the light cone expansion of Brandt and
Preparata.

If one is not interested in differentiability at the origin one might as well be content with the original formula (4.5) and define a bilocal operator by

$$E(x, \xi) = \frac{: A(x+\xi) A(x-\xi):}{h(\xi^2)} \tag{4.20}$$

which is finite on the light cone.

There is one serious defect of the given treatment. The main question is how h, the leading light cone singularity, can be read off from the short distance expansion. Precisely this question has not been answered. For in principle it is conceivable that none of the leading terms of the Wilson expansion carries the leading light cone singularity. How could one exclude such a pathological behaviour? A simple way out would be to set up a Wilson expansion for the bilocal operator (4.20)

$$E(x, \xi) = \sum_{j=1}^{N} \sum_{r} L_j(\xi^2) \, \xi^{\mu_1} \cdots \xi^{\mu_r} \, B_{j \mu_1 \cdots \mu_r}(x) + R_N(x, \xi)$$

with

$$\lim_{\xi \to 0} \frac{R_N}{\xi^{N+1}} = 0$$

for suitable N. We now require in addition that $E(x, \xi)$ does not decrease faster than any power of ξ for $\xi \to 0$, if η is on the light cone. Under this condition at least one of the coefficients, say L_k, does not vanish in the limit $\xi^2 \to 0$. The term involving L_k determines the function h in the following way. The Wilson expansion of the original operator product is

$$: A(x+\xi) A(x-\xi): = h(\xi^2) E(x, \xi) =$$

$$= \sum \ell_j(\xi^2) \, \xi^{\mu_1} \cdots \xi^{\mu_r} \, B_{j \mu_1 \cdots \mu_r}(x)$$

where $\ell_j(\xi^2) = h(\xi^2) L_j(\xi^2)$

The terms with the strongest behaviour near the light cone are given by

$$L_j(0) \neq 0$$

of which there is at least one. Hence the leading light cone singularity is represented by any such term and we may set

$$h(\tfrac{\xi^2}{\zeta^2}) = \ell_j(\tfrac{\xi^2}{\zeta^2}) \quad \text{if} \quad L_j(0) \neq 0$$

The final rule is what one would have expected: The leading light cone singularity of the operator product is given by those terms of the Wilson expansion which behave strongest near the light cone.

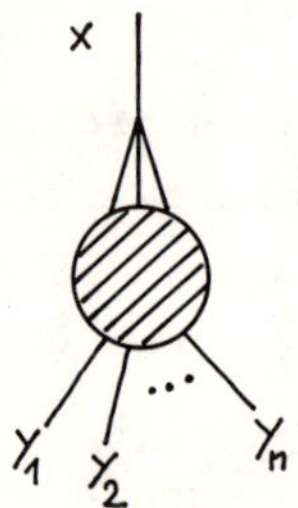

Fig. 1   DIAGRAMS OF TYPE I

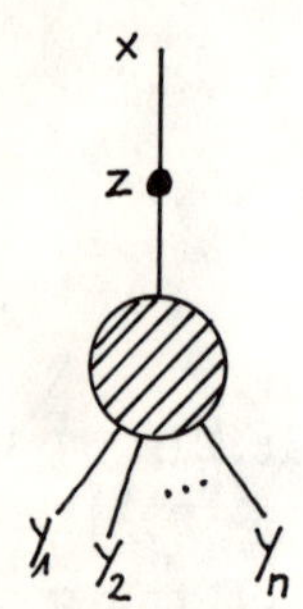

Fig. 2   DIAGRAMS OF TYPE II

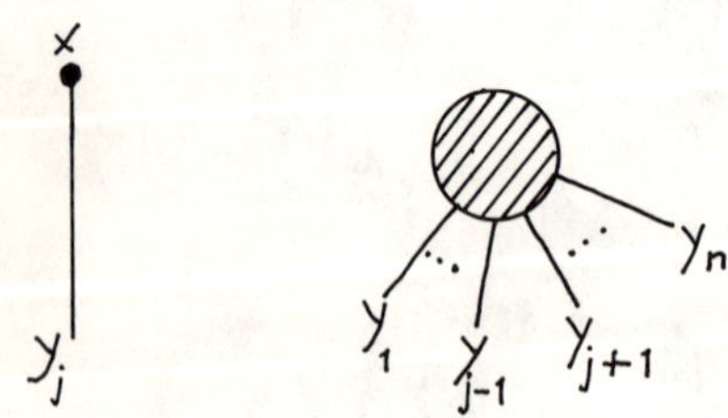

Fig. 3   DIAGRAMS OF TYPE III

REFERENCES

1. K. Wilson, Cornell Report 1964.
   K. Wilson, Phys. Rev. $\underline{179}$, 1499 (1969).
2. K. Wilson and W. Zimmermann, Comm.Math.Phys. $\underline{24}$, 871 (1972).
3. P. Dirac, Proc. Cambridge Phil. Soc. $\underline{30}$, 150 (1934).
   W. Heisenberg, Z. Phys. $\underline{9o}$, 209 (1934).
4. L. Garding and A. Wightman, Arkiv för Fys. $\underline{28}$ (1965).
5. P. Otterson and W. Zimmermann, Comm. Math. Phys. $\underline{24}$, 107, (1972)
6. W. Zimmermann, to be published.
7. G. Dell'Antonio, NYU-preprint 1972.
8. R. Brandt and Ng, NYU-preprint 1972.
9. W. Zimmermann, Brandeis Summer School Lectures Vol I, 1970 and NYU-preprints.
10. For an alternative method see
    H. Epstein, V. Glaser, preprint CERN Th 1344 and 1400 (1971)
    R. Stora, Lecture Notes, Les Houches Summer School 1971
    H. Rouet, R. Stora CNRS-preprints 1971.
11. N.N. Bogoliubov, D.V. Shirkov, Introduction to the Theory of Quantized Fields, Interscience Publ., New York (1960).
12. J. Lowenstein, Phys. Rev. $\underline{D4}$, 2281 (1971).
13. J. Lowenstein, Comm. Math. Phys. $\underline{24}$, 1 (1971).
14. B. Schroer, J. Lowenstein, Phys. Rev. July 1972.
15. R. Brandt, Ann.Phys. $\underline{44}$, 221 (1967) and $\underline{52}$, 122 (1969).
16. G. Dell'Antonio and W. Zimmermann, in preparation.
17. R. Brandt and G. Preparata, Nucl. Phys. $\underline{B27}$, 541 (1971).
18. Y. Frishman, Ann.Phys. $\underline{66}$, 373 (1971)
19. H. Fritzsch and M. Gell-Mann, Proceedings of the Tel-Aviv Conference 1971.

APPLICATION OF THE NORMAL-PRODUCT
ALGORITHM (N.P.A.) TO ZERO MASS LIMITS,
BROKEN SYMMETRIES, AND GAUGE FIELDS

B. Schroer

Institut für Theoretische Physik
Freie Universität Berlin

I. INTRODUCTION

One of the recent motivations for being interested in zero mass
field theories comes from Wilson's short distance expansion for operator
products[1]. This expansion, which generalizes the ETC framework, leads
to singular coefficient functions. Its usefulness is related to Wilson's
suggestion that the leading singularity is that of a zero mass theory,
i.e. the most singular coefficient does not depend on the particle masses. This picture is particular relevant if one tries to understand the
Bjorken scaling law in a field theoretical framework. In the following
we will show that there is an appropriate method, namely the N.P.A.,
which allows one to give a simple discussion of zero mass theories and
the scale invariant (Gell-Mann Low) limit.

There is another field theoretical topic for which this algorithm permits to analyse a very complex situation. This is the area of
the gauge field theories which presently enjoy great popularity as unified renormalizable theories of weak + electrodynamic interactions.
According to my best knowledge the use of NPA techniques for proving
gauge invariance is essential: there has been no other convincing (or
better rigorous since convincing has a somewhat subjective interpretation) treatment of the very delicate gauge problems. Here we will
discuss the gauge problems in two theories: the abelian vector gluon
theory (and its QED limit) and the abelian broken symmetry theory
of Higgs.

The material is organized in the following way:

In the next section (II) we will introduce the N.P.A. in the massive
Thirring model and derive the Callan-Symanzik and the Renormalization-
Group equations. They will be used to construct (in this case known)
the finite zero mass limit. In the section (III) we give a NPA treatment
of broken and spontaneously broken symmetries and we focus in particular on the NPA treatment of the σ-model. The zero mass (infrared)
aspects for the spontaneously broken case will be commented on in the
case of the σ-model. Some aspects of the abelian Higgs model are treated in the IV. section after a brief discussion of the "Gauge Invariance
Criterion" for the vector gluon model.

The last section contains some remarks on the scale invariant limit
which bridge the connection to Mack's and Todorov's talk. We elaborate
in particular on some "paradoxical" aspects of the vector gluon model
regarding the anomalies of the axial current and the scale limit.

The material we selected for this series of lectures has,
at least in parts, already appeared in preprint form. The section II
is taken from a recent preprint by Gomes and Lowenstein[2]. The main
emphasis in this section is pedagogical. The section III on broken
and spontaneously broken symmetries is inspired by B.W.Lee's[3] and
K. Symanzik's[4] work in this topic. Our NPA method however, simplifies
and shortens the already existing treatment considerably. As already
indicated, the power of this method comes to full blossom in the discussion of gauge invariance section IV and V. This material (without
the treatment of the Higgs model) as well as the "paradox" of the
current in the scale limit (in a vector-gluon theory) was already
presented at the "Colloquium on Renormalization of Yang-Mills Fields
and Applications to Particle Physics" (19 - 23 June 1972).

II.　　　THE N.P.A. DERIVED IN THE CONTEXT OF THE MASSIVE THIRRING

　　　　MODEL

The NPA can be introduced from any renormalizable model[5,6].
From a pedagogical point of view it is advantageous to set up the
algorithm in a simple model and then to carry it over to other models
by analogy. As a simple model in which we can illustrate the power of
the method, we take the massive Thirring model ($S \neq 1$!).

We follow the treatment of Gomes and Lowenstein[2]. The formal(classical) Lagrangian is

$$\mathcal{L} = \frac{i}{2}\,\overline{\Psi}\,\gamma^{\mu}\,\overleftrightarrow{\partial_{\mu}}\,\Psi \;-\; m\,\overline{\Psi}\,\Psi \;-\; \frac{1}{2}g\,\overline{\Psi}\gamma^{\mu}\Psi\,\overline{\Psi}\gamma_{\mu}\Psi \qquad (2.1)$$

In two dimensional space-time g is dimensionless and the γ's have the representation:

$$\gamma^{0} = \begin{pmatrix} 0 & 1 \\ 1 & 0 \end{pmatrix}, \quad \gamma^{1} = \begin{pmatrix} 0 & 1 \\ -1 & 0 \end{pmatrix}, \quad \gamma^{5} = \gamma^{0}\gamma^{1} = \begin{pmatrix} -1 & 0 \\ 0 & 1 \end{pmatrix} \qquad (2.2)$$

The canonical dimension of the field is $\frac{1}{2}$.

Let
$$X = \overset{n}{\underset{1}{\widetilde{\Pi}}}\,\Psi(x_{i})\,\overset{n}{\underset{1}{\widetilde{\Pi}}}\,\overline{\Psi}(y_{i}) \qquad (2.3)$$

Then the formal (i.e.divergent) Feynman rules are given by the formula:

$$\langle 0|TX|0\rangle = \langle\Phi_{0}|TX_{0}\,e^{i\int d^{2}x\,:\,\mathcal{L}_{I}^{(0)}(x):}\,|\Phi_{0}\rangle_{\otimes} \qquad (2.4)$$

$$|\Phi_{0}\rangle = \text{ Fock-vacuum, } \Psi_{0}, X_{0} \text{ free fields}$$

$$\mathcal{L}_{I}(x) = -\frac{1}{2}g\,\overline{\Psi}\gamma^{\mu}\Psi\,\overline{\Psi}\gamma_{\mu}\Psi$$

$$\otimes \;=\; \text{omission of vacuum bubbles.}$$

Apart from ultraviolett divergences, the Green's functions computed with (2.4) do not have their poles at the physical mass m. Both problems are solved according to BPHZ (Bogoliubov, Parasiuk, Hepp, Zimmermann) by adding counterterms to the Lagrangian

$$\mathcal{L} \rightarrow \mathcal{L}_{eff} = \frac{i}{2}(1+b)\,\overline{\Psi}\gamma^{\mu}\overleftrightarrow{\partial_{\mu}}\Psi - (m-a)\overline{\Psi}\Psi - \frac{1}{2}(g-c)\overline{\Psi}\gamma^{\mu}\Psi\,\overline{\Psi}\gamma_{\mu}\Psi$$
$$= \frac{i}{2}\,\overline{\Psi}\gamma^{\mu}\overleftrightarrow{\partial_{\mu}}\Psi - m\,\overline{\Psi}\Psi + \mathcal{L}_{I\,eff} \qquad (2.5)$$

and applying the BPHZ "Finite Part" prescription which acts on the Feynman integrands and makes them absolutely convergent for any sub-integration:

$$\langle 0|TX|0\rangle = \text{F.P.}\,\langle\Phi_{0}|TX_{0}\,e^{i\int \mathcal{L}_{I\,eff}(\Psi_{0},\overline{\Psi}_{0})d^{2}x}\,|\Phi_{0}\rangle_{\otimes} \qquad (2.6)$$

For the explicit definition of the F.P. we refer to Zimmermann[5].

Two general remarks are in order:

1) In the transition from $\mathcal{L} \to \mathcal{L}_{eff}$ one not only adds the mass, wave function- and coupling constant-renormalization counterterms (a, b, c), but in order to obtain full flexibility of normalization, one must also add any new term of dimension ≤ 2 (in 4 dim.space-time ≤ 4) which is consistent with the symmetries of the Lagrangian even if it was not already contained in $\mathcal{L}$. This means that we would have to add $\bar{\Psi}\Psi\bar{\Psi}\Psi,\ \bar{\Psi}\gamma_5\Psi\bar{\Psi}\gamma_5\Psi$ etc. with new coupling constants. Because of Fermi statistics these terms all give $\Psi_1\Psi_2\Psi_1^+\Psi_2^+$ which is proportional to the $\bar{\Psi}\gamma_\mu\Psi\bar{\Psi}\gamma^\mu\Psi$ term and hence they can be incorporated into the c-counterterm.

2) In (2.4) and (2.6) we should write the interaction Hamiltonians instead of the Lagrangian. They are identical for couplings not involving derivatives. For derivative couplings, for example, the minimal electromagnetic coupling of a scalar particle, the noncovariant terms in $\mathcal{H}_{I\ eff}$ compensate the noncovariant terms in the naive time ordered products [7]. The resulting Green functions can be demonstrated to be the same as taking the formula (2.4,6) with the stipulation that time ordered contractions involving derivatives are to be interpreted to mean the derivative acting on the time ordered propagators of the scalar fields. For gauge theories one needs furthermore the introduction of ghosts (Stückelnberg ghosts and Faqeev-Popov ghosts) into $\mathcal{L}_{eff}$. This will be explained in section IV.

The constants in front of the counterterms are additional arbitrary parameters which are only fixed in terms of the physical parameters after enforcing normalization conditions. Usually one determines them imposing the mass shell normalization conditions:

$$\text{Def.}\quad (2\pi)^2\ \delta\left(\sum p_i + \sum q_i\right)\ \Gamma^{(2N)}(p_1\cdots p_N; q_1\cdots q_N) \tag{2.7}$$

$$= \int \prod_1^N dx_i\, dy_i\ e^{i\sum_1^N p_i x_i + q_i y_i}\ \langle 0|TX|0\rangle^{prop}$$

The "vertices" $\Gamma^{(2N)}$ are normalized:

$$\Gamma^{(2)}(p,-p)\big|_{\not p=m} = \Gamma^{(2)}_{(0)}(p,-p)\big|_{\not p=m} = 0 \tag{2.8a}$$

$$\gamma^\mu \frac{\partial}{\partial p^\mu} \Gamma^{(2)}(p,-p)\big|_{\not p=m} = \gamma^\mu \frac{\partial}{\partial p^\mu} \Gamma^{(2)}_0(p,-p)\big|_{\not p=m} = i\gamma^\mu\gamma_\mu = 4i \tag{2.8b}$$

con-
stant $\Gamma^{(4)}(p_1\,p_2;\,p_3 p_4)\Big|_{\text{s.p.}} = \Gamma^{(4)}\text{ lowest order}\Big|_{\text{s.p.}} = -ig\cdot\gamma\text{-matrices}$
part of

$$\tag{2.8c}$$

The symmetry point on the mass shell is s.p. $= p_i\,p_j = \dfrac{m^2}{3}(4\delta_{ij}-1)$.
Only the p_i independent part (constant part) of $\Gamma^{(4)}$ is normalized.

For many discussions one is forced to take the intermediate
renormalization at a point p= μ :

$$\Gamma^{(2)}(p_1-p)\Big|_{p=\mu} = i(\mu-m)$$

$$\tag{2.8b'}$$

con-
stant $\Gamma^{(4)}(p_1 p_2; p_3 p_4)\Big|_{p_i p_j = \frac{\mu^2}{3}(4\delta_{ij}-1)} = -ig\cdot\gamma - \text{matrices}$
part of

If zero mass particles are involved it is preferable to take a negative
μ^2 away from the region where Γ's have cuts.

The N.P.A. is an extension of the BPHZ formula (2.6) to
Green functions involving local functions of the basic fields.

Let $\mathcal{O}$ be a local function of ψ and $\bar\psi$, for example $\bar\psi\psi$.
Then we define

$$\langle 0|TN[\mathcal{O}](x)\,X|0\rangle = \text{F.P.}\,\langle\phi_0|T:\mathcal{O}[\psi_0\bar\psi_0]:X_0\,e^{i\int\mathcal{L}_{I\text{eff}}(x)\,d^2x}|\phi_0\rangle_\otimes$$

$$\tag{2.9}$$

Clearly the Feynman rules have the form:

Fig. 1

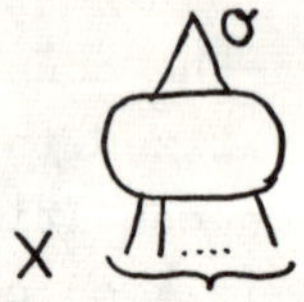

Inside the bubble we have just the ordinary renormalization parts for
which we already defined the F.P. The new renormalization parts are
those which involve the new vertex $\mathcal{O}$. For those subgraphs γ we
choose a divergence degree (i.e. a formula which tells us the number.
of Taylor subtractions):

$$\delta(\gamma) = 2 - \tfrac{1}{2}n - (2-\delta_\mathcal{O})$$

with $\delta_\mathcal{O} \gtrsim \dim\mathcal{O}$

Reminder: For renormalization parts with n external legs connected
to the remaining graph we have the well known Dyson-BPH formula

$$\delta(\psi) = 2 - n \cdot \frac{1}{2}$$

According to our choice of δ we will have different Green functions.
We label those different objects by placing an index on N: $N_\delta(\sigma)$.
The minimal δ which is equal to the canonical dimension of σ is ge-
nerally omitted. The so defined Green functions have several properties.
Without proof we state:[5,6]

1.   Dimensional rule:

$$\partial_\mu \langle TN_\delta[\sigma](x)X \rangle = \langle TN_{\delta+1}[\partial_\mu \sigma](x)X \rangle \tag{2.10}$$

2.   Validity of field equation:

$$[(1+b)i\partial\!\!\!/-m+a]\langle T_\psi(x)X \rangle = i(g-c)\langle TN_{\frac{3}{2}}[\bar\Psi \gamma^\mu \Psi \gamma_\mu \Psi] \tag{2.11}$$
$$(x)X \rangle - i\sum \delta(x-y_i)\langle 0|TX_{\hat{y_i}}|0 \rangle$$

and inside a normal product: [29]

$$\langle TN_\delta[B\{(1+b)i\partial\!\!\!/-m+a\}\Psi](x)X \rangle = (g-c)\langle TN_\delta[B\bar\Psi \gamma^\mu \Psi$$
$$\gamma_\mu \Psi](x)X \rangle - i\sum \delta(x-y_i)\langle B(x)X_{\hat{y_i}} \rangle \tag{2.12}$$

Here B is a $\bar\Psi$ (with γ-matrices) or a spatial derivative thereof.

3.   Zimmermann's Identity:

$$N_\delta[\sigma] = N[\sigma] + \sum r_i N_\delta[\sigma_i] \tag{2.13}$$

here the σ_i are all the other operators which have the same transformation properties as σ and can. dim. $\sigma_i \leqq \delta$.

This relation is to be understood inside a Green function. If one ex-
cepts the validity of this formula one can easily derive a formula for
the r_i.
Let us explain this in the special case $\sigma = \bar\Psi\Psi$ and $\delta = 2$, $N_1[\bar\Psi\Psi]$
$\equiv N[\bar\Psi\Psi]$

$$N_2[\bar\Psi\Psi] = N[\bar\Psi\Psi] + r_2 N_2[\tfrac{i}{2}\bar\Psi \gamma^\mu \overleftrightarrow{\partial_\mu} \Psi] \tag{2.14}$$
$$+ r_1 N_2[i\partial^\mu \bar\Psi \gamma_\mu \Psi] + r_3 N_2[\tfrac{1}{2}\bar\Psi \gamma^\mu \Psi \bar\Psi \gamma_\mu \Psi]$$

Suppose we compute r_3. Let us sandwich the above relation inside a
proper Green function with 4 fields.

$$\langle 0|TN[\bar\Psi\Psi](0)\tilde{\bar\Psi}(p_1)\tilde{\Psi}(p_2)\tilde{\bar\Psi}(p_3)\tilde{\Psi}(p_4)|0 \rangle^{prop} + \tag{2.15}$$

$$+ r_3 \langle 0|TN_2[\tfrac{1}{2}\bar\Psi\gamma^\mu\Psi\bar\Psi\gamma_\mu\Psi](0)\tilde{\bar\Psi}(p_1)\tilde\Psi(p_2)\tilde{\bar\Psi}(p_3)\tilde\Psi(p_4)|0\rangle$$

$$+ \text{ other } N_2 - \text{terms} = 0.$$

The second proper Green function (vertex function) can be decomposed into the "elementary" part:

Fig.2

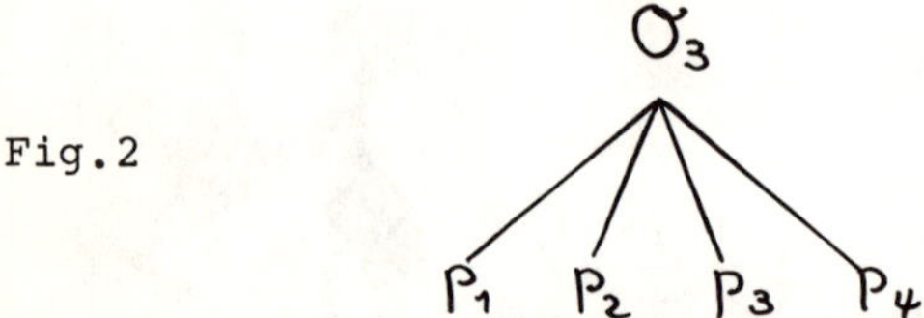

and the rest involving at least one loop integration

Fig.3

The F.P. prescription in the definition of the normal product is a Taylor subtraction scheme (at momentum zero) for the loop parts. With $\delta(G) = 2-2 = 0$ we know that the subtracted expression vanishes if all momenta are equal to zero.

$$r_3 \langle TN_2[\tfrac{1}{2}\bar\Psi\gamma_\mu\Psi\bar\Psi\gamma^\mu\Psi](0)\tilde{\bar\Psi}(0)\tilde\Psi(0)\tilde{\bar\Psi}(0)\tilde\Psi(0)|0\rangle^{\text{prop}} \tag{2.16}$$

$$= r_3 \langle \ldots\ldots \rangle^{\text{el.part}} = r_3\cdot\text{kinematical }(\gamma\text{-dep.})\text{ factor.}$$

The other N_2-terms have <u>no</u> elementary part in this special configuration and they vanish at 0 because the N_2 prescription subtracts out the constant term.

Hence:

$$-\langle TN[\bar\Psi\Psi](0)\tilde{\bar\Psi}(0)\tilde\Psi(0)\tilde{\bar\Psi}(0)\tilde\Psi(0)\rangle^{\text{prop}} = r_3\cdot\text{kin.factor.}$$

This is an explicit (finite!) formula for r_3. By using Green's functions involving derivatives of two fields one can play the same game for the other $r_i's$.

We obtain:

$$r_3 = \frac{i}{16} \, \delta_{\alpha_1 \alpha_2 \alpha_3 \alpha_4} \langle TN[\bar{\Psi}\Psi](0) \, \tilde{\Psi}_{\alpha_1}(0) \tilde{\bar{\Psi}}_{\alpha_2}(0) \tilde{\bar{\Psi}}_{\alpha_3}(0) \tilde{\Psi}_{\alpha_4}(0) \rangle^{\text{prop}} \qquad (2.17a)$$

$$r_1 = \frac{i}{8} \, \text{tr} \, \gamma^\mu \Big(\frac{\partial}{\partial p^\mu} + \frac{\partial}{\partial q^\mu} \Big) \langle TN[\bar{\Psi}\Psi](0) \, \tilde{\bar{\Psi}}(p) \tilde{\Psi}(q) \rangle^{\text{prop}} \Big|_{p=q=0} \qquad (2.17b)$$

$$r_2 = \frac{i}{4} \, \text{tr} \, \gamma^\mu \frac{\partial}{\partial p^\mu} \langle TN[\bar{\Psi}\Psi](0) \, \tilde{\bar{\Psi}}(p) \tilde{\Psi}(-p) \rangle^{\text{prop}} \Big|_{p=0} \qquad (2.17c)$$

The first application of the N.P.A. [i.e rules 1-3] is the derivation of Ward identities:

a)   Vector current W.I.: $j_\mu = (1+b) N_1[\bar{\Psi}\gamma_\mu \Psi]$

$$\partial^\mu \langle 0|T j_\mu(x) X|0\rangle = \partial^\mu (1+b) \langle 0|TN_1[\bar{\Psi}\gamma_\mu \Psi](x) X|0\rangle \qquad (2.18)$$

$$= (1+b) \langle TN_2[\partial^\mu \bar{\Psi}\gamma_\mu \Psi](x) X|0\rangle$$

$$= -i \langle TN_2[\bar{\Psi}\{(1+b)i\vec{\partial} - m + a\}\Psi](x) X\rangle$$

$$+i \langle TN_2[\bar{\Psi}\{-i\overleftarrow{\partial}(1+b) - m + a\}\Psi](x) X\rangle$$

$$\text{(field eqn.)} \qquad = \sum_1^N (\delta(x-y_i) - \delta(x-x_i)) \langle TX\rangle$$

b)   Axial current W.I.: $j\mu_5 = (1+b) N_1[\bar{\Psi}\gamma_\mu \gamma^5 \Psi]$

$$\partial^\mu \langle T j_{\mu 5}(x) X\rangle = (1+b) \partial^\mu \langle TN_1[\bar{\Psi}\gamma_\mu \gamma^5 \Psi](x) X\rangle$$

$$= i \langle TN_2[\bar{\Psi}\gamma^5 \{i(1+b)\vec{\partial} - m + a\}\Psi](x) X\rangle$$

$$+i \langle TN_2[\bar{\Psi}\{-i\overleftarrow{\partial}(1+b) - m + a\}\gamma^5 \Psi](x) X\rangle$$

$$+2i(m-a) \langle TN_2[\bar{\Psi}\gamma^5 \Psi] X\rangle$$

$$\text{(field eqn.)} = -\sum_1^N (\delta(x-x_i)\gamma_i^5 + \delta(x-y_i)\gamma_i^{5T}) \langle TX\rangle$$

$$+ 2i(m-a) \langle TN_2[\bar{\Psi}\gamma^5 \Psi] X\rangle \qquad (2.19)$$

According to rule 3 we may write the last term as:

$$N_2[\bar{\Psi}\gamma^5 \Psi] = N_1[\bar{\Psi}\gamma^5 \Psi] + s N_2[\partial^\mu \bar{\Psi}\gamma_\mu \gamma^5 \Psi] \qquad (2.20)$$

We therefore obtain the Ward identity:

$$\partial^\mu \langle T\hat{j}_{\mu 5}(x) X \rangle = 2i\,(m-a) \langle T N_1[\bar\Psi \gamma^5 \Psi] X \rangle$$
$$- \sum_1^N (\delta(x-x_i)\gamma_i^5 + \delta(x-y_i)\gamma_i^{5T}) \langle TX \rangle \qquad (2.21)$$

where
$$\hat{j}_{\mu 5} = (1 + b - s)\, N_1[\bar\Psi \gamma_\mu \gamma^5 \Psi]$$

Since with $\hat{j}_{\mu 5}$ the δ-function terms in the Ward identity have the coefficient 1, it is reasonable (from the point of view of asymptotic chiral invariance) to call $\hat{j}_{\mu 5}$ the γ_5-current. The γ_5-partner of j_μ has an "anomalous" δ-function term (equal-time commutators), a feature well known from the exact solution [8].

We now come to an important application of the N.P.A., namely the derivation of the C.S.- and renormalization group equation. Those equations result if we combine Zimmermann's identity with the renormalized version of Schwinger's variational principle. We define the "renormalized action" to be the operator:

$$A = \int d^2x\, N_2[\mathcal{L}_{eff}](x) \qquad (2.22)$$

If we now perform an infinitesimal change of the parameter ε $(=m_1 g_1 \ldots)$ of the Lagrangian we obtain for the corresponding change of the renormalized Green's function:

$$\frac{\partial}{\partial \varepsilon} \langle TX \rangle = i \langle T \frac{\partial A}{\partial \varepsilon} X \rangle \qquad (2.23)$$

Here $\frac{\partial A}{\partial \varepsilon}$ means the differentiation with respect to the explicit ε-dependence of the coefficients in $\mathcal{L}_{eff}$. For example:

$$\frac{\partial}{\partial m} \langle TX \rangle = \left\{ \left(\frac{\partial a}{\partial m} - 1\right)\Delta_1 + \frac{\partial b}{\partial m}\Delta_2 + \frac{\partial c}{\partial m}\Delta_3 \right\} \langle TX \rangle$$
$$\frac{\partial}{\partial \mu} \langle TX \rangle = \left\{ \frac{\partial a}{\partial \mu}\Delta_1 + \frac{\partial b}{\partial \mu}\Delta_2 + \frac{\partial c}{\partial \mu}\Delta_3 \right\} \langle TX \rangle$$
$$\frac{\partial}{\partial g} \langle TX \rangle = \left\{ \frac{\partial a}{\partial g}\Delta_1 + \frac{\partial b}{\partial g}\Delta_2 + \left(\frac{\partial c}{\partial g} - 1\right)\Delta_3 \right\} \langle TX \rangle \qquad (2.24)$$

Here the action of a Δ_i on a Green's function means the insertion of an integrated normal product:

$$\Delta_1 = i \int d^2x\, N_2[\bar\Psi \Psi](x) \qquad (2.25)$$

$$\Delta_2 = -\tfrac{1}{2} \int d^2x \, N_2 [\bar\Psi \gamma^\mu \overleftrightarrow{\partial_\mu} \Psi] (x)$$

$$\Delta_3 = \tfrac{i}{2} \int d^2x \, N_2 [\bar\Psi \gamma^\mu \Psi \bar\Psi \gamma_\mu \Psi] (x)$$

Since $\bar\Psi\Psi$ has canonical dimension one, the Δ_0 defined by

$$\Delta_0 = i \int d^2x \, N_1 [\bar\Psi \Psi] (x)$$ also exists and it fulfills the Zimmermann relation:

$$\Delta_0 \langle TX \rangle = \{ \Delta_1 + r_2 \Delta_2 + r_3 \Delta_3 \} \langle TX \rangle \tag{2.26}$$

We can easily find one more relation between the 3 Δ_i's by considering
the so called "counting identities" of the model. Let $2N$ be the
number of external lines of a graph, L the number of internal lines,
V_2 the number of 2-vertices and V_4 the number of 4-vertices. Since
every 2-vertex has two lines and every 4-vertex 4 lines we find:

$$2L + 2N = 2V_2 + 4V_4 \tag{2.27}$$

Taking the set of all graphs with (V_4-1) 4-vertices and inserting one
additional 4-vertex we evidently obtain V_4 times the set of all graphs
with V_4 4-vertices. An analogous argument holds for the 2-vertices.
Hence (2.27) can also be written as:

$$(2L+2N)\langle TX \rangle = \{ 2(a\Delta_1 + b\Delta_2) + 4(c-g)\Delta_3 \} \langle TX \rangle \tag{2.28}$$

The insertion of $m\Delta_1 - \Delta_2$ into $\langle TX \rangle$ is graphically:

$$\{ m\Delta_1 - \Delta_2 \} \langle TX \rangle = \sum \left\{ \vcenter{\hbox{⬭}} - \tfrac{1}{2} \vcenter{\hbox{⬭}} - \tfrac{1}{2} \vcenter{\hbox{⬭}} \right\}$$

$$= \sum \vcenter{\hbox{⬭}}$$

i.e. this process just counts the number of all lines (for one particle
irreducible parts the number of internal lines only):

$$(L + 2N) \langle TX \rangle = (m \cdot \Delta_1 - \Delta_2) \langle TX \rangle \tag{2.29}$$

Taking (2.28) and (2.29) together we obtain the "external leg counting
identity":

$$-2N \langle TX \rangle = \{ 2(a-m)\Delta_1 + 2(b+1)\Delta_2 + 4(c-g)\Delta_3 \} \langle TX \rangle \tag{2.30}$$

Hence we have altogether 5 relations which relate the 3 Δ_i's to other quantities. It is clear that $2N$, Δ_o and the three differentiations must be connected by two relations between them. Then two relations can easily be demonstrated[6] to be any linear combination of the Callan-Symanzik equation:

$$m\frac{\partial}{\partial m} + \mu\frac{\partial}{\partial\mu} + \beta\left(\frac{m}{\mu},g\right)\frac{\partial}{\partial g} + 2N\gamma\left(\frac{m}{\mu},g\right)\langle TX\rangle \qquad (2.31)$$

$$= \alpha\cdot m\,\Delta_o\,G\,\langle T\,X\rangle$$

and the renormalization group equation

$$\mu\frac{\partial}{\partial\mu} + \sigma\left(\frac{m}{\mu},g\right)\frac{\partial}{\partial g} + 2N\tau\left(\frac{m}{\mu},g\right)\langle TX\rangle = 0 \qquad (2.32)$$

The $\beta,\gamma,$ and α as well as the σ and τ are related to the a, b, and c by a system of equations which has a unique solution in perturbation theory.

We now consider the C-S equations for the composite objects $N_1[\bar\Psi\gamma_\mu\Psi]$ and $N_1[\bar\Psi\gamma_\mu\gamma^5\Psi]$. The only change in the derivation is due to the new counting identity:

$$(2.33)$$

$$(-2N-2)\;\langle T\,N_1[\bar\Psi\gamma_\mu\Psi]\,X\rangle = \left\{2(a-m)\Delta_1 + 2(b+1)\Delta_2 + 4(c-g)\Delta_3\right\}\langle T\,N_1X\rangle$$

and the Zimmermann identity:

$$\Delta_1\langle T\,N_1\,X\rangle = \left\{\Delta_o - r_2\Delta_2 - r_3\Delta_3\right\}\langle T\,N_1X\rangle \qquad (2.34)$$

$$- t\;\langle T\,N_1\,X\rangle$$

The additional last term results from the observation that subgraphs, in which Δ_1 and N_1 appear inside the <u>same</u> renormalization part, lead to a new subtraction:

Fig.4

Since $N_1[\bar\Psi\gamma_\mu\gamma^5\Psi] = \varepsilon_\mu^\nu\,N_1[\bar\Psi\gamma_\nu\Psi]$ as a property of the two-dimensional Dirac matrices, we obtain the same C-S equation for both objects.

$$\left\{ m\frac{\partial}{\partial m} + \mu\frac{\partial}{\partial \mu} + \beta\frac{\partial}{\partial g} + (2N+2)\gamma \right\} \langle TN_1 X \rangle = \alpha m(\Delta_0 - t)\langle TN_1 X \rangle \tag{2.35}$$

Taking the divergence for the vector current on both sides we obtain:

$$\left\{ m\frac{\partial}{\partial m} + \mu\frac{\partial}{\partial \mu} + \beta\frac{\partial}{\partial g} + (2N+2)\gamma \right\} \frac{1}{1+b} \sum_i (\delta(x-x_i) - \delta(y-y_i))$$
$$\cdot \langle TX \rangle = \alpha m(\Delta_0 - t)\frac{1}{1+b}\sum_i (\delta(x-x_i) - \delta(y-y_i))\langle TX \rangle \tag{2.36}$$

hence (using the C-S equations for $\langle T\ X \rangle$)

$$-\frac{1}{(1+b)^2}\left(m\frac{\partial}{\partial m} + \mu\frac{\partial}{\partial \mu} + \beta\frac{\partial}{\partial g} + 2\gamma \right)(1+b) = -\alpha m t\frac{1}{1+b} \tag{2.37}$$
$$\alpha m t = -\frac{1}{1+b}\left(\beta\frac{\partial}{\partial g} + 2\gamma \right)(1+b)$$

Since γ is the "would be" anomalous asymptotic dimension
of the field (the dimension for the scale invariant "multilated" C-S
equation: $\beta = 0$) the "would be" anomalous dimension of the conserved
current is equal to the canonical dimension since the extra 2γ in
(2.35) cancels with the t on the right hand side. This observation,
which resulted by using the Ward identity and the C.-S. equations for
the field, is model independent.

Taking the divergence of the C.-S. equation for the N_1
$[\bar{\Psi}\gamma_\mu\gamma^5\Psi]$ and using the C.-S. equation for the N $[\bar{\Psi}\gamma^5\Psi]$ one obtains
(by comparing the δ-function terms on both sides) one more piece of
information, namely:

$$\left(\beta\frac{\partial}{\partial g} + 2\gamma \right)(1+b) + \alpha m t(1+b) - 2(1+b)\beta\frac{\partial}{\partial g}\left(\frac{m-a}{1+b}s \right) = 0 \tag{2.38}$$

together with (2.37) and the nonvanishing of $\frac{m-a}{1+b}$ s in lowest perturbation theoretical order we obtain $\beta = 0$, the expected statement on the
asymptotic scale invariance. This statement is a property of the particular model and can be traced back to the algebraic relation $\gamma_\mu\gamma^5$
$= \varepsilon_\mu{}^\nu\gamma_\nu$ which led to the same C.-S. equations for the
$N_1[\bar{\Psi}\gamma_\mu\Psi]$ and the $N_1[\bar{\Psi}\gamma_\mu\gamma^5\Psi]$. The asymptotic scale invariance
is by no means a property of all renormalizable two dimensional models
(but rather involves an additional assumption concerning the existence
of a Gell-Mann Low eigenvalue, see below).

Let us now study the $m \rightarrow 0$ technique.

First we consider the fields.

<u>Statement:</u>
$$ m \frac{\partial}{\partial m} \langle T\, X \rangle = \mathcal{O}(m \log^x m) $$

This statement follows by taking the difference of the C.-S. and the R.G. equation. One needs in addition the Weinberg theorem on power counting which gives (modulo logs) a zero power for

$$ \langle TX \rangle \,, \; \Delta_o \langle TX \rangle \,, \; \langle TN_1[\ ]X \rangle \,, \; \Delta_o \langle TN_1[\ ]X \rangle $$

and all dimensionless normalization constants as b, c, α, β, γ, ... (where the mass power of BPH normalization constants is equal to its dimension).

Since the $\langle T\, X \rangle \sim \log^x m$, the above statement tells us that the logs are in fact absent[30]. We call the theory obtained in the limit $m \rightarrow 0$ the Gell-Mann Low "preasymptotic theory". The Gell-Mann Low limiting theory is obtained at a zero for the function β. In the case of the Thirring model the preasymptotic theory is already equal to the limiting theory.

For any quantity given in terms of normal products of the basic fields, we can figure out whether it exists in the preasymptotic theory. Let us look for example at the vector current $j_\mu(x) = (1+b)\, N[\bar{\Psi} \gamma_\mu \Psi]$.
It can easily be shown that b is infrared divergent and fulfills the differential equation $m \dfrac{\partial\, b}{\partial\, m} = -\alpha\, m\,(1+b)\,t$ (this follows by subtracting from (2.37) the renormalization group equation for b).
It is only the product of this singular factor with $N_1[\bar{\Psi}\gamma_\mu\Psi]$ which approaches a finite limit: This results from:

$$ m \frac{\partial}{\partial\, m} \langle T\, j_\mu\, X \rangle \sim 0 \; (m \log^x m). $$

Similar consideration hold for the properly normalized axial current and the energy momentum tensor.
In order to obtain a complete picture of the Thirring model as the limit of the massive 4 fermion coupling one would have to establish the field equations (including the correct space-time limiting process) which lead according to an argument of Johnson[8] to the correct Green functions. Unfortunately this discussion has been carried out only for the nonexceptional euclidean points[31]. The existence of the limiting distributions in the physical points is still an open problem. Hence it has not been established on a firm mathematical basis that the preasymptotic theory is equivalent (after redefining the coupling constant in an appropriate way) say to the Klaiber[8] solution of the Thirring model.

III   <u>N.P.A. AND BROKEN SYMMETRIES</u>

A second powerful application of the N.P.A. is the broken-symmetry problem. Let us again be concrete and take a specific model, namely the mesonic σ- model.

The most general effective Lagrangian (in BPHZ sense) which is isospin and parity-invariant and only involves the π and σ field is:

$$\mathcal{L}_{eff} = \tfrac{1}{2}(1-b)(\partial_\mu \sigma)^2 + \tfrac{1}{2}(1-d)(\partial_\mu \vec{\pi})^2 - \tfrac{1}{2}(\mu_\pi^2 + a_\pi)\vec{\pi}^2$$

$$- \tfrac{1}{2}(\mu_\sigma^2 + a_\sigma)\sigma^2 + \tfrac{1}{3}\lambda_1 \sigma^3 + \tfrac{1}{2}\lambda_2 \sigma \vec{\pi}^2 + \tfrac{1}{4}g_1 \sigma^4 \qquad (3.1)$$

$$+ \tfrac{1}{4}g_2 \vec{\pi}^4 + \tfrac{1}{2}g_3 \vec{\pi}^2 \sigma^2$$

This Lagrangian leads to the field equations for the renormalized (i.e. finite, but not properly normalized) Green functions

$$\left(X = \prod_1^n \sigma(x_i) \prod_1^m \vec{\pi}(y_i)\right)$$

$$\left\{(1-b)\partial_x^2 + \mu_\sigma^2 + a_\sigma\right\} \langle T\sigma(x)X \rangle$$

$$= \langle TN_3[\lambda_1 \sigma^2 + \tfrac{1}{2}\lambda_2 \vec{\pi}^2 + g_1 \sigma^3 + g_3 \sigma \vec{\pi}^2](x)X \rangle \qquad (3.2a)$$

$$- i\sum_i \delta(x-x_i) \langle TX_{\widehat{\sigma(x_i)}} \rangle$$

$$\left\{(1-d)\partial_x^2 + \mu_\pi^2 + a_\pi\right\} \langle T\vec{\pi}_a(x)X \rangle$$

$$= \langle TN_3[\lambda_2 \sigma \vec{\pi}_a + g_2 \vec{\pi}^2 \vec{\pi}_a + g_3 \sigma^2 \vec{\pi}_a](x)X \rangle \qquad (3.2b)$$

$$- i\sum_i \delta(x-y_i)\, \delta_{ab} \langle TX_{\widehat{\pi_e(y_i)}} \rangle$$

We want to find restrictions among the 9 parameters of the Lagrangian such that there exists an axial current of the form:

$$\vec{j}_{\mu 5}(x) = c_1 N_3[\vec{\pi}\,\partial_\mu \sigma](x) + c_2 N_3[(\partial_\mu \vec{\pi})\sigma](x) + c_3 \partial_\mu \vec{\pi}$$

and $j_{\mu 5}$ fulfills the Ward identity:

$$\partial^\mu \langle T\, j_{\mu 5 a}(x) X \rangle = i \sum_i \delta(x - x_i) \langle TX\, (\sigma(x_i) \to \pi_a(x_i)) \rangle$$

$$- i \sum_i \delta(x - y_i) \langle TX\, (\pi_b(y_i) \to \delta_{ab}\, \sigma(y_i)) \rangle$$

$$- i \sum_i \delta(x - y_i)\, \sigma_{ab}\, c\, \langle TX\, \underbrace{}_{\pi_b(y)} \rangle \qquad (3.4)$$

$$+ \alpha\, \langle T \pi_a(x)\, X \rangle$$

Corresponding to asymptotic chiral invariance. It is easy to see that the Lagrangian (3.1) leads to a Ward identity having the δ-function structure of (3.4) if we choose: $c_1 = - (1-b)$; $(1-d) = c_2$.

The other N_3 terms which follow from (3.3) match the wanted structure (3.4) if and only if

$$c_1 = - c_2 \qquad (3.5)$$

$$g_2 = g_1 = g = g_3$$

$$\lambda_1 = \lambda_2 = \frac{c_3}{1-d}\, g$$

$$\mu_\sigma^2 + a_\sigma = \mu_\pi^2 + a_\pi - 2 \left(\frac{c_3}{1-d}\right)^2 g \,, \quad \alpha = \frac{c_3}{1-d}\, (-\mu_\pi^2 - a_\pi)$$

We see that the effective Lagrangian we obtained by a systematic construction is the same as the Lagrangian obtained from the following intuitive device, (private communication A. Rouet and R. Stora).

1.   Start from the symmetric Lagrangian

$$\mathcal{L} = \tfrac{1}{2}(1-b) \left[(\partial_\mu s)^2 + (\partial_\mu \vec{\pi})^2 \right] - (m^2 + a)(\vec{\pi}^2 + s^2) \qquad (3.6)$$
$$- \tfrac{1}{4} g (\vec{\pi}^2 + s^2)^2$$

and add a linear term $c \cdot s(x)$

2. Perform a "translation" $s \longrightarrow \sigma + \sigma_0$ with $\sigma_0 = \langle s \rangle$ and rewrite the Lagrangian in terms of σ but compensate the linear terms in σ and drop the constant term.

Note that in the BPHZ renormalization framework self-closing tadpole-like loops are always zero. Hence if one compensates linear terms in the Lagrangian there will be no graphs from trilinear terms for which σ vanishes into the vacuum. After the renormalization has been performed it is convenient to undo the translation and introduce an operator:

$$N_4 \left[\mathcal{L}(\vec{\pi}, s) \right] + cs \qquad (3.7)$$

It is this nice symmetric looking Lagrangian which is conveniently used in manipulation involving the renormalized action integral or the construction of the renormalized energy, momentum tensor, Noether currents and field equations.

In order to avoid infinite resummations, the BPHZ renormalization computation should however be done in the translated σ-form as explained in 2.

If the breaking is stronger than the PCAC (for example a breaking of a symmetry by a mass matrix), then the $\mathcal{L}_{eff}$ leading to a soft symmetry breaking current ($\partial^\mu j_\mu \sim N_\delta$ with $\delta < 4$) has a completely unsymmetric appearance. Even the highest polynomial terms in $\mathcal{L}_{eff}$ do not show the symmetry. The realization of this fact, namely that in BPHZ the softness of symmetry breaking divergencies of currents is not formally seen by looking at the Lagrangian $\mathcal{L}_{eff}$, but rather resides in the renormalized Ward identities, is due to Lowenstein[6].

Of special interest is the case of a conserved current. There are two solutions to this problem: $\frac{c_3}{1-d} = 0$ the symmetry-limit, or $\frac{c_3}{1-d} \neq 0$ and $\mu_\pi^2 + a_\pi = 0$, the spontaneously broken symmetry limit. Note that for the latter case our renormalization Lagrangian $\mathcal{L}_{eff}$ is unsymmetric (because of summing up of breaking effects) whereas the $N_4 \left[\mathcal{L}_{eff} \right]$ used for application of the action principle (equation of motion, Noether currents) can be chosen symmetric by doing the inverse translation.

The spontaneously broken case illustrates also the very delicate infrared problem which we also have to face in the broken gauge theories of the section V: the vanishing mass of the Goldstone boson forces one to renormalize at a fixed space like point $p^2 = -\mu^2$. By writing $g \longrightarrow g + f$ and $c_3 \longrightarrow c_3 + h$ one can make all graphs involving two or four σ-lines (normalized at $p^2 = -\mu^2$) infrared finite. But then the question arises of whether these infrared divergent counter-terms d, f and h which also appear in the two point and four point graphs of the π's automatically take care of the infrared divergence of those graphs. We have checked in lowest order that this indeed so. It seems that one should be able to prove this statement in general by showing for $m_\pi \longrightarrow 0$

$$m_\pi \frac{\partial}{\partial m_\pi} \langle TX \rangle = O(m_\pi \log^x m_\pi) \qquad (3.8)$$

By counting the number of variational equations, counting relations and Zimmermann identities one sees that there are 3 more than there exist independent Δ_i's in $\mathcal{L}_{eff}$. Hence there should be 3 relations. We expect that one of these relations (together with Weinberg's power counting) will lead to (3.8).

IV   GAUGE THEORIES AND THE NPA

Lagrangians involving vector mesons can only be renormalized (if at all) by quantizing them in an indefinite metric Hilbertspace. The selection of the physical observables (which decouple from the ghosts and generate a positive metric space) is done by a gauge principle. For zero mass vector fields the trouble with quantizations is particularly evident if one starts with the classical Lagrangian:

$$\mathcal{L} = -\tfrac{1}{4}(\partial_\mu A_\nu - \partial_\nu A_\mu)^2 + \dots$$

Since $\mathcal{L}$ is gauge invariant, we do not find a unique extremal point in the variational principle. This problem is related to the trouble with canonical quantization. In order to have an action principle we pick a particular gauge by adding to $\mathcal{L}$ the term $\tfrac{1}{2}\alpha(\partial_\mu A^\mu)^2$. For nonabelian gauge theories this is still insufficient for quantization. As Fadeev and Popov showed, one has to introduce in addition scalar fermion ghosts[9]. In these lectures we stay with the abelian gauge theories.

Following these instructions for the case of an abelian gauge field coupled to a spinor particle one obtains[5]:

$$\mathcal{L}_{eff} = (1+d)\tfrac{i}{2}\bar{\Psi}\gamma^\mu \overleftrightarrow{\partial}_\mu \Psi - (M-c)\bar{\Psi}\Psi - \tfrac{1-b}{4}(\partial_\mu A_\nu \qquad (4.1)$$
$$- \partial_\nu A_\mu)^2 + \tfrac{1}{2}(m^2+a)A_\mu A^\mu + (e+f)\bar{\Psi}\gamma^\mu \Psi A_\mu - \tfrac{1}{2\alpha}(\partial_\mu A^\mu)^2$$

According to the general BPHZ construction explained in section II, one should add a counterterm: $h(A_\mu A^\mu)^2$. Anticipating the result that gauge invariance requires $h=0$, we omit this term. It is convenient to write α as $1/2\alpha = (m^2+a)/m_o^2$ with m_o = ghostmass.

$$\mathcal{L}_o = \tfrac{i}{2}\bar{\Psi}\gamma^\mu \overleftrightarrow{\partial}_\mu \Psi - M\bar{\Psi}\Psi - \tfrac{1}{4}(\partial_\mu A_\nu - \partial_\nu A_\mu)^2 + \tfrac{1}{2}m^2 A_\mu A^\mu \qquad (4.2)$$

$$- \frac{1}{2} \frac{m^2}{m_o^2} \left(\partial_\mu A^\mu \right)^2$$

The meson propagator following from this $\mathcal{L}_o$ is

$$\frac{-i}{k^2-m^2} \left(g_{\mu\nu} - \frac{k_\mu k_\nu}{k^2} \right) - \frac{i}{k^2-m_o^2} \frac{m_o^2}{m^2} \frac{k_\mu k_\nu}{k^2}$$

Note that the pole at $k^2=0$ is only superficially there: its residuum cancels. The renormalized field equations for the Green functions are derived from $\mathcal{L}_{eff}$ (Euler-Lagrange equation + contact terms) similarly to the derivation in section II. In the same spirit one derives the Ward identity from the field equations:

$$(1+d)\partial^\mu \langle T N_3 [\bar{\Psi} \gamma_\mu \Psi](x) X \rangle = \sum_i [\delta(x-y_i) - \delta(x-x_i)] \langle T X \rangle \tag{4.3}$$

here $\quad X = \prod_i^n \Psi(x_i) \prod_i^n \bar{\Psi}(y_i) \prod_i^m A_{\mu_i}(z_i)$

From the Ward identity one obtains the formula of the "First Gauge Criterion" in the following way:

Attach a ghost line ---- (longitudinal part of full vector-meson propagator) to a an arbitrary graph of $\langle T \ X \rangle$. The ghost line hooks up either directly to an external meson line or to a vertex:

Fig.5

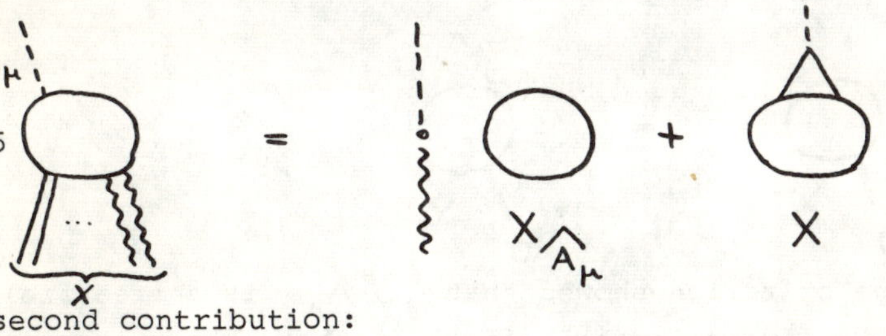

The second contribution:

$$-i(e+f) \frac{m_o^2}{m^2+a} \int d^4x' \Delta_F(x-x', m_o^2) \partial'_\mu \langle T N_3 [\bar{\Psi} \gamma^\mu \Psi](x') X \rangle \tag{4.4}$$

can be rewritten with the help of the Ward identity. In this way we see that the ghost can hook up only to an external meson line or to an external fermion line. In the connected part the contraction to an external meson line does not contribute and we obtain:[*]

---

[*] The adoption of any type of normalization conditions (on shell or off shell) will lead to $f = e \, d$

$$\langle 0 | T \partial_\mu A^\mu(x) X \rangle_c = i \, \frac{e+f}{1+d} \, \frac{m_0^2}{m^2+a} \sum_i \left(\Delta_F(x-x_i, m_0^2) \right.$$

$$\left. - \Delta_F(x-x_i, m_0^2) \right) \langle TX \rangle_c \qquad (4.5)$$

The 1st gauge criterion expresses the decoupling of the real ghost from gauge invariant fields. We call $\mathcal{O}(x)$ a gauge invariant field if it fulfills the relation:

$$\langle T \partial_\mu A^\mu(x)\, \mathcal{O}(y)\, X \rangle = -i \, \frac{m_0^2}{m^2+a} \sum_i \partial_{\nu_i} \Delta_F(x-z_i) \langle T\mathcal{O}(y) X_{\hat{z}_i} \rangle$$

$$+ i \, \frac{e+f}{1+d} \, \frac{m_0^2}{m^2+a} \sum \left(\Delta_F(x-x_i) - \Delta_F(x-y_i) \right) \langle T\mathcal{O}(y) X \rangle \qquad (4.6)$$

i.e. the $\mathcal{O}(y)$ does not lead to any additional possibilities for hooking up a ghost line. It is easy to see that all normal products belonging to classical gauge invariant fields or functions of fields (the δ-index is irrelevant) pass this 1st gauge criterion. A truely gauge invariant $\mathcal{O}(x)$ should also be unchanged if we change the virtual ghost lines, i.e. it should be genuinly independent of α or m_0. In order to achieve this independence as a consequence of the 1st gauge criterion we must attach a virtual ghost line to graphs i.e.

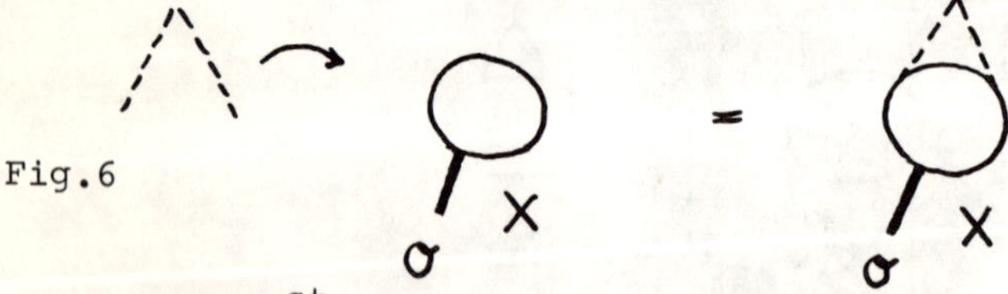

Fig.6

Since the 1st gauge criterion showed that $\partial^\mu A_\mu$ is a free field of mass m_0^2 (take the LSZ limits in the X-fields (4.5)!), the only reasonable way to formulate the relation between the m_0 change and the attachment of a "ghost fork" is : the "2nd gauge criterion":

$$m_0^2 \, \frac{\partial}{\partial m_0^2} \langle T\mathcal{O}(y) X \rangle = \frac{i}{2} \, \frac{m^2+a}{m_0^2} \int d^4x \, \langle \hat{T} : (\partial^\mu A(x))^2 : \mathcal{O}(y) X \rangle \qquad (4.7)$$

The time ordering of the Wick product on the right hand side is defined graphically as the sum over all external line attachments of the "ghost" fork" <u>excluding</u> the external bubbles i.e. for the connected part we have Fig.7:

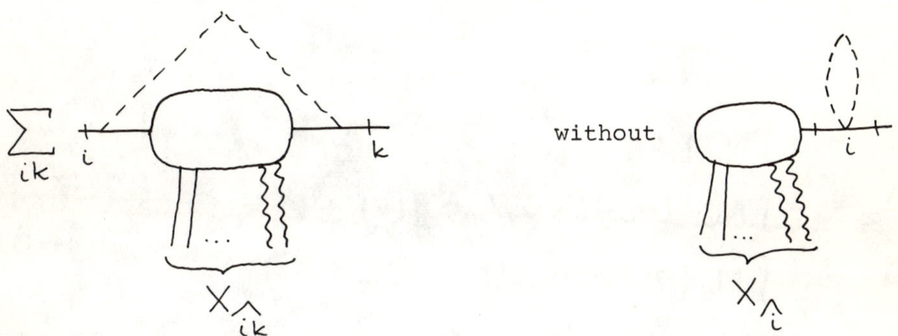

In the Green's function language this can be best understood by starting from

$$\langle T : \partial_\mu A^\mu (x+\xi) \partial_\nu A^\nu (x-\xi) : X \rangle = \langle T \partial_\mu A^\mu (x+\xi) \partial_\nu A^\nu (x-\xi) X \rangle$$

$$(4.8)$$

$$- \langle T \partial_\mu A^\mu (x+\xi) \partial_\nu A^\nu (x-\xi) \rangle \langle T X \rangle$$

According to (4.5) the $\partial_\mu A^\mu$ lines become attached to the external ψ-lines and (4.8) develop a logarithmic singularity for $\xi \to 0$ due to the "external fork bubbles". In order to define the time ordering of the Wick product in the limit $\xi \to 0$ we subtract these offending contributions:

Def.
$$\langle \hat{T} : (\partial_\mu A^\mu (x))^2 : X \rangle = \lim_{\xi \to 0} \left\{ \langle T : \partial_\mu A^\mu (x+\xi) \partial_\nu A^\nu (x-\xi) : X \rangle \right.$$

$$(4.9)$$

$$\left. - \omega(\xi^2) \sum_i (\delta(x-x_i) + \delta(x-y_i)) \langle T X \rangle \right\}$$

with
$$\omega(\xi^2) = - e^2 \left(\frac{m_o}{m^2+a}\right)^2 \Delta_F^2 (\xi)$$

The correct factors in (4.7) follow by <u>formally</u> applying Schwinger's action principle to the infinitesimal m_o^2 change. The <u>renormalized</u> Schwinger's action principle gives instead of (4.7) the relation:

$$\frac{\partial}{\partial m_0^2} \langle T O(y) X \rangle = \Big\{ \frac{\partial a}{\partial m_0^2} \Delta_1 + \frac{\partial b}{\partial m_0^2} \Delta_2 + \frac{\partial c}{\partial m_0^2} \Delta_3 + \frac{\partial d}{\partial m_0^2} \Delta_4$$
$$+ \frac{\partial f}{\partial m_0^2} \Delta_5 - \frac{\partial}{\partial m_0^2} \left(\frac{m^2 + a}{m_0^2} \right) \Delta_6 \Big\} \langle T O(y) X \rangle \tag{4.10}$$

Here the Δ_i's are the integrated parts of $\mathcal{L}_{\text{eff}}$

$$\Delta_1 = \frac{i}{2} \int N_4 [A_\mu A^\mu](x) \, d^4x,$$
$$\Delta_2 = \frac{i}{4} \int N_4 [(\partial_\mu A_\nu - \partial_\nu A_\mu)^2](x) \, d^4x, \tag{4.11}$$
$$\Delta_3 = i \int N_4 [\bar\Psi \Psi](x) \, d^4x, \quad \text{etc.}$$

In order to derive the wanted relation (4.7) from the already established relation (4.8), we write down Zimmermann's identity for the connection of the Wick product[*] with the N_4:

$$\frac{i}{2} \int d^4x : (\partial^\mu A_\mu(x))^2 : = \sum_{i=1}^{7} r_i \Delta_i \tag{4.12}$$

Here Δ_7 is a new object which did not make its appearance in the Lagrangian:

$$\Delta_7 = \frac{i}{4!} \int N_4 [(A_\mu A^\mu)^2](x) \, d^4x \tag{4.13}$$

As explained in the section II, the r_i are determined from the normalization of Δ_i-renormalization parts.

$$r_1 = \frac{1}{8} \int d^4x \, \langle \hat T : (\partial^\nu A_\nu)^2(x) : A_\mu(0) \tilde A^\mu(0) \rangle^{prop}$$

$$r_2 = \frac{1}{8} \int d^4x \left[\frac{\partial}{\partial k^2} \langle \hat T : (\partial^\nu A_\nu)^2(x) : A_\mu(0) \tilde A^\mu(-k) \rangle^{prop} \right]_{k=0}$$

$$r_3 = \frac{1}{8} \, tr \int d^4x \, \langle \hat T : (\partial^\nu A_\nu)^2(x) : \Psi(0) \tilde{\bar\Psi}(0) \rangle^{prop} \tag{4.14}$$

---

[*]   More precisely we have to show that

$$\int \langle T : (\partial_\mu A^\mu (x))^2 : X \rangle \, d^4x = \sum_1^7 r_i \Delta_i \langle T X \rangle$$

The existence of such a relation is a straightforward consequence of the Zimmermann identity for bilocal products[5].

$$r_4 = \frac{1}{32}\, \mathrm{tr}\, \gamma^\mu \int d^4x \left[\frac{\partial}{\partial p^\mu} \langle \hat{T} : (\partial^\nu A_\nu)^2(x) : \Psi(0) \tilde{\bar{\Psi}}(-p) \rangle^{prop} \right]_{p=0}$$

$$r_5 = \frac{1}{32}\, \mathrm{tr}\, \gamma^\mu \int d^4x \langle \hat{T} : (\partial^\nu A_\nu)^2(x) : \Psi(0) \tilde{\bar{\Psi}}(0) \tilde{A}_\mu(0) \rangle^{prop}$$

$$r_6 = 1 \; , \quad r_7 = \frac{1}{16} \int d^4x \langle \hat{T} : (\partial^\nu A_\nu)^2(x) : A_\lambda(0) \tilde{A}^\lambda(0) \tilde{A}_\mu(0) \tilde{A}^\mu(0) \rangle^{prop}$$

prop = proper part (one-particle irreducible).

prop'= proper part without the zero order.

r_6 must be one since the N.P. with the higher index and that with the lower index enter the Zimmermann identity with the same factor. Because of the transversality of A_μ in a proper Green function (due to amputation, A_μ can be replaced by e j_μ) we see that $r_1 = 0 = r_2 = r_7$. Use of the Ward identity leads to $r_5 = e\, r_4$. Equating the action formula (4.8) with the Zimmermann identity yields differential equations for the BPHZ constants in the counterterms. They have the solution:

$$a = a_0$$
$$b = b_0$$
$$c = c_0 + \int_0^{m_0^2} dm_0'^2 \, \frac{m^2 + a}{m_0'^4} \, r_3(m_0') \qquad\qquad (4.15)$$
$$d = d_0 + \int_0^{m_0^2} dm_0'^2 \, \frac{m^2 + a}{m_0'^4} \, r_4(m_0')$$
$$f = ed$$

The a_0, b_0, ... are independent of the ghost mass.

Now we begin to see the delicacy of gauge invariance. Since the m_0 dependence of the BPHZ renormalization parameters is fixed, we are in general not able to formulate m_0^2 independent normalization conditions. <u>Only on mass shell</u> the residuum of the propagator pole and the on shell vertex may be prescribed in a m_0^2 independent manner. In order to see this let us for example consider the m_0-change of the vertex function

$$m_0^2 \frac{\partial}{\partial m_0^2} \Gamma_\mu(p,p',k) = \frac{i}{2} \frac{m^2+a}{m_0^2} \int d^4x \langle \hat{T} : (\partial_\mu A^\mu(x))^2 : \Psi(0) \tilde{\bar{\Psi}}(p') \tilde{A}_\mu(k) \rangle^{prop}$$

$$(4.16)$$

A straightforward consideration shows[*], that the right hand side has the following graphical representation:

---

[*] The reader can find a more elaborate version of this argument in forthcoming lecture notes of J.H. Lowenstein (Maryland 1972).

Fig.8

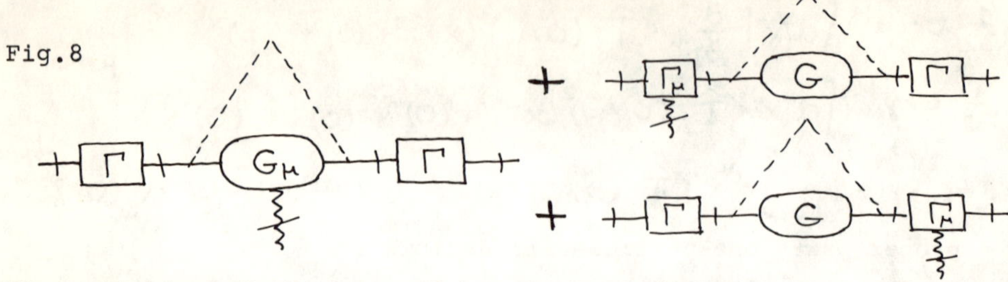

The vanishing of this expression on mass shell is a consequence of the fact that each term contains at least one factor $\Gamma^{(2)}$ which vanishes on shell. A similar consideration for the mass- and wave function normalization. Hence the mass shell normalizations are <u>independent</u> of m_o. If we had chosen a m_o-independent normalization condition at $p = 0$ (intermediate renormalization) or $p^2 = -\mu^2$, only the c_o comes out to be independent of m_o. Hence in this case the <u>normalization condition</u> for the wave function and the coupling constant must be chosen <u>m_o -dependent</u>. The m_o dependence is determined if we for example choose in the Landau gauge ($m_o = 0$):

$$\Gamma^{(2)}(p, -p)\Big|_{\not{p}=\mu} = i\,(\mu - M) \qquad \text{Landau gauge}$$

then

$$\Gamma^{(2)}(p,-p)\Big|_{\not{p}=\mu} = i\,(\mu - m)\,F(m_o) \qquad \text{in general gauge} \qquad (4.17)$$

$$\Gamma^{(3)}_\mu(p,-p;0)\Big|_{\not{p}=\mu} = \begin{cases} ie\gamma_\mu & \text{Landau gauge} \\ ie\gamma_\mu\,G(m_o) & \text{general gauge} \end{cases} \qquad (4.18)$$

With the m_o - dependent choice of normalization we now have:

$$m_o^2\,\frac{\partial}{\partial m_o^2}\,\langle TX \rangle = \frac{i}{2}\,\frac{m^2+a}{m_o^2}\int d^4x\,\langle \hat{T}:(\partial^\mu A_\mu)^2: X \rangle \qquad (4,19)$$

From (4.18) it becomes clear that the 2nd gauge criterion (4.6) expresses the fact that $\mathcal{O}$ does not introduce any additional attachments of "ghost forks". It is easy to see that $F_{\mu\nu}$ fulfills this criterion. However $N_3[\bar{\Psi}\Gamma\Psi](x)$ which passes the 1st criterion does not fulfill (4.6). Fortunately one can repair this defect by a multiplicative finite renormalization:

$$N_3[\bar{\Psi}\Gamma\Psi](x) \;\longrightarrow\; (1 + \alpha_\Gamma(m_o))\,N_3[\bar{\Psi}\Gamma\Psi] \qquad (4.20)$$

$$= \hat{N}_3[\bar{\Psi}\Gamma\Psi]$$

where $\alpha_\Gamma(0) = 0$.

For N_4 normal products a multiplicative change is not enough, one also has to perform an additive change for example ($\check{F}_{\mu\nu} = \varepsilon_{\mu\nu\kappa\lambda} F^{\kappa\lambda}$)

$$\hat{N}_4 [F_{\mu\nu} \check{F}^{\mu\nu}] = N_4 [\check{F}_{\mu\nu} F^{\mu\nu}] + \beta(m_0) N_4 [\partial^\mu (\bar\Psi \gamma_\mu \gamma^5 \Psi)]$$

For the proof of these statements we refer to the original publication[10]. For any [N] which passes the 1st gauge test, there is a corresponding [N] which fulfills the 2nd criterion. By taking the LSZ limit on (4.5) one easily establishes

$$m_0^2 \frac{\partial}{\partial m_0^2} \qquad S = 0 \qquad\qquad (4.21)$$

Among the many interesting relations one can derive with the help of the NPA and the 2nd gauge criterion, there is one of particular importance which we need for a discussion in the last section. This is the famous "anomaly" of the axial current[11]. Let us first consider the divergence of the (gauge dependent) object $N_3 [\bar\Psi \gamma_\mu \gamma^5 \Psi]$:

$$(1+d)\ \partial^\mu \langle T N_3 [\bar\Psi \gamma_\mu \gamma^5 \Psi](x) X \rangle = (1+d) \langle T N_4 [\partial^\mu \{\bar\Psi \gamma_\mu \gamma^5 \Psi\}(x)\} X \rangle$$

(field eqn.)
$$= 2i (M-c) \langle T N_4 [\bar\Psi \gamma^5 \Psi](x) X \rangle$$
$$- \sum (\delta(x-x_i) \gamma_{x_i}^5 + \delta(x-y_i) \gamma_{y_i}^{5T}) \langle T X \rangle$$

(Zimmermann's identity) $= 2i (M-c) \langle T N_3 [\bar\Psi \gamma^5 \Psi](x) X \rangle \qquad (4.22)$

$$+ r \langle T N_4 [F\check{F}](x) X \rangle$$

$$+ s \langle T N_4 [\partial^\mu \bar\Psi \gamma_\mu \gamma^5 \Psi](x)) + \delta\text{-fct.-terms}$$

It is not difficult (however somewhat time consuming) to see[10] that the gauge invariant version of this relation takes the form:

$$\partial^\mu \langle T j_{\mu 5}(x) X \rangle = 2i M \langle T j_5(x) X \rangle + r \langle T \hat{N}_4 [F\check{F}](x) X \rangle$$

$$- \sum (\delta(x-x_i) \gamma_{x_i}^5 + \delta(x-y_i) \gamma_{y_i}^{5T}) \langle T X \rangle \qquad (4.23)$$

Here: $j_{\mu 5}(x) = (1 + b - s - \beta r) N_3 [\bar\Psi \gamma_\mu \gamma^5 \Psi](x)$

$$j_5(x) = (1 - \frac{c}{M}) N_3 [\bar\Psi \gamma^5 \Psi](x) \qquad\qquad (4.24)$$

The normalization of $j_{\mu 5}$ is uniquely fixed by the requirement that the δ-functions appear with the coefficient one. BPH techniques lead to

$$r^{(2)} = \frac{e^2}{(4\pi)^2} \qquad , \quad r^{(4)} = 0$$

Adler and Bardeen have given an argument based on regulators that $r = r^{(2)}$. Recently we succeeded in finding a NPA proof of the Adler Bardeen statement[24].

V. N P A TREATMENT OF THE HIGGS MODEL

The starting point for the Higgs model is the QED of a scalar charged boson. The BPHZ $\mathcal{L}_{eff}$ (in the sense of the Gell-Mann Low perturbation formula (2.6) using the covariant contractions) contains a $(\phi^+\phi)^2$-counterterm and is given by:

$$\mathcal{L}_{eff} = -\tfrac{1}{4}(1-b)(\partial_\mu A_\nu - \partial_\nu A_\mu)^2 - \tfrac{1}{2\alpha}(\partial^\mu A_\mu)^2$$
$$+ (1+d)\,(D_\mu \phi)^\dagger (D^\mu \phi) - \mu^2 \phi^\dagger \phi - h(\phi^\dagger \phi)^2$$
$$D_\mu = \partial_\mu - ie A_\mu$$

(5.1)

Here we have already anticipated that as a result of the Ward identity

$$(1+d)\, i\, \partial^\mu \langle T N_3 [\phi^\dagger \{ \overset{\leftrightarrow}{\partial}_\mu - 2i\, \tfrac{e+f}{1+d} A_\mu \} \phi](x)\, X \rangle$$
$$= -\sum_i (\delta(x-x_i) - \delta(x-y_i)) \langle T X \rangle$$

(5.2a)

and the field equation:

(5.2b)

$$(1-b)\, \partial_\mu \partial^\mu A_\nu + \tfrac{1}{\alpha} \partial_\nu \partial^\mu A_\mu = i\,(e+f)\, N_3 [\phi^\dagger \{ \overset{\leftrightarrow}{\partial}_\mu - 2i\, \tfrac{e+f}{1+d} A_\mu \} \phi](x)$$

the normalization properties of the proper parts at p=0 lead to $e+f = e(1+d)$ as in the spinor case. It is completely straightforward to formulate the differential equation for the counterterms d, μ^2 and h which lead to the 2nd gauge criterion.

The Higgs model results if we take $\mu^2 < 0$ [13]. In this case the classical field strength ϕ which gives the lowest energy is not zero, but rather $\phi_{cl} = e^{ic} \cdot (-\mu^2/2h)^{1/2}$. By a global gauge transformation one can obtain c = 0. It is plausible that the corresponding quantum statement is that the field develops a nonvanishing vacuum expectation value: $\langle \phi \rangle = (-\mu^2/2h)^{1/2} = 2^{-\frac{1}{2}} v$

Introducing: $\phi = 2^{-\frac{1}{2}}(v + \psi + i\chi)$ with $\langle \psi \rangle = 0 = \langle \chi \rangle$

into the Lagrangian

we obtain

$$\mathcal{L} = \mathcal{L}_0(A_\mu) + \frac{1}{2}\frac{e^2 v^2}{1+d} A_\mu^2 + \frac{1+d}{2}\left[(\partial_\mu \psi)^2 + (\partial_\mu \chi)^2\right]$$

$$- ev\, A_\mu \partial^\mu \chi - h v^2 \psi^2 + \mathcal{L}_{int} \tag{5.3}$$

where in $\mathcal{L}_{int}$ we have collected all the tri- and quadrilinear inter-action terms. Note that due to the relation: $-\frac{\mu^2}{2 h} = \frac{v}{2}$, the linear contribution in ψ vanishes.

The bilinear terms describe a system of a massive vectormeson, a zero mass (Goldstone) boson χ and a massive ψ particle. This bilinear part can also be written as:

$$-\frac{1}{4}(1-b)(\partial_\mu B_\nu - \partial_\nu B_\mu)^2 + \frac{1}{2}\frac{e^2 v^2}{1+d} B_\mu^2 - \frac{1}{2\alpha}\left[\partial^\mu\left(B_\mu + \frac{1+d}{ev}\partial_\mu \chi\right)\right]^2 \tag{5.4}$$

$$+ \frac{1}{2}(1+d)(\partial_\mu \psi)^2 - h v^2 \psi^2$$

with $\qquad B_\mu = A_\mu - \frac{1+d}{e\,v}\partial_\mu \chi$

Apart from the α-term we have a system of two decoupled physical fields B_μ and ψ . So if we start an iteration procedure on this bilinear Lagrangian and if we would find quantities $\mathcal{O}$ which are α independent (in an analogous sense as the gauge invariant quantities of the previously discussed unbroken gauge theory), then we are assured that such quantities are free of ghosts i.e. they generate a positive definite Hilbertspace. In this case the Goldstone boson and the zero mass vectormeson ghost (dipole ghost) contained in the free propagators (without the wave function renormalization):

$$\langle T \tilde{A}_\mu(p) A_\nu(0)\rangle = -i\left(g_{\mu\nu} - \frac{p_\mu p_\nu}{p^2}\right)\frac{1}{p^2 - (ev)^2} - i\alpha \frac{p_\mu p_\nu}{p^4}$$

$$\langle T \tilde{A}_\mu(p) \chi(0)\rangle = \alpha ev\, \frac{p_\mu}{p^4} \tag{5.5}$$

$$\langle T \tilde{\chi}(p) \chi(0)\rangle = \frac{i}{p^2} - i\alpha (ev)^2 \frac{1}{p^4}$$

simultaneously decouple from the physical quantities. It is an important question whether this intuitive picture of Higgs can really be
established in renormalized perturbation theory. For the discussion of
this problem, the NPA is extremely helpful. The effective Higgs model
Lagrangian in the BPHZ sense is the most general renormalizable Lagrangian whose free part is given by (5.4) (leaving out the wave function
renormalization) which leads to a conserved current for the A_μ field.
The result can be worked out in the same way as for the σ model.
We will write it down giving its (straightforward) derivation:

$$\mathcal{L}_{eff} = -\frac{1}{4}(1-b)(\partial_\mu A_\nu - \partial_\nu A_\mu)^2 + \frac{1}{2}\frac{\hat{e}^2 v^2}{1+d} A_\mu^2 - \frac{1}{2\alpha}(\partial_\mu A^\mu)^2$$

$$+ \frac{1}{2}(1+c)(\partial_\mu \psi)^2 + \frac{1}{2}(1+d)(\partial_\mu \chi)^2 - \left(\frac{1+c}{1+d}\right)^2 h v^2 \psi^2$$

$$- ev A_\mu \partial^\mu \chi + \hat{e}(\chi \overleftrightarrow{\partial_\mu} \psi) A^\mu + \frac{1}{2}\frac{\hat{e}^2}{1+c} A^\mu A_\mu \psi^2$$

$$+ \frac{1}{2}\frac{\hat{e}^2}{1+d} A_\mu A^\mu \psi^2 + \frac{\hat{e}^2 v}{1+d} A_\mu A^\mu \psi - \frac{h}{4}\chi^4$$

$$- \frac{1}{2}\frac{1+c}{1+d} h \psi^2 \chi^2 - \frac{1}{4}\left(\frac{1+c}{1+d}\right)^2 h \psi^4 - \frac{1+c}{1+d} v h \psi \chi^2$$

$$- \left(\frac{1+c}{1+d}\right)^2 h v \psi^3 \quad \text{with} \quad \hat{e} = e + f \tag{5.6}$$

Anticipating the relation : $e(1+d) = e(1+c) = \hat{e}$ as a result from the
Ward identity and the normalization properties of vertex functions at
$p=0$ we may simplify the above Lagrangian. In that case the conserved
current is:

$$j_\mu(x) = (1+d) N_3 \left[\chi \overleftrightarrow{\partial_\mu} \psi + e A_\mu(\psi^2 + \chi^2) + 2ev A_\mu \psi \right.$$
$$\left. - v \partial_\mu \chi + ev^2 A_\mu \right](x) \tag{5.7}$$

From the effective field equations (Euler-Lagrange equations of $\mathcal{L}_{eff}$)
one immediately derives the following Ward identity:

$$\partial^\mu \langle T j_\mu(x) X \rangle_c = i \sum_i \delta(x-x_i) M_i \langle TX \rangle_c + iv \, \delta(x-x_i)\delta_{\chi,X} \tag{5.8}$$

here $\quad X = \prod_{i=1}^{n} \phi(x_i) \prod_{i=1}^{m} A_{\mu_i}(y_i)$

$$\tag{5.9}$$

where $\quad \phi = \begin{pmatrix} \psi \\ \chi \end{pmatrix}$ and $M = \begin{pmatrix} 0 & -1 \\ 1 & 0 \end{pmatrix}$

In order to study the implications of the Ward identity (5.8) for the BPHZ renormalization and gauge invariance it is very helpful to introduce the generalized vertex functions of this model. They are identical with the proper part (one-particle irreducible) of the Green functions. Their explicit construction is (due to the appearance of trilinear coupling) a bit tedious, their most elegant construction is via functional Legendre transform. This technique, which is due to Jona-Lasinio[14], consists in introducing a generating functional for the connected Green functions:

$$G_c \{J, \eta^\mu\} = \frac{i^n}{n!} \frac{i^m}{m!} \int \langle T \, \phi(x_1) \dots \phi(x_n) A_{\mu_1}(y_1) \dots A_{\mu_m}(y_m) \rangle_c$$

$$\tag{5.10}$$

$$\times \prod_i J(x_i) \prod_i \eta_{\mu_i}(y_i) \; d^\nu x_i \; d^\nu y_i$$

$$J = \begin{pmatrix} J_1 \\ J_2 \end{pmatrix}$$

Defining "dressed" source functions

$$i \phi(x) = \frac{\delta G_c}{\delta J(x)}, \quad i A^\mu(x) = \frac{\delta G_c}{\delta \eta_\mu(x)} \tag{5.11}$$

the vertex functions are introduced via the Legendre transformation:

$$\Gamma \{\phi, A^\mu\} = G_c \{J, \eta^\mu\} - i \int (\phi(x) - J(x) + \eta^\mu A_\mu(x)) \, d^\nu x \tag{5.12}$$

The vector-vertex functions used for the formulation of the Ward identity are given by

$$\Gamma_\mu (x; \phi, A^\nu) = G_{c,\mu}(x; J; \eta^\nu) \tag{5.13}$$

Then the Ward identity (5.8) which in the functional form reads:

$$\partial^\mu G_{c,\mu}(x; J; \eta^\mu) = i J(x) M \frac{\delta G_c}{\delta J(x)} - v J_2(x) \tag{5.14}$$

goes over into:

$$\partial^\mu \Gamma_\mu (x; \phi; A^\nu) = - \frac{\delta \Gamma}{\delta \phi(x)} M i \phi(x) - v i \frac{\delta \Gamma}{\delta \phi_2(x)} \tag{5.15}$$

Specializing at first to functions involving at most 3 ϕ's on the left hand side, we obtain (for n=3)

$$- i p^\mu \Gamma_\mu(P, P_1, P_2, P_3) = i \{ M_1 \Gamma(P_1 + P, P_2, P_3) + M_2 \Gamma(P_1, P_2 + P, P_3) \tag{5.16}$$

$$+ M_3 \Gamma(P_1, P_2, P_3 + P) - v \Gamma_\chi(P, P_1, P_2, P_3) \}$$

where the index χ indicates that the first argument is a χ-variable. Taking $p \rightarrow 0$ and observing that there are no one-particle singularities in the Γ_μ we obtain (by taking specific components):

$$\Gamma^{[3,0]} - 2\Gamma^{[1,2]} - v\,\Gamma^{[2,2]} = 0 \tag{5.17a}$$

$$3\Gamma^{[1,2]} - v\,\Gamma^{[0,4]} = 0 \tag{5.17b}$$

and from the corresponding relations with two instead of three momenta:

$$\Gamma^{[0,2]} = 0 \tag{5.18a}$$

$$\Gamma^{[2,0]} - v\,\Gamma^{[1,2]} = 0 \tag{5.18b}$$

Specializing (5.17) to 4 momenta we obtain for $p \rightarrow 0$:

$$3\Gamma^{[2,2]} - \Gamma^{[0,4]} - v\,\Gamma^{[1,4]} = 0 \tag{5.19a}$$

$$\Gamma^{[4,0]} - 3\Gamma^{[2,2]} - v\,\Gamma^{[3,2]} = 0 \tag{5.19b}$$

The BPHZ Lagrangian (5.6) with $c = d = \frac{f}{e}$ leads to ($\underline{\quad} = \psi$ line, $--- = \chi$ line, $\sim\!\sim = A_\mu$ -line)

$$\Gamma^{[0,2]}(0,0) = 0 \quad , \quad \Gamma^{[2,0]}(0) = i h v^2$$

$$\Gamma^{[3,0]}(0,0,0) = -6\,i h v = \quad$$

$$\Gamma^{[1,2]}(0,0,0) = -2\,i h v = \quad$$

$$\Gamma^{[4,0]}(0) = -6\,i h = \quad$$

$$\Gamma^{[2,2]}(0) = -2\,i h = \quad$$

$$\Gamma^{[0,4]}(0) = -6\,i h = \quad$$

We see that the result of the Ward identity (5.15) reproduces the structure of the A_μ independent terms in $\mathcal{L}_{eff}$ if we put

$$\Gamma^{[3,2]}(0) = 0 = \Gamma^{[1,4]}(0).$$

However, the vanishing of these not primitively divergent quantities is not a consequence of the Ward identity (5.15) alone but rather has to be viewed as a statement following from combining the Ward identity with the structure of $\mathcal{L}_{eff}$.

By using the normalization relation (follows from (5.6))

$$\partial^\mu_{p_1} \Gamma^{[1,1]}_\mu (p, p_1, p_2)\Big|_0 = \frac{1}{e} \partial^\mu_{p_1} \langle T A_\mu(0) \tilde{\Psi}(p_1) \chi(p_2) \rangle^{prop}\Big|_0 \tag{5.20}$$

$$= 4 \ (d+1)$$

The equation resulting by differentiation in p^μ and $p \to 0$ of (5.16) for n=2 instead of n=3 is:

$$- i\partial^\mu_{p_1} \Gamma^{[1,1]}_\mu\Big|_{p=0} = -i\partial^\mu_{p_1}\partial^p_\mu \Gamma^{[0,2]}\Big|_{p=0} + i\partial^\mu_{p_1}\partial^p_\mu \Gamma^{[2,0]}\Big|_{p=0} \tag{5.21}$$

$$+ iv\, \partial^\mu_{p_1}\partial^p_\mu \Gamma^{[1,2]}\Big|_{p=0}$$

At $p_1 = 0$ this gives the relation between the wave function renormalization and trilinear coupling constant renormalization of the effective Lagrangian only if the last term vanishes. Hence this relation between the counterterms of the effective Lagrangian is not a consequence of the Ward identity (5.15) as it would be the case for the unbroken gauge theory. In order to obtain the relations between the terms having two A_μ's and the ones with one A_μ one considers the Ward identity involving

$$\frac{\delta}{\delta A^\nu(x)} \Gamma_\mu\Big|_{A=0}$$

Note that in this case this is no v-term corresponding to the last term in (5.17).

The formulation of the 1st gauge criterion results from the question of how a ghost line $\partial^\mu A_\mu = \ldots\ldots$ can be attached to a graph. In the connected part of the Green function this line can only attach to a vertex. Again, by using the Ward identity one can throw the internal vertex attachments onto the external Ψ and χ lines:

$$\langle T \partial^\mu A_\mu(x) X \rangle_c = ie \sum_i \Delta_F(x-x_i, a) M_i \langle T X \rangle_c$$

$$+ iev\, \Delta_F(x-y)\, \delta_{x,x} \tag{5.22}$$

For the formulation of the 2nd gauge criterion we have to establish the formula:

$$\frac{d}{d\alpha} \langle T X \rangle = \frac{i}{2\alpha^2} \int \langle T : (\partial_\mu A^\mu)^2 : (x) X \rangle\, d^4x \tag{5.23}$$

Again as in the unbroken case, the time ordered Wick-product does not include graphs of the type indicated in Fig.7.

The method was explained in the symmetric case: one compares the formula from the renormalized variation principle with Zimmermann's identity:

$$\frac{i}{2} \int : (\partial_\mu A^\mu(x))^2 : d^4x = \sum_1^{16} r_i \Delta_i \qquad (5.24)$$

here the Δ_i's are the 16 possible integrated NP's of degree 4 (omitting the $(A_\mu A^\mu)^2$-term whose r-vanishing coefficient can be established easily). Let us first concentrate on those terms which only involve the fields Ψ and χ without derivatives.

$$\Delta_1 = \frac{i}{2} \int N_4[\Psi^2](x)\, d^4x \;, \qquad \Delta_2 = \frac{i}{3!} \int N_4[\Psi^3](x)\, d^4x$$

$$\Delta_3 = \frac{i}{2} \int N_4[\Psi\chi^2](x)\, d^4x \;, \qquad \Delta_4 = \frac{i}{4!} \int N_4[\Psi^4](x)\, d^4x$$

$$\Delta_5 = \frac{i}{4} \int N_4[\Psi^2\chi^2](x)\, d^4x,$$

$$\qquad (5.25)$$

$$\Delta_6 = \frac{i}{4!} \int N_4[\chi^4](x)\, d^4x \;, \qquad \Delta_7 = \frac{i}{2} \int N_4[\chi^2](x)\, d^4x$$

The coefficients r_i are computed according to the rules explained in the 2nd section, for example:

$$r_1 = \frac{1}{4} \quad \left\langle T \int d^4x : (\partial_\mu A^\mu(x))^2 \Psi(o)\Psi(o) \right\rangle^{prop}$$

Since the coefficients of the Δ_i's in the action

$$A = \int N_4[\mathcal{L}_{eff}](x)\, d^4x \qquad \text{are parameters which are related}$$

to each other [for instance in the case of the 7 Δ_i's (5.27) the 7 Lagrangian coefficients are given in terms only two parameters h and v] , the differential equation for these coefficients could lead to a contradiction if there would not be as many relations between the r_i's. So in our special example there should be 6 relations among the 7 r_i's. The Ward identity for

$$\left\langle T \int d^4x' : (\partial_\mu A^\mu(x'))^2 j_\mu(x) \; x \right\rangle^{prop}$$

should give the analoga of relations (5.17a - 5.19b) (with $\Gamma^{[1,4]} = \Gamma^{[3,2]} = 0$ in (5.19)):

$$\tau_2 - 2\tau_3 - \upsilon\tau_5 = 0 \tag{5.26a}$$

$$3\tau_3 - \upsilon\tau_6 = 0 \tag{5.26b}$$

$$\tau_7 = 0 \tag{5.26c}$$

$$2\tau_1 - \upsilon\tau_2 = 0 \tag{5.26d}$$

$$3\tau_5 - \tau_6 = 0 \tag{5.26e}$$

$$\tau_4 - 3\tau_5 = 0 \tag{5.26f}$$

We have not been able to establish these relations. We hope to complete the NPA treatment of the Higgs model in a future publication.

VI   <u>PROPERTIES OF THE SCALE INVARIANT (GELL-MANN LOW) LIMIT</u>

The Callan-Symanzik equation and the renormalization group equation can be derived for any renormalizable coupling. They always result from the fact that the number of relations between the $\Delta_i's$ appearing in $\mathcal{L}_{eff}$ (Schwingers action principle, counting identities and Zimmermann's identity) exceed the number of $\Delta_i'S$ by at least two.

Letting $m \longrightarrow 0$ (preasymptotic theory), and postulating the existence of a Gell-Mann Low limiting theory, one obtains scale invariant Green functions $\langle T \ X \rangle$. Is this limit also conformally invariant? The easiest way to settle this problem is to study the renormalized energy-momentum tensor (for an A^4 coupling)

$$\Theta_{\mu\nu} = N_4 \left[\Theta_{\mu\nu}^{eff} \right] \tag{6.1}$$

where $\Theta_{\mu\nu}^{eff}$ is the classical symmetric energy-momentum tensor computed with $\mathcal{L}_{eff}$. We define an "improved" tensor[16] by adding a completely transversal contribution which does not influence the connection with the Poincaré generators:

396

$$\hat{\Theta}_{\mu\nu} = \Theta_{\mu\nu} + f \left(\partial_\mu \partial_\nu - g_{\mu\nu} \partial_k \partial^k \right) N_2 [A^2] \qquad (6.2a)$$

A straightforward but somewhat involved computation[17] shows that the requirement

$$\hat{\Theta}^\mu_{\ \mu} = 0 \qquad (6.2b)$$

becomes completely equivalent with the defining equation for the Gell-Mann Low functions σ and τ [at the G.-L eigenvalue σ (g = g_o)] in term of the BPHZ renormalization functions a, b and c of the A^4 coupling if and only if the parameter f in (6.2) is chosen to be

$$f = - \frac{1+b}{6} \left(1 + 2\tau_o \right)$$

The vanishing of this trace leads to conformal invariance and since the Gell-Mann Low limiting theory is (as was explained at this conference by G. Mack) the scale invariant asymptote of a massive theory, this scale invariant limit will also be conformal invariant. Hence there is no difference between a Gell-Mann Low asymptote of a Lagrangian QFT and the Migdal-Polyakov[18] "bootstrap" approach based on conformal invariance. Note however that this M.-P. approach has only been formulated for renormalizable trilinear coupling as for example the A^3 coupling in 6 dimensional space-time.

Of special interest is the behaviour of composite fields (i.e.N.P.'s of basic fields) in the preasymptotic theory and their scale dimensions in the scale invariant Gell-Mann Low limiting theory. Consider for example the composite field $N_2 [A^2] (x) = N [A^2] (x)$. The Callan-Symanzik equation for the composite Green's function: $\langle T \ N [A^2](x) \ X \rangle$ are different from those of $\langle T \ X \rangle$. For the latter the N.P.A. method explained in section II leads [6] to the C.S. equations (with on shell normalization):

$$\left\{ m \frac{\partial}{\partial m} + \beta \frac{\partial}{\partial g} + N \gamma \right\} \ \langle T \ X \rangle = \rho \Delta_o \langle T \ X \rangle \qquad (6.3)$$

$$\text{with} \quad \Delta_o = \tfrac{i}{2} \int N_2 [A^2] (x) \ d^4x$$

The composite Green function has a different counting identity which can be obtained by substituting $N \to N + 2$. The changed Zimmermann identity for $\Delta_o \langle T \ N [A^2](x) \ X \rangle$ is:

$$(\Delta_1 + r_2 \Delta_2 + r_3 \Delta_3) \ \langle T \ N [A^2] (x) \ X \rangle = \Delta_o \langle T \ N [A^2](x) \ X \rangle$$

$$(6.4)$$

$$+ \text{ correction term}$$

with
$$\Delta_1 = \tfrac{i}{2} \int N_4 [A^2](x)\, d^4x$$

$$\Delta_2 = \tfrac{i}{2} \int N_4 [(\partial A)^2]\,(x)\, d^4x, \quad \Delta_3 = \tfrac{i}{4} \int N_4 [A^4](x)\, d^4x$$

The correction term is proportional to $\langle T\ N[A^2](x)\ X\rangle$ and originates from the fact that in addition to the old renormalization parts there exists a new renormalization part for which the $N[A^2]$ and the Δ_0 resp. the Δ_1 are in the <u>same</u> subgraph. For this subgraph which has two lines connecting to the rest of the graph the Δ_0 and the Δ_1 subtraction scheme is different:

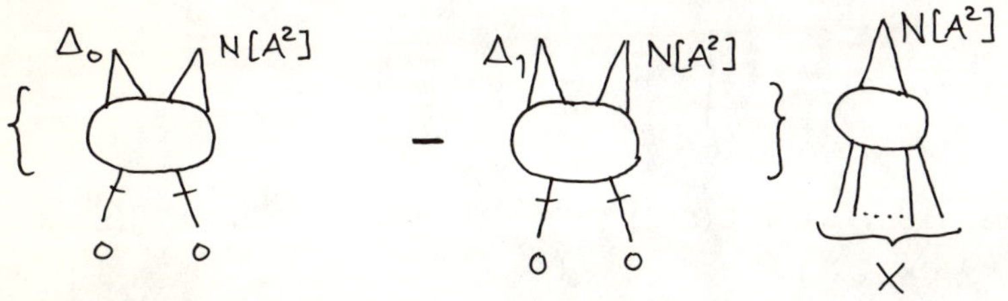

Fig.9

Hence the new Zimmermann identity is of the form:

$$\wp(\Delta_1 + r_2\Delta_2 + r_3\Delta_3)\ \langle T\ N[A](x)\ X\rangle = \wp\Delta_0 \langle T\ N[A^2](x)X\rangle$$
$$- u\ \langle T\ N[A^2](x)\ X\rangle$$

Again u can be computed from normalization properties for N.P.'s and we obtain

$$u = \tfrac{1}{2}\wp\Delta_0\ \langle T\ N[A^2](o)\ \widetilde{A}(o)\ \widetilde{A}(o)\rangle$$

Therefore the new C.S. equations are:

$$(D + (N+2)\gamma)\ \langle T\ N[A^2](x)\ X\rangle = (\wp\Delta_0 - u)\langle T\ N[A^2](x)\ X\rangle$$

$$(6.5)$$

where we use the abbreviation D for the C-S differential operator. In the case of μ-normalization we obtain the C.S. equation (6.5) with

$$D \longrightarrow D_\mu = m\frac{\partial}{\partial m} + \mu\frac{\partial}{\partial \mu} + \beta\frac{\partial}{\partial g}$$,as well as the R.-G. equation:

398

$$\left\{ \mu \frac{\partial}{\partial \mu} + \sigma \frac{\partial}{\partial g} + (N+2)\tau \right\} \left\langle T \, N \left[A^2 \right] (x) \, X \right\rangle = 0 \tag{6.6}$$

Defining: $\qquad \hat{N} \left[A^2 \right] (x) = (1+t) \, N \left[A^2 \right] (x)$

we see that the difference of equation (6.5) and (6.6) leads[*] to

$$m \frac{\partial}{\partial m} \left\langle T \, \hat{N} \left[A^2 \right] (x) \, X \right\rangle \sim 0 \, (m \, \log^x m) \tag{6.7}$$

provided t fulfills:

$$m \frac{\partial}{\partial m} (1+t) = (1+t) u \tag{6.8}$$

i.e. $\qquad 1+t = \exp. \displaystyle\int_0^m u \, (m') \, dm'$

This leads to a dimension: $N+2+ 2(N+2)\tau + 2u$ for $\left\langle T \, N \left[A^2 \right](x) \, X \right\rangle$ and hence:

$$\dim N \left[A^2 \right] = 2 + 4\tau + 2u \tag{6.9}$$

The situation becomes more interesting if we look at the objects of canonical dimension 4. There are three scalar objects:

$$Y_1 = N \left[(\partial_\mu A)^2 \right] (x), \quad Y_2 = N \left[A \partial^2 A \right] (x), \quad Y_3 = N \left[A^4 \right] (x) \tag{6.10}$$

Again the change in the counting identity is taken care of by $N \rightarrow N+2$ resp. $N \rightarrow N+4$. The new Zimmermann's identity is:

$$\left\{ \Delta_1 + r_2 \Delta_2 + r_3 \Delta_3 \right\} \left\langle T \, Y_i (x) \, X \right\rangle = \Delta_0 \left\langle T \, Y_i (x) \, X \right\rangle$$
$$+ \text{correction}$$

It is easy to see that the correction which originates from renormalization subgraphs containing both the Δ_1 (resp. Δ_0) and the Y_i will lead to a mixing of the Y_i's. Hence our new C.S. equation will be a matrix equation of the form:

---

[*] For the Green function and the coefficients we use the Weinberg power counting similar to section II.

$$\{D + M\gamma + \mathcal{U}\} \langle T\, \underset{\sim}{Y}(x)\, X\rangle = \rho\Delta_o \langle T\, \underset{\sim}{Y}(x)\, X\rangle \qquad (6.11)$$

$$\text{with } M = \begin{pmatrix} N+2 & & O \\ & N+2 & \\ O & & N+4 \end{pmatrix}$$

and $\mathcal{U} = (u_{ik})$ consisting of Green function of $\rho\Delta_o\, \underset{\sim}{Y}(x)$ at zero momenta.

The Ansatz $\underset{\sim}{\hat{Y}}(x) = (1 + T)\, \underset{\sim}{Y}(x)$ $\qquad (6.12)$

will lead to μ-normalized C.S. equations whose difference with the R.G. equations again gives

$$m\,\frac{\partial}{\partial m}\,\underset{\sim}{\hat{Y}} \sim O\,(m\,\log^x m) \qquad (6.13)$$

provided

$$m\,\frac{\partial}{\partial m}\,(1 + T) = \mathcal{U}(1 + T) \qquad (6.14)$$

Depending on whether the matrix:

$$S + \mathcal{U} \text{ with } S = \begin{pmatrix} 4+4\gamma & & O \\ & 4+4\gamma & \\ O & & 4+8\gamma \end{pmatrix} \qquad (6.15)$$

is diagonalizable or not, the $\underset{\sim}{Y}$ split into fields belonging to irreducible one-dimensional representations of the dilatation group or (in the nondiagonalizable case in which we only achieve a Jordan form) the $\underset{\sim}{Y}$ forms a non completely reducible "Dell'Antonio column"[23]. In order to see which of the possibilities actually occur, one has to enter a detailed discussion. We have not attempted to do this. Note that even without going into any detailed investigation we can say that certain objects for example the energy-momentum tensor must decouple from other tensor fields of dimension 4. This does however not mean that the other objects also decouple from the energy-momentum tensor. In other words, the energy-momentum tensor may turn out to be the first member of a Dell'Antonio column leading to an infinitesimal dilatation of the form:

$$\begin{pmatrix} \lambda_1 & 1 & \\ & \ddots & \ddots \\ O & & \end{pmatrix}\begin{pmatrix} T_{\mu\nu}(o) \\ \vdots \end{pmatrix} \qquad (6.16)$$

The A^4 theory is a good illustration of what happens in the general case.
It is particularly interesting to look at the scale limit of a theory
which develops axial current analogy. Consider the axial current in
the vector-gluon model of section IV. The C.S. equation of
$N_3 [\bar{\Psi} \gamma_\mu \gamma^5 \Psi]$ mixes with that of $N_3 [\overset{\vee}{F}_{\mu\nu} \; A^\vee]$. Hence two
suitable linear combinations $Y_i(x)$ $i = 1,2$ exist in the preasymptotic
zero mass limit. They are therefore candidates for composite fields of
the Gell-Mann Low limiting theory. In the following we will however show
that the uniquely defined gauge invariant axial current does <u>n o t</u>
exist in the Gell-Mann Low limiting theory.For this we look back at the
Ward identity(4.22) for the gauge invariant axial current (4.23). Let
us study its gauge invariant 3-point function

$$\langle j_{\mu 5}(x) \; F_{\mu_1 \nu_1}(y) \; F_{\mu_2 \nu_2}(z) \rangle \tag{6.17}$$

Since in the Gell-Mann Low limiting theory $F_{\mu\nu}$ must have
canonical dimension (because of $\partial^\nu F_{\mu\nu} = j_\nu$), its two point
function is identical to a free two point function. Therefore according to a well known theorem $F_{\mu\nu}(x)$ has to be a free field (and
therefore $j_\mu = 0$) on the positive definite gauge invariant factorspace
generated cyclically from the vacuum by applying gauge invariant operators. As a special consequence for the 3-point function (6.4) we have
the validity of the free Maxwell equation in y and z. But this gives
us very stringent information on this 3-point function. Consider for
example the triple commutator

$$\langle [[j_{\mu 5}(x) \; F_{\mu_1 \nu_1}(y)] \; F_{\mu_2 \nu_2}(z)] \rangle \tag{6.18}$$

It is a function which satisfies the free field equation in y and z and
reduces (because of causality) on the hypersurface $y_o = x_o$ to a
$\delta^{(3)}(x-y)$ function and its derivatives: for $z_o = x_o$ we obtain a derivation of $\delta^{(3)}(x-z)$. Note that the coefficients of these δ-functions
must be finite as a consequence of a theorem of Malgrange[20]. Therefore
we obtain for (6.5) as a solution of a Cauchy problem

$$P_{\mu\mu_1\nu_1\mu_2\nu_2}(\frac{\partial}{\partial y}, \frac{\partial}{\partial z}) \quad iD(x-y) \; iD(x-z)$$

One easily goes back to the original Wightman function and obtains:

$$\langle j_{\mu 5}(x) \; F_{\mu_1\nu_1}(y) \; F_{\mu_2\nu_2}(z) \rangle = P_{\mu\mu_1\nu_1\mu_2\nu_2}(\frac{\partial}{\partial y}, \frac{\partial}{\partial z}) iD^{(+)}(x-y)$$
$$iD^{(+)}(x-z) \tag{6.19}$$

Hence $j_{\mu 5}$ must have an integer dimension and the only acceptable possibility is dim j_μ = can. dim = 3. In that case a trivial but somewhat lengthy computation leads to a Wightman function which is identical (this is to be expected) to the Wightman function we obtain from the composite field:

$$j_{\mu 5}(x) = c : \overset{\vee}{F}_{0\mu\nu}(x)\, A^{\overset{\vee}{}}_0 :$$

Such a current is evidently not gauge invariant unless $c = 0$. This in turn leads to contradiction with:

$$\left\langle \partial^\mu T\, j_{\mu 5}(x)\, F_{\mu_1\nu_1}(y)\, F_{\mu_2\nu_2}(z) \right\rangle = r \left\langle T\, N_4\!\left[F_{\mu\nu}\,\overset{\vee}{F}^{\mu\nu}\right](x) \right.$$

$$\left. F_{\mu_1\nu_1}(y)\, F_{\mu_2\nu_2}(z) \right\rangle$$

$$(6.20)$$

In order to obtain (6.7) we used the vanishing of the $2\,i\,m\;j_5$ in the Gell-Mann Low limit. Note that the Green's function on the right hand side can not vanish because the kinematical factor $\left\langle T\, N_4[\overset{\vee}{F}F]\,(0)\, \widetilde{F}_{\mu_1\nu_1}(0)\right.$ $\left. F_{\mu_2\nu_2}(0)\right\rangle$ is universal, i.e. independent of the coupling constant. Because of the Adler-Bardeen statement r cannot develop a Gell-Mann Low zero, which leads to the promised contradiction.

Perhaps the value of such paradoxa resulting from the Gell-Mann Low theory of vector gluons is very limited by the observation of Adler[21] that the scale invariant limit is approached in such a weak fashion that it seems to be questionable whether a Gell-Mann Low (scale)-limit can be separated (as a unitary field theory) from the correction terms[22].

A reasonable way out without giving up the idea of a Gell-Mann Low limiting theory altogether is to say that there exists a limiting field theory for the basic fields ψ and A_μ but that certain composite fields do not have limits or in the language of Mack and Symanzik[25] that the Bethe-Salpeter equations for the "would be" composite objects (in particular the two point function bootstrap equation) have no solution.

The picture emerging from this Gell-Mann Low limit seems to be very attractive[26] from the point of view of Bjorken scaling.

If $j_\mu(x)$ can be shown to be zero in the big indefinite Hilbertspace (and not only on the gauge invariant factor space), then Ψ is a free field and the limiting SU_3 currents (to which no vector gluon is coupled if we assume a "strong" isoscalar vector gluon) would be free currents. Since known mathematical methods only allow to make statements on the gauge invariant charge zero sector[27] and do not permit to conclude that $j_\mu(x)$ just creates gauge excitations (which could be gauged away in the big Hilbertspace) one would be inclined to expect that a formulation of the theory in positive definite Hilbertspace based in terms of bilocal integrals[28] is more appropriate to derive the desired result. We will come back to this interesting problem in a future publication.

<u>ACKNOWLEDGEMENT</u>

I am indebted for clarifying discussions to Profs. H.J. Lowenstein, G. Mack, R. Stora, and A.J. Swieca. I also would like to thank the organizers of this summer school, in particular Prof. W. Rühl, for their hospitality.

REFERENCES

1. K. Wilson, Phys. Rev. $\underline{D3}$, 1818 (1971)

2. M. Gomes and J.H. Lowenstein, University of Pittsburgh
    preprint NYO-3829-90. See also
    A.H. Mueller and T.L. Trueman, Phys.Rev. $\underline{D4}$, (1971) 1635

3. B.W. Lee, Nucl. Phys. $\underline{B9}$, 649 (1969)

4. K. Symanzik, Commun. Math. Phys. $\underline{16}$, 48 (1970)

5. W. Zimmermann, Brandeis Lectures (Cambridge, 1970)
    compare also NYU Technical Report 10, 11/72 where the same
    material is presented in a more accessible form

6. J.H. Lowenstein, Phys.Rev. $\underline{D4}$, 228 (1971)
    Commun. Math. Phys. $\underline{24}$, 1 (1971)

7. P.T. Matthews, Phys. Rev. $\underline{75}$, 1270 (1949)

8. K. Johnson, Nuovo Cim. $\underline{20}$, 773 (1962)
    B. Klaiber, Boulder Lectures in Theoretical Physics 1967,
    Gordon and Breach, New York 1968

9. L.D. Fadeev and V.N. Popov, Phys. Letters $\underline{25B}$, (1967)

10. J.H. Lowenstein and B. Schroer to be published in
    Phys.Rev. D, July 1972

11. S.L. Adler, Lectures on Elementary Particles and Quantum Field
    Theory, Volume 1 and literature quoted there

12. S.L. Adler and W.A. Bardeen, Phys.Rev. $\underline{182}$, 1517 (1969)

13. P. Higgs, Phys.Rev. $\underline{145}$, 1156 (1966)
    B.W. Lee, Phys.Rev. $\underline{D5}$, 823 (1972)

14. G. Jona-Lasinio, Nuovo Cim. $\underline{34}$, 1790 (1964)

15. G. 't Hooft and M. Veltman "Combinatorics of gauge fields",
    University of Utrecht preprint, 1972

16. S. Coleman and R. Jackiw, Ann.of Phys. $\underline{67}$, 552 (1971)

17.   B. Schroer, Nuovo Cim. Lett.<u>2</u>, 887 (1971)

M. Hortaçsu, R. Seiler and B. Schroer, Phys. Rev. <u>D5</u>, 2519 (1972

18.   A.M. Polyakov, JETP Lett. <u>12</u>, 538 (1971)

A.A. Migdal, Phys.Letters <u>37B</u>, 98 (1972) ibid.386
The program has meanwhile been carried through for the A^3-coupling in 6-space-time dimension by G. Mack and K. Symanzik, DESY preprint 72/20 (1972)
Compare also the lecture notes (of this conference) by
G. Mack and J.T. Todorov

19.   After we presented this discussion at the "Colloquium on Renormalization of Yang-Mills Fields and Application to Particle Physics"(June 19-23, 1972), we received a preprint entitled "Constraints on Anomalies" by S.L. Adler, Curtis G. Callan Jr., D.J. Gross, and Roman Jackiw. The conclusions in this paper are similar, the main difference lies in the assumption. Whereas we based our discussion on the <u>uniquely defined</u> (by normalization of δ-terms) axial current and its anomalous Ward identity (which we established in renormalized perturbation theory) in a vector gluon model, the quoted authors base their discussion on the validity of the short distance formula of Crewthers.

20.   B. Malgrange, see A.S. Wightman, Les Houches Lectures 1960, page 291

21.   S.L. Adler "Short Distance Behaviour of Quantum Electrodynamics and an Eigenvalue Condition for α", to be published in Phys.Rev.

22.   A conformal invariant Gell-Mann Low limiting theory can be extracted if the scale limit can be interchanged with the integrations involved in Bethe-Salpeter equations. For the case of corrections which have a $\lambda \to \infty$ power-behaviour, this interchange is the basis of the Migdal-Polyakov bootstrap. For gauge theories as Q E D one can however show that the correction terms drop off only logarithmically. In this case the interchange of the scale limit with the Bethe-Salpeter integrations become questionable. In other words one does not know whether the "asymptote" is again a unitary and analytic quantum field theory. I thank Prof.Mack for a

discussion of this point.

23.   G. Dell'Antonio, NYU preprint 1972
The concept of Dell'Antonio columns is related to the
"associated" eigenstates of the dilatation (associated homo-
geneous function) introduced in I.M. Gel'fand and G.E.
Shilov "Generalized Functions", Vol. 1, Academic Press Inc.
New York and London 1964.

24.   J.H. Lowenstein and B. Schroer, NYU preprint 1972
For an alternative derivation using more conventional field
theoretic methods see A. Zee, Institute for Advanced Study
report.

25.   G. Mack and K. Symanzik, DESY preprint 72/19.

26.   B. Schroer talk presented at the topical conference on
Outlook for Broken Conformal Symmetry, Frascati 4/5 May,
1972, to be published.

27.   F. Strocci, comment section of Phys.Rev. $\underline{D5}$, August 1972.

28.   Line integrals over the vector potential were already
introduced in the $30'^S$ by W. Heisenberg.

29.   Recently M. Gomes (University of Pittsburgh Thesis 1972)
has shown that if B = polynomial in the field with degree $\geqslant 2$,
the equations of motion develop anomalies. See also
Y.M.P. Lam. Phys. Rev. $\underline{D6}$ (Oct. 15, 1972) and
M. Gomes and J.H. Lowenstein "Linear Relations among
Normal-Product Fields" University of Pittsburgh preprint 1972.

30.   A. Sirlin has already pointed out that the Callan-Symanzik
equations lead to a significant simplification of the work
of Kinoshita, T.D. Lee and Nauenberg on mass singularities:
A. Sirlin, Phys.Rev. $\underline{D5}$, 2132 (1972)
T. Kinoshita, J. Math. Phys. $\underline{3}$, 650 (1962)
T.D. Lee and M. Nauenberg, Phys. Rev. $\underline{133}$ B,1549 (1964)

31.   These concepts have been introduced by K. Symanzik, Commun.
Math.Phys. $\underline{23}$, 49 (1971). The situation for the nonexceptional
euclidean points in the massive Thirring model has been
investigated by M. Gomes, University of Pittsburgh Thesis,
1972.

Lecture Notes in Physics

Bisher erschienen / Already published